PRIMARY NEURAL SUBSTRATES OF
LEARNING AND BEHAVIORAL CHANGE

PRIMARY NEURAL SUBSTRATES OF LEARNING AND BEHAVIORAL CHANGE

Edited by

DANIEL L. ALKON and JOSEPH FARLEY

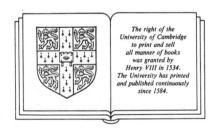

The right of the
University of Cambridge
to print and sell
all manner of books
was granted by
Henry VIII in 1534.
The University has printed
and published continuously
since 1584.

CAMBRIDGE UNIVERSITY PRESS

Cambridge

London   New York   New Rochelle

Melbourne   Sydney

Published by the Press Syndicate of the University of Cambridge
The Pitt Building, Trumpington Street, Cambridge CB2 1RP
32 East 57th Street, New York, NY 10022, USA
296 Beaconsfield Parade, Middle Park, Melbourne 3206, Australia

First published 1984

Printed in the United States of America

Library of Congress Cataloging in Publication Data
Main entry under title:
Primary neural substrates of learning and behavioral change.
Includes index.
1. Learning – Physiological aspects.   2. Paired-
association learning – Physiological aspects.
3. Animal behavior.   I. Alkon, Daniel L.   II. Farley,
Joseph. [DNLM: 1. Learning – Physiology.   2. Behavior –
Physiology.   3. Neurons – Physiology.   WL 102.5 P952]
QP408.P74   1984   591.1'88   83-7681
ISBN 0 521 25472 8

55,994

Contents

Contributors

T. W. ABRAMS *Center for Neurobiology and Behavior, College of Physicians and Surgeons, Columbia University, New York, New York 10032*

DANIEL L. ALKON *Section on Neural Systems, Laboratory of Biophysics, IRP, NINCDS, National Institutes of Health at the Marine Biological Laboratory, Woods Hole, Massachusetts 02543*

JACK D. BARCHAS *Department of Psychiatry and Behavioral Sciences, Stanford University Medical Center, Stanford, California 94305*

S. R. BARRY *Department of Neurology, University of Michigan Medical Center, Ann Arbor, Michigan 48109*

J. S. CAMARDO *Center for Neurobiology and Behavior, College of Physicians and Surgeons, Columbia University and New York State Psychiatric Institute, New York, New York 10032*

T. J. CAREW *Department of Psychiatry and Center for Neurobiology and Behavior, College of Physicians and Surgeons, Columbia University and New York State Psychiatric Institute, New York, New York 10032*

GREGORY A. CLARK *Department of Psychology, Stanford University, Stanford, California 94305*

DAVID H. COHEN *Department of Neurobiology and Behavior, State University of New York, Graduate Biology Building, Stony Brook, New York 11794*

JOHN A. CONNOR *Bell Laboratories, 600 Mountain Avenue, Murray Hill, New Jersey 07974*

TERRY CROW *Department of Physiology, School of Medicine, University of Pittsburgh, Pittsburgh, Pennsylvania 15261*

NELSON DONEGAN *Department of Psychology, Stanford University, Stanford, California 94305*

JOACHIM ERBER *Fachbereich Biologie, Freie Universität Berlin, FB23, WE5, Grünewaldstrasse 34, D-1000 Berlin 41, FRG West Germany*

JOSEPH FARLEY *Department of Psychology, Princeton University, Princeton, New Jersey 08540*

DEBRA FORTHMAN QUICK *Department of Psychology, University of California, Los Angeles, California 90024*

JOHN GARCIA *Department of Psychology and Mental Retardation Research Center of Neuropsychiatric Institute, University of California, Los Angeles, California 90024*

ALAN GELPERIN *Department of Biology, Princeton University, Princeton, New Jersey 08544*

I. GORMEZANO *Department of Psychology, University of Iowa, Iowa City, Iowa 52242*

R. D. HAWKINS *Department of Psychiatry and Center for Neurobiology and Behavior, College of Physicians and Surgeons, Columbia University and New York State Psychiatric Institute, New York, New York 10032*

PHILIP E. HOCKBERGER *Department of Molecular Physics, Bell Laboratories, 600 Mountain Avenue, Murray Hill, New Jersey 07974*

E. R. KANDEL *Department of Physiology and Psychiatry and Center for Neurobiology and Behavior, College of Physicians and Surgeons, Columbia University and New York State Psychiatric Institute, New York, New York 10032*

RONALD E. KETTNER *Department on Psychology, Stanford University, Stanford, California 94305*

DAVID G. LAVOND *Department of Psychology, Stanford University, Stanford, California 94305*

DAVID A. MCCORMICK *Department of Psychology, Stanford, University, Stanford, California 94305*

JOHN MADDEN IV *Department of Psychiatry and Behavioral Sciences, Stanford University, Stanford, California 94305*

MICHAEL D. MAUK *Department of Psychology, Stanford University, Stanford, California 94305*

RANDOLF MENZEL *Institut Tierphysiologie, Freie Universität Berlin, Grünewaldstrasse 34, D-1000 Berlin 41, FRG West Germany*

JOSEPH T. NEARY *Laboratory of Biophysics, National Institutes of Health at the Marine Biological Laboratory, Woods Hole, Massachusetts 02543*

MICHAEL M. PATTERSON *College of Osteopathic Medicine and Department of Psychology, Ohio University, Athens, Ohio 45701*

HOWARD RASMUSSEN *Departments of Physiology, Cell Biology, and Internal Medicine, Yale University School of Medicine, 333 Cedar Street, New Haven, Connecticut 06510*

ROBERT A. RESCORLA *Department of Psychology, University of Pennsylvania, Philadelphia, Pennsylvania 19104*

ANTHONY G. ROMANO *College of Osteopathic Medicine and Department of Psychology, Ohio University, Athens, Ohio 45701*

JERRY W. RUDY *Department of Psychology, University of Colorado, Boulder, Colorado 80309*

CHRISTIE L. SAHLEY *Department of Psychology, Yale University, New Haven, Connecticut 06520*

A. S. SIEGELBAUM *Department of Pharmacology and Center for Neurobiology and Behavior, College of Physicians and Surgeons, Columbia University and New York State Psychiatric Institute, New York, New York 10032*

JOSEPH E. STEINMETZ *College of Osteopathic Medicine and Department of Psychology, Ohio University, Athens, Ohio 45701*

RICHARD F. THOMPSON *Department of Psychology, Stanford University, Stanford, California 90024*

BRENT WHITE *Department of Psychology, Centre College, Danville, Kentucky 40422*

S. J. WIELAND *Deaprtment of Biology, Princeton University, Princeton, New Jersey 08544*

C. D. WOODY *Laboratory of Neurophysiology, University of California, Los Angeles, California 90024*

Preface

Scientific observation and discovery often involve the translation of intuitive understanding into a communicable language. In the process of translation, intuition is verified by rigorously defining the phenomenon of interest and by vastly extending the breadth and number of observations that originally suggested an intuitive insight.

Investigators of learning have developed a need for a common language. Over the last 75 years, students of this subject have asked their questions within diverse disciplines, including behavioral psychology, neurophysiology, and biochemistry. They have focused their attention on animals ranging from humans, cats, and rabbits to bees and grasshoppers to squid and snails. A language that will encompass such diversity cannot be based on superficial approximations at equivalence. We cannot equate the learning of complex abstractions such as mathematical operations with conditioning of an eye blink reflex. Nor do we expect associative learning of a snail to be readily describable as "classical conditioning."

We convey more information and cause less confusion by describing features that characterize the phenomenon of interest. Does the learning involved require a fixed temporal relationship between two distinct sensory stimuli? Is such learning specific to the stimuli presented during training? Has a neural change been in fact correlated with acquisition and retention of learning, or is such a correlation an attractive hypothesis? Has a change of a potassium channel really been implicated in causing learning, and if it has, does it have both a short- and long-lasting time course?

With time, identity of natural phenomena in widely different species may be established. For the present, however, empirical descriptions in terms understandable to the nonspecialist are usually most valuable. It is our purpose in presenting the chapters herein to provide a substantial and authoritative source of such empirical descriptions as well as initial attempts at conceptual syntheses derived from the data available. From the contributions included, it is clear that we are entering an exciting period in learning research. Until 2 to 3 year ago, there had been no evidence for any cellular mechanisms of associative learning. The promise of the current findings is that they will lead to hypotheses for such mechanisms in several species and thereby permit eventual formulation of principles that relate to neural systems in general as they encode and record the occurrence of temporally associated stimuli during an animal's experience.

December 1983   Daniel L. Alkon
Woods Hole, Massachusetts

PART I

BEHAVIOR

Introduction to Part I

JOSEPH FARLEY

The neurosciences are beginning to come of age, and the prospects for relating behavior to the functioning of the nervous system are perhaps more exciting now than at any other time in the recent past. The area of research concerned with the neural bases of learning and memory, phenomena that until very recently have simply resisted analysis at any level more molecular than that of systems neurophysiology, is beginning to yield insights into how the acquisition, storage, and retrieval of information are accomplished within the nervous system and how these processes influence the behavior of intact organisms.

Part I consists of three chapters concerned with (1) the behavioral characteristics of associative learning in vertebrates, (2) the theoretical constructs that have guided the development of contemporary learning theory, and (3) the prospects for a productive translation of these theories into models and heuristics of use to the neurophysiologist, biochemist, and cell biologist.

Chapter 1 appropriately notes that associative learning and classical conditioning are not synonymous. The latter is an experimental paradigm, with its own vocabulary and set of empirical findings. These only partially overlap with our intuitive appreciation of the characteristics of associative learning, derived largely from nineteenth-century British epistemology. In skillful hands, classical conditioning preparations have provided fundamental insights into the behavioral processes implicated in vertebrate associative learning and more complex forms of knowledge acquisition in which associative learning participates. In addition, classical conditioning preparations are characterized by three distinct advantages for the neurophysiologist: (1) the degree of experimental control that can be exercised over the events to be learned about; (2) the ability to index reliably and precisely the product of learning and memory, behavior; and (3) the capacity for tracing the flow of information from the initial stage of sensation and perception through to motor output.

Chapter 2 reminds us that the scope and generality of theories about learning derived from conditioning experiments is broadened if we choose to conceptualize much of the learning involved in classical/Pavlovian conditioning as learning about relations among events. The findings that the author of this chapter and others (Tolman and Brunswick 1935; Dickinson 1980; Mackintosh 1975; Kamin 1969; Bitterman et al. 1980) have reviewed concerning the phenomena of blocking, overshadowing, higher-order conditioning, and contingency learning suggest that even within the seemingly rarified confines of the conditioning experiment, animals are usefully conceptualized as engaged in discerning the "causal texture" of their environments. Temporal contiguity among events is only one such relation involved in causal detection that has been studied within model system preparations (see Part II), and a clear goal for future research is to begin an exploration of the many other interevent relations that now form the focus of contemporary behavioral research, such as stimulus similarity, event correlations, and the distinctions between procedural and declarative knowledge.

Chapter 3 strikes the theme that both the generality of the laws of learning across phylogeny, as well as their species-specific uniqueness, derive from evolutionary selection pressures. The similarities of gustatory aversion–learning phenomena in species as evolutionarily divergent as man and mollusk are argued to reflect the evolution of homologous programs within nervous systems for regulating feeding in animals that secure their food in similar ways. The differences between the nature of the associative learning underlying conditioned fear and flavor aversions are equally striking, however.

Collectively, the three contributions remind us that even at the level of behavioral discourse, associative learning is no simple matter and a variety of phenomenally quite distinct forms of "associations" can be characterized. Whether common cellular mechanisms are involved is impossible to answer at the present time (Parts II and III). Finally, we must always keep in mind the complexity of the behavioral phenomena to be explained, so that in our reductions to more molecular levels of discourse and scientific inquiry, the phenomena that originally formed the focus of our investigation have not slipped away.

REFERENCES

Bitterman, M. E., Lolordo, V. M., and Overmier, J. B. 1980. *Animal Learning.* New York: Plenum Press.

Dickinson, A. 1980. *Contemporary Animal Learning Theory.* Cambridge: Cambridge University Press.

Kamin, L. J. 1969. Predictability, surprise, attention and conditioning. In *Punishment and Aversive Behavior* (B. A. Campbell and R. M. Church, eds.), pp. 279–96. New York: Appleton-Century Crofts.

Mackintosh, N. J. 1975. A theory of attention: Variations in the associability of stimulus with reinforcement. *Psychol. Rev. 82:*276–98.

Tolman, E. C., and Brunswick, E. 1935. The organism and the causal texture of the environment. *Psychol. Rev. 42:*43–77.

1 · The study of associative learning with CS–CR paradigms

I. GORMEZANO

INTRODUCTION

In a contemporary context, our classical conditioning research with rabbit preparations, in particular the nictitating membrane response (NMR), can be viewed from at least four perspectives: (1) as laboratory models of associative learning (see Gormezano 1972; Gormezano and Kehoe 1981); (2) as laboratory models of behavioral adaptation (see Gormezano 1965; Gormezano and Coleman 1973); (3) as a source of axioms for theories regarding the mediation of extended sequences of goal-directed activity (Gormezano 1980); and (4) as model preparations for the study of the underlying neural processes of learning (see Thompson 1976 and Chapter 4). Nevertheless, our original motive for developing these rabbit preparations was to obtain classical conditioning data of a quality suitable for the study of associative learning processes. In some quarters, "associative learning" and "conditioning" are used interchangeably, but we have commented in detail upon those theoretical and empirical considerations that militate against the comfortable but simpleminded identification of the empirical laws of classical conditioning with the doctrines of philosophical associationism, in particular, the law of contiguity (Gormezano and Kehoe 1981; Gormezano et al. 1982). Moreover, the persistent confusion and controversy as to what constitute Pavlovian conditioning and "appropriate controls" have served to cloud the application of associationism to conditioning and vice versa. Accordingly, a substantial portion of our presentation will be addressed to these broader conceptual and methodological issues. Reference to our experimental work will serve to illustrate how we have employed associative doctrine to guide heuristically our experimental and theoretical efforts.

CONDITIONING AND ASSOCIATIVE LEARNING

Classical conditioning paradigms defined

Whereas a broad but still objective definition of "learning" provides that a change in behavior must be a result of "experience" and be relatively permanent, associative doctrine provides the additional stipulation that the "experience" consist of a temporal conjunction of two events. Accordingly, as noted very early by Lashley (1916), it is the full experimental control of the temporal conjunction between the conditioned stimulus (CS) and the unconditioned stimulus (UCS), as well as the specified reaction (the unconditioned response, UCR), that makes classical conditioning preparations particularly attractive vehicles for the study of associative learning. In particular, the basic classical conditioning paradigm has the unique capabilities of specifying the stimulus antecedents to the target response and thus of providing the *basis* for an unequivocal determination of associations. In the fully specified basic classical conditioning paradigm there is a set of experimental operations involving a UCS that reliably produces a UCR and a CS that has been shown by test not initially to produce a response resembling the UCR. The CS and UCS are then presented repeatedly to the organism in a specified order and temporal spacing, and a response similar to the UCR develops to the CS that is called the conditioned response (CR); that is to say, CS–CR functions are obtained. Although various temporal arrangements of the CS and UCS characterize the classical conditioning paradigm, what distinguishes it from instrumental conditioning is that (1) presentation or omission of the UCS is independent of CR occurrence, and (2) the definition of a CR is restricted to the selection of the target response from among those effector systems elicited as UCRs by the UCS.

Regrettably, the requirement that CS and UCS occurrences be independent of the CR has become the hallmark of the classical conditioning paradigm to the exclusion of the second requirement of selection of a target response from among those effector systems elicited as UCRs. The tendency to emphasize the stimulus presentation procedure of classical conditioning, although ignoring selection of the target response, can be traced to the couching of the procedural distinction between classical and instrumental conditioning in terms of their respective stimulus–stimulus and response–stimulus contingencies (see Skinner 1938). Moreover, the observation that in the vast majority of classical conditioning preparations the topography of CRs are not *identical* to the UCRs has been used to dismiss target response selection as a necessary part of the definition of classical conditioning. However, an objective criterion for specifying a CR is obtained by requiring that the response to the CS appear in the *same* effector system as the UCR. Although such a criterion admits to even wider differences between the CR and UCR (e.g., decelerative heart rate CRs and accelerative heart UCRs), the original stimulus antecedent (i.e., the UCS) for the target response is known and under experimental control. Of course, such a definition of the CR is not meant to deny the logical possibility that CS and UCS presentations may result in the acquisition of a response to the CS in an effector system not elicited by the UCS. However, if such a response were acquired, it should be distinguished from a CR because the original stimulus antecedents for the response remains to be specified.

Other conditioning paradigms

The tendency to specify a classical conditioning procedure simply in terms of the response-independent presentation of the CS and UCS has led to the aggregation of a broad set of learning phenomena under the heading of "classical conditioning." Specifically, the term "classical conditioning" has been extended gradually from the historic, basic CS–CR paradigm to include conditioned stimulus–instrumental response (CS–IR), discriminative instrumental approach, and "autoshaping" procedures (see Gormezano and Kehoe 1975). The CS–IR paradigms include most notably the conditioned suppression procedure, as well as other "transfer of control" or "classical–instrumental transfer" procedures (Gormezano and Moore 1969). In CS–IR procedures, the stimulus pairings of classical conditioning are carried out with a CS and biological significant event (e.g., shock), but without any measurement of a target response. Then, in the test phase, the CS is presented during ongoing instrumental responding and the CS's facilitatory or disruptive effect on instrumental behavior is measured. Accordingly, it may be said that CS–IR functions are obtained in these paradigms that purport to indirectly assess classical conditioning processes. Some discriminative instrumental approach procedures have also been designated as "Pavlovian," simply because an explicit cue (CS) is presented and a food (grain) hopper, designated the UCS, is made available at a fixed time following the onset of the cue (e.g., Holland and Rescorla 1975; Longo et al. 1964). In these paradigms, approach behavior is necessary and, by definition, instrumental to actual receipt of the reinforcing event. Yet the approach activity, usually measured grossly by stabilimetric devices, has been mistakenly identified as a "UCR." Thus, while the development of anticipatory approach activity ("CR") to the discriminative stimulus presented prior to food availability is undeniably associative in nature, it is also undeniably an instrumentally conditioned response. Generally, the autoshaping procedure consists of response-independent presentation of a lighted manipulandum (e.g., lighted key) as a CS and activation of a food magazine as the UCS, with the target response being contact with the manipulandum (e.g., key pecking). Although the target response (key pecking) is not an instrumental response, it also does not appear in the constellation of UCRs (Woodruff and Williams 1976). Accordingly, the acquisition of a response in an effector system not affected by the UCS would qualify autoshaping as a "new" associative learning procedure arising only from the stimulus presentations of the CS–CR procedures but with an acquired response having stimulus antecedents that are to a degree separate from the UCS and remaining to be specified.

Although all of the above paradigms have an associative character, the class of behaviors conventionally employed in each of them affects their suitability for investigations of not only basic associative processes but also the underlying neural substrate. Specifically, they lack the unique capabilities of CS–CR paradigms to exercise absolute control over the timing and sequencing of stimulus events and to identify the stimulus antecedents to the target response from the outset of training. Moreover, with CS–CR paradigms the class of behaviors studied have an anatomically defined set of movements or secretions, mediated by a relatively small group of muscles and/or glands. Consequently, CS–CR procedures have allowed for the potential identification of neural final common pathway(s) for behavior (e.g., Cegavske et al. 1976; Gray et al. 1981). Since in CS–CR paradigms the target response system is elicited by the UCS, it has been possible to identify effector pathways outside the conditioning situation, which permits the observation of changes in the activity of those pathways from the start of conditioning (see Thompson 1976). On the other hand, in the other associative paradigms, we have referenced, it is much more difficult to identify the final common pathways for the target behavior or to observe changes in them brought about by training, since the target response is usually outcome-defined (e.g., press the bar) and thus allows for a wide variety of different body movements to yield the required outcome. The allowable variation makes it difficult, if not impossible, to clearly identify a final common pathway for the movements that make up the behavior. Furthermore, since there is usually no known unconditioned stimulus for the target behaviors, their pathway cannot be identified outside the learning situation.

Control methodology

The associative nature of a given preparation has come to be determined not only by the mere use of contiguous occurrence of the CS and UCS, but also in terms of a set of control operations intended to estimate the contribution of other possible processes to responding. If behavior in a classical conditioning preparation were governed strictly by an associative process, then the single observation of the designated target response to a CS after CS–UCS pairing would be sufficient to indicate the establishment of a CR. Not surprisingly, this ideal case has never been achieved, for not all responses observed in connection with a CS uniquely result from prior CS–UCS pairing. At a minimum, all response systems show some level of baseline activity, often raised by UCS presentations, that will produce an accidental coincidence of the CS and target response. Moreover, the likelihood of a response to the CS may be systematically affected by (1) alpha responses, which are unconditioned responses to the CS in the same effector system as the target response; and (2) "pseudo-conditioned" and "sensitized" responses, which can be established on the basis of prior UCS-alone presentations. By employing a control group given CS-alone presentations, a detailing of the latency, duration, amplitude, and course of habituation of the alpha response can provide a basis for eliminating it from consideration as a CR. In particular, alpha responses are usually of a shorter latency than CRs; hence, if a sufficiently long CS–UCS interval is employed, both the alpha response and the CR can be observed in the interval and scored accordingly. Further complexities can arise in distinguishing alpha responses from CRs (e.g., sensitization) and the details of dealing with such complexities can be found elsewhere (see Gormezano 1966; Gormezano and Moore 1969).

Once the alpha response has been eliminated from consideration, the possible contribution of pseudo-CRs to CR measurement must be assessed. If the UCS is presented one or more times prior to the presentation of a CS (particularly if the UCS is noxious), the procedure may frequently result in the occurrence of a response to the CS, labeled a pseudo-CR, that may be quite indistinguishable from a CR. The mechanism producing the pseudo-CR is not known at present; however, it is traditionally treated as separate from those responses acquired by classical conditioning because of its occurrence in the absence of previous CS–UCS pairings. Although the UCS-alone procedure provides the clearest demonstration of the phenomenon, it does not permit an assessment of the strength of a pseudo-CR on a trial-by-trial basis for comparison with the acquisition of a CR. To provide such a trial basis for comparison, a single control procedure has been employed for the assessment of both pseudo-CRs and baseline responses. Known as the "unpaired" control, this procedure involves presenting CS-alone and UCS-alone trials the same number of times as the experimental group is given CS–UCS pairings. However, the stimuli are presented in a random fashion with any CS–UCS interval being variable and far exceeding the maximum CS–UCS interval believed to be effective in establishing a CR for the response system under observation. Under such a procedure, an examination of the responses occurring on CS trials (excluding responses in the alpha latency range) yields a summative measure of pseudo-conditioned and baseline responses, or possibly some synergistic action of both stimuli.

Use of the unpaired control procedure to estimate nonassociative contributions to responding is based on the assumptions that (1) temporal contiguity of the CS and UCS is the necessary (and perhaps sufficient) condition for CR acquisition, and (2) any responding produced by the unpaired procedure is nonassociative in nature, since the randomized sequencing of CS and UCS presentations and extremely large intervals between stimulus events would preclude the operation of any CS–UCS contiguity effects. However, associative theory and its control methodology in classical conditioning was challenged by the assertion that associative learning could be reconceptualized in molar terms, in which the statistical relation between the CS and UCS is the fundamental determinant of response acquisition (Prokasy 1965; Rescorla 1967). In particular, a contingency hypothesis was proposed (Rescorla 1967) that focused on the degree to which the CS "predicts" or carries "information" about the UCS, as specified in terms of the probability of UCS occurrence in the presence and absence of the CS. Furthermore, the contingency hypothesis assumed that if the probability of a UCS is greater in the presence of the CS than its absence, a "positive contingency" would prevail and "excitatory" associative effects would accrue to the CS; and conversely, "inhibitory" associative effects would presumably accrue if the probability of a UCS were higher in the absence of the CS than in its presence to yield a "negative contingency." Therefore, an unpaired control condition, in which there is a perfect negative contingency, was expected to be inhibitory in its consequences. Hence, to provide an associatively neutral condition against which to assess inhibitory conditioning as well as excitatory conditioning, Rescorla (1967) proposed the "truly random control," which was variously specified in terms of independent programming of the CS and UCS (p. 74) or, more precisely, in terms of equal probabilities of UCS occurrence in the "presence" and "absence" of the CS (p. 76). Presumably, the "truly random" control would produce no associative effects since the CS is "irrelevant" to the UCS. In contrast, under the pairings hypothesis, the "chance" CS–UCS pairings occurring with the "truly random" control would be expected to produce positive associative effects.

While both CS–UCS "pairing" and "contingency" describe features of the sequence of stimulus durations and interstimulus intervals in a classical conditioning session, the precise specification of "pairing"/"unpairing" or the value of a "contingency" for a given preparation cannot be arrived at through a priori arguments but relies on extensive *empirical* knowledge. At a minimum,

"CS–UCS pairing" denotes the range of CS–UCS intervals that are efficacious in producing response acquisition, and, by the same token, the "explicitly unpaired" procedure also relies on a delineation of the effective range of CS–UCS intervals for a given preparation. Although contingency hypotheses are couched in terms of the statistical relations between the CS and UCS, the *operational* implementation of the truly random control must acknowledge CS–UCS interval effects to identify what constitutes the effective "presence" or "absence" of the CS at the time of UCS occurrence. Yet Rescorla's (1967) "CS–UCS contingency" hypothesis, rather than explicitly acknowledge CS–UCS interval effects, proposed that the statistical regularities among a sequence of CSs and UCSs are the determinants of responding; and, accordingly, the "truly random" control would provide an associatively neutral condition.

Subsequent to Rescorla's (1967) proposal, research with truly random control procedures has revealed that substantial conditioning ("excitation") can be obtained depending on the placement of "chance CS–US pairings" within the overall schedule of events (Benedict and Ayres 1972) and on the number of "chance CS–US pairings" (Ayres et al. 1975; Kremer and Kamin 1971; Quinsey 1971). Since these effects demonstrated the potent consequences of occasional "CS–US pairings," they provided substantial support for associative accounts based on trial-by-trial CS and UCS interstimulus relations rather than accounts based on the overall statistical properties of the event sequence. Moreover, as formulated, the contingency hypothesis requires the organism to possess a rather complex computational apparatus to calculate the CS/UCS contingency. Accordingly, since the organism cannot be supplied directly with this contingency, the scope of the hypothesis is largely limited to asymptotic predictions of performance after sufficient time and events have interceded to permit the organism to calculate the CS/UCS contingency. Rescorla and his associates (e.g., Holland and Rescorla 1975; Rescorla 1973; Rescorla and Cunningham 1978), apparently recognizing the difficulties involved in the use of the "truly random" control, have reverted to the more traditional unpaired control procedure. Unfortunately, many investigators, perhaps unaware of the shift, have persisted in the use of the truly random control despite the attendant methodological and conceptual difficulties.

MECHANISMS OF CONDITIONING

CS trace hypotheses

Although associative doctrine has guided classical conditioning research, investigations in the 1930s, 1940s, and 1950s revealed that the empirical laws of classical conditioning with CS–CR paradigms depart in important ways from a simple law of contiguity. In brief, it has been found that CR acquisition can reliably occur when the CS and UCS are temporally separated. This fundamental challenge to the law of contiguity was originally raised by Pavlov's discovery of "trace conditioning." A second challenge to a contiguity principle has been the observation that strict simultaneity between a CS and UCS yields little or no responding, whereas asynchronous presentation of the CS and UCS produces the most reliable CR acquisition (e.g., Bernstein 1934; Kimble 1947; McAllister 1953). In addition to the problems of trace conditioning and stimulus asynchrony, the topographic characteristics of the behavioral CR have placed a burden on the conversion of associative doctrine to a scientific theory. Specifically, it has been long recognized that the form of the CR differs in many respects from the UCR (e.g., Hilgard 1936a,b). Most noticeably, the CR is initiated in advance of the UCS, rather than coinciding with the time of UCS application. Accordingly, we will consider these challenges to the law of contiguity as revealed by the characteristics of anticipatory CRs and the effects of stimulus asynchrony and CS properties on conditioning of the rabbit's nictitating membrane response (NMR). We shall also consider how CS trace accounts have been developed to meet these empirical challenges to contiguity.

The anticipatory response. The most ubiquitous feature of the acquired behavior in both classical and instrumental conditioning is its forward movement to stimuli antedating the reinforcing event. Although the classical conditioning paradigm is regarded as providing an ideal vehicle for demonstrating the occurrence of anticipatory responses, detailed measurements of their quantitative characteristics have often been obscured by base rate, alpha, and pseudoconditioned responses. However, contributions from these nonassociative sources have been negligible (2–4 percent) in all of our rabbit conditioning preparations – namely, eyelid (Schneiderman et al. 1962), nictitating membrane (Gormezano et al. 1962), eyeball retraction (Deaux and Gormezano 1963), jaw movement (Smith et al. 1966), and heart rate response (Schneiderman et al. 1966). Accordingly, it has been possible to observe clearly the emergence of anticipatory CRs. In the NMR preparation, in particular, examination of CR latency data invariably reveals that the first CRs are initiated at or just before the time of UCS onset, but then as conditioning progresses, CR initiation shifts to progressively earlier portions of the CS–UCS interval. The systematic decrease in CR latency, which has constituted a primary quantitative feature of the anticipatory CR in all of our investigations, led us to examine CS trace accounts of classical conditioning (see Hull 1952; Pavlov 1927) as a source of insight into the processes governing the emergence of the anticipatory CR and for deducing the systematic decrease in CR latency.

Simply expressed, unless one is willing to credit animals and humans with precognizance, anticipatory CRs must be tied to an interaction between prior training history and some aspect of the stimulus conditions immediately antedating the CR. Pavlov (1927, pp. 39–40), in attempting to provide stimulus antecedents to the "empty" interval between CS offset and UCS onset, under his "trace conditioning" procedure, proposed that the CS left a "trace," that is, perseverative representation in the central nervous system. Accordingly, he postulated that the portion of the trace immediately antedating the CR was the effective instigator of the response. However, more recent CS trace accounts, in attempting to address the apparent necessity of temporal asynchrony for conditions of CS–CR preparations and observed maxima, have postulated that CS onset initiates a molar stimulus trace that rises in intensity to a maximum some time after CS onset, after which the trace gradually decays back to a null value (Anderson 1959; Gormezano 1972; Hull 1943, 1952).

Associative strength is presumed to accrue at the point of contiguity between the CS trace and UCS–UCR occurrence, and the increment in associative strength on each trial is presumed to be a direct function of the intensity of the CS trace contiguous with the UCS–UCR. Hence, to explain the occurrence of anticipatory CRs, these trace formulations assume that anticipatory CRs result from generalization along the intensity dimension from the point of CS trace and UCS–UCR contiguity to earlier portions of the trace. Moreover, as associative strength at the point of contiguity increases, generalization would be expected to extend further along the intensity dimension and, accordingly, the CS trace. Thus, consistent with expectations from the CS trace hypothesis, the first CRs occur near the UCS after which CR latency decreases toward the onset of the CS (see Gormezano 1972; Gormezano and Kehoe 1981).

Stimulus asynchrony. According to the more recent CS trace accounts of conditioning (Anderson 1959; Gormezano 1972; Hull 1943, 1952), the largest increments in associative strength would result from training at those forward CS–UCS intervals for which the CS trace is at a high or maximum intensity at the time of the UCS–UCR. At shorter or longer CS–UCS intervals, the CS trace is presumed to be too weak to produce appreciable increments in associative strength (or to be altogether absent). Consequently, the form of the empirical interstimulus interval–CR frequency function is postulated to reflect the variation in the intensity of the CS trace over time. In research with the rabbit NMR preparation, the trace hypothesis has served as a heuristic guide in generating a considerable body of data that, in turn, has provided indirect but converging evidence for a stimulus trace (Gormezano 1972; Gormezano and Kehoe 1981). For example, systematic manipulations of the CS–UCS interstimulus interval (ISI) in an extensive series of rabbit NMR studies (Gormezano 1972; Kehoe 1979; Schneiderman 1966; Schneiderman and Gormezano 1964; Smith 1968; Smith et al. 1969) have typically yielded concave-shaped functions between CR frequency and ISI, as shown in Figure 1.1 for the trace conditioning study of Smith et al. (1969).

It should be noted that although the "stimulus trace" has carried sensory and neurophysiological connotations, the form and characteristics of the "trace" are actually behavioral constructs defined in a given preparation under a given set of conditions (Gormezano 1972; Gormezano and Kehoe 1981). A case-by-case behavioral anchoring of the "trace" has been necessary since the divergence in ISI–CR frequency functions dashed early hopes that their form would be invariant over species and response systems (see Gormezano and Moore 1969, pp. 135–138; Hall 1976, p. 110). For example, within the rabbit preparations, the ISI–CR frequency functions for the jaw movement and heart rate response systems are substantially broader than, and may even have different optimal values from, the function for the NMR preparation (Gormezano 1972; Schneiderman 1972). Since these divergences in ISI effects occur in the same species, it is clear then that the neural counterparts of the "CS trace" construct must lie only partly in the sensory system.

CS manipulations. Our initial attempts to manipulate the CS trace were guided by Hull's (1943) conceptual appeal to the rate of neural firings as the

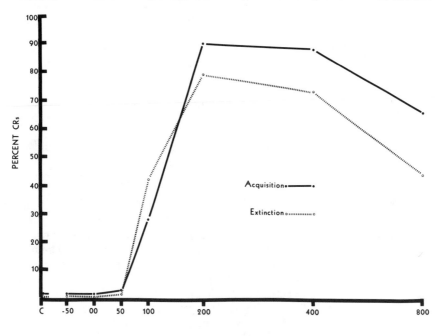

INTERSTIMULUS INTERVAL IN MSEC

Figure 1.1. The mean percentage of NMRs in blocks of 21 test trials (days) in acquisition and in blocks of 20 trials in extinction as a function of CS–UCS interval (msec). On each of eight daily acquisition sessions, the seven paired groups ($n = 12$) received 80 CS–UCS pairings at CS–UCS intervals of 800, 400, 200, 100, 50, 0, and -50 msec. In addition, a control (C) group ($n = 12$) received unpaired presentations of 80 CS-alone and 80 UCS-alone trials. The CS was a 50-msec, 92-dB, 1000-Hz tone, and the UCS was a 50-msec, 4-mA, 60-Hz shock.

physical basis for the intensity of the neural trace. Accordingly, a series of experiments were conducted using manipulations of the CS that might be expected to affect the frequency of neural firings. For instance, through activation of on–off fibers, intermittent stimuli should generate a greater number of neural firings contiguous with the UCR to produce more effective conditioning. By the same token, the offset of a stimulus should, by itself, produce a brief train of neural firings and hence a weaker trace relative to the onset of a delay CS. In agreement with trace expectations, a pulsed 800-sec tone CS (50 msec on, 50 msec off) produced faster CR acquisition than a constant CS, and the offset of a tone stimulus present during the intertrial interval produced slower conditioning than the same tone in a delay conditioning procedure (Gormezano 1972, pp. 156–158). Similarly, trace conditioning, in which the CS is brief, produces slower CR acquisition than a delay conditioning procedure (Schneiderman 1966). In further agreement with the trace hypothesis, CR acquisition has been observed to be a direct function of the intensity of a tone

CS over the range of 65 to 86 dB (Gormezano 1972 pp. 157–159; Scavio and Gormezano 1974).

In an interesting assessment of trace formulations, Patterson (1970) manipulated the time course of the trace by using electrical stimulation of the rabbit's inferior colliculus as a CS. Such a stimulus, in bypassing a portion of the afferent system, should reduce the initial recruitment time of the CS trace and thereby foreshorten the minimal ISI necessary for conditioning. Consistent with the trace accounts, Patterson (1970) obtained a substantial level of responding at an ISI of 50 msec, a value that had yielded no evidence of NMR conditioning when a tone CS was used (Smith et al. 1969) and that could not be attributed to a dynamogenic effect, since the intensity of the intracranial CS (38 μA) was one that had been demonstrated to support a lower level of responding than the tone CS. Another means of behaviorally anchoring the form of the trace has involved determining the NMR's ISI–UCR amplitude (reflex modification) function (Ison and Leonard 1971; Thompson 1976). The function is obtained by measuring UCR amplitudes to threshold-intensity UCSs preceded by a tone or light CS at various ISIs and with the total number of "CS – UCS" trials restricted to preclude the appearance of CRs. In general, it has been found that UCR amplitude varies as a function of the ISI, and, in terms of the CS trace hypothesis, the effects of the CS on the UCR can be construed as reflecting the inherent dynamogenic effects of CS trace intensity at the time of UCR occurrence. In particular, Thompson (1976) has obtained an ISI–UCR amplitude function that closely corresponds to the concave ISI–CR frequency function depicted in Figure 1.1.

Mediated associations

Over a broad range of parameters, our NMR preparation has consistently revealed a tightly bound range of CS–UCS intervals for conditioning with maxima between 200 and 400 msec and levels of responding declining to negligible levels as the ISI approaches 3–4 sec (see Gormezano et al. 1982). Accordingly, as a model preparation with a clearly delineated contiguity gradient, our NMR preparation has readily permitted us to identify classical conditioning mechanisms (e.g., higher-order conditioning, sensory conditioning) that could operate to yield CRs to distal components of a series of stimuli at intervals well beyond the bounds of a CS–UCS contiguity gradient. Specifically, an extensive series of NMR serial compound conditioning studies have provided us with support for associative mediational processes that could serve to extend CR acquisition to component CSs lying far beyond the bounds of the empirically determined single-CS ISI–CR frequency function (see Gormezano and Kehoe 1980). Briefly, our studies revealed that substantial acquisition of the rabbit's NMR can be obtained to a stimulus (CS1) temporally remote from the UCS when it is placed in a reinforced serial compound (CS1–CS2–UCS) in which the second stimulus (CS2) is more contiguous with the UCS.

In our initial serial compound conditioning experiment (Kehoe et al. 1979, experiment 1), the CS2–UCS interval was fixed at an optimal value of 350 msec, while the CS1–UCS interval was varied up to a value of 2750 msec, an

ISI that, by itself, yielded an asymptotic level of CRs of less than 25 percent and yet in the serial compound (CS1–CS2–UCS) the level of responding to CS1 rose over the course of 6 days (360 trials) to a maximum level near 80 percent CRs. Moreover, control experiments (experiments 2 and 3) revealed that the 80 percent level of CS1 responding could not be attributed to direct conditioning brought about by CS1–UCS presentations, cross-modal generalization from CS2–UCS training, and/or any synergistic effect of mixed CS1–UCS and CS2–UCS trials. Another experiment (experiment 4) revealed a similar pattern of CS1 responding in serial compounds at CS1–UCS intervals ranging over the values of 4750, 8750, and 18,750 msec. Thus, even when CS1 was located at CS–UCS intervals many times longer than values that produce even slight evidence of acquisition of NM CRs to a single CS (i.e., 4 sec), a relatively simple two-element serial compound promoted the acquisition of CRs to CSs at a CS–UCS interval of almost 19 sec. Subsequently, another study was undertaken to provide a direct demonstration that CR acquisition to CS1 in a serial compound resulted from mediated associations from CS2 based on CS1–CS2 and CS2–UCS training inherent in the serial compound training (Kehoe pers. comm.). In the investigation, a group that received both CS1–CS2 and CS2–UCS pairings (group *P–P*) showed CR acquisition to a level that was substantially higher than that of any of the three control groups – namely, group *U–P*, which received unpaired CS1/CS2 presentations and CS2–UCS pairings; group *P–U*, which received CS1–CS2 pairings and unpaired CS2/UCS presentations; and group *U–U*, which received all stimuli in unpaired presentations.

Having demonstrated that separate CS1–CS2 and CS2–UCS pairings could reliably produce CR acquisition to CS1, we examined the effects of manipulating the CS1–CS2 interval. Each of the 16 days of acquisition training consisted of 30 CS2–UCS trials interspersed with 30 CS1–CS2 trials with the CS2–UCS interval being fixed at 400 msec, while the CS1–CS2 interval was manipulated at CS1–CS2 intervals of 400, 1400, 2400, 4400, and 8400 msec, respectively. The groups were designated by their respective CS1–CS2 "trace intervals" between the offset of CS1 and the onset of CS2 of 0, 1, 2, 4, and 8 sec, respectively. To control for cross-modal generalization from CS2 to CS1 and any nonassociative contributions arising from the UCS, a group labeled *"UP"* received 30 unpaired presentations each of CS1 and CS2 interspersed among CS2–UCS trials. In addition, all groups received two daily nonreinforced test trials each to CS1 and CS2. The experiment revealed, as seen in parts *A* and *C* of Figure 1.2, that CS1 responding was a negative, decelerated function of the CS1–CS2 trace interval, whereas CS2 responding, as shown in part *B*, was relatively unaffected by trace interval. Thus, responding to CS1 was highest at the 0 trace interval and rapidly declined as the CS1– CS2 trace interval increased up to 2 sec, and then, at longer intervals, responding stabilized at a level substantially higher than that of the unpaired control. Accordingly, the study would appear to indicate that CR acquisition to CS1 in a mediated (second-order) CS1–CS2 association follows a "contiguity gradient" that is similar but more extended than that seen in the ISI–CR frequency functions to single CSs for the rabbit NMR preparation (see Gormezano 1972; Schneiderman and Gormezano 1964).

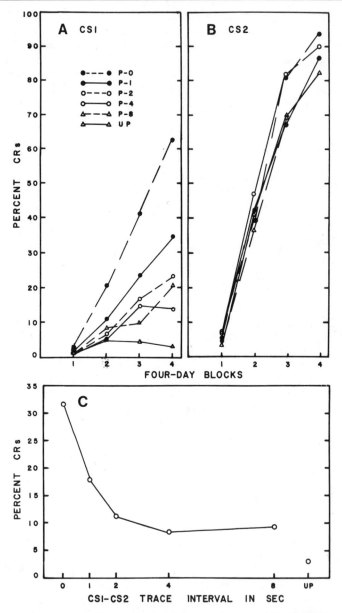

Figure 1.2. The mean percentage of NM CRs as a function of CS1–CS2 trace interval (msec) to CS1 (*A*) and CS2 test trials (*B*) in 4-day blocks and to CS1 test trials across all days of training (*C*). On each of 16 daily acquisition sessions, the five paired groups (*n* = 12) received 30 CS2–UCS trials, at a CS–UCS interval of 400 msec, interspersed with 30 CS1–CS2 trials manipulated across groups at values of 400, 1400, 2400, 4400, and 8400 msec. An unpaired (*UP*) control group received 30 unpaired presentations each of CS1 and CS2 interspersed among 30 paired CS–UCS trials.

CLASSICAL CONDITIONING AND PSYCHOPHARMACOLOGY

Component analysis of drug effects

The earliest research regarding the behavioral effects of pharmacological agents was conducted by Pavlov (1927, p. 35), who examined acquisition of the overt responses produced by apomorphine to the cues arising from the antecedent injection procedure. Despite Pavlov's pioneering efforts, drug research with CS–CR procedures has been rare compared to the numerous experiments using instrumental procedures. Yet because of the experimenter's control over presentation of the CS and UCS, the specificity of the target response, and its well-developed control methodology, classical conditioning would appear to have a great deal to offer investigators interested in detailing the behavioral effects of drug interventions. Accordingly, we have begun to use the NMR preparation to delineate the mode of action of purported psychotropic drugs by examining their possible effects on three components of conditioning: (1) sensory processing of stimuli, (2) learning processes, and (3) motor functioning.

It is important to note that the above three-component framework is orthogonal to the long-standing but deceptive distinction between "learning" and "performance" frequently invoked in current drug research. On one hand, the learning–performance distinction has been useful as an acknowledgment that behavior at any given moment is determined in a multiple fashion by current, relatively temporary organismic states that may mask or reveal the more permanent consequences of past experience. On the other hand, the learning–performance distinction has fostered the illusion that it is easy to ascertain the relatively irreversible and reversible effects of any given variable. In fact, not all behavior theories acknowledge a learning–performance distinction (see Brown 1961, pp. 99; Guthrie 1959), and those that do acknowledge the distinction also recognize that many variables can enter both the learning and performance process. Thus, in Hull's (1943, 1952) theory, for example, CS intensity is a performance variable in that it has a dynamogenic (potentiating) effect on current behavior; however, CS intensity is also an associative variable that determines, in part, the increment in associative strength on each reinforced trial. Similarly, it is possible that a given drug could have multiple temporary and permanent effects that interact with each other. Accordingly, it is our expectation that by first delineating the component process(es) affected by a drug, the later determination of the relatively temporary or permanent nature of the drug effects will be easier to determine.

Locus of drug effects

Our procedures for assessing drug effects was developed in connection with our investigation of LSD. In brief, the strategy has been one of progressive refinement in the localization of the drug's effect. The initial assessment took the form of a dose–response curve determined for the effects of LSD on the acquisition of CRs under daily doses of 0 (saline), 1, 10, 30, 100, and 300 nmol/kg intravenously (Gimpl et al. 1979). Each dosage group received 10 days of

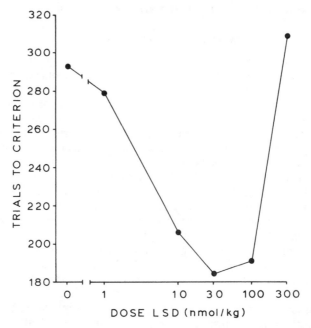

Figure 1.3. Effects of LSD on rate of acquisition of NM CRs. Data are expressed as mean number of trials preceding the first occurrence of 10 consecutive CRs as a function of LSD dosage. The zero dosage represents vehicle controls. Each point is the mean of 12 rabbits, and the "*" indicates significant mean differences from vehicle control ($p < .05$) as calculated by the method of Tukey.

acquisition training involving 30 tone–shock pairings intermixed with 30 light–shock pairings per day. A parallel set of six unpaired control groups were given the same drug dosages under acquisition training consisting of 30 tone-alone, 30 light-alone, and 60 shock-alone trials presented in a restricted random unpaired fashion. As seen in Figure 1.3, the paired groups revealed a biphasic dose–response curve in which dosages of 1 to 100 nmol/kg significantly enhanced CR acquisition, with the maximal enhancement occurring at 30 nmol. However, the 300-nmol/kg dose retarded initial CR acquisition relative to the vehicle control. On the other hand, for the unpaired groups, the level of responding to either the tone or light CS was negligible (2–4 percent) across all dosage levels. Thus, the enhancing effect of LSD on CR acquisition could not be attributed to such nonassociative factors as an elevation in the base rate of responding, sensitization, or pseudoconditioning. Furthermore, since no differences were apparent between responding to the tone versus light CS in either the paired or unpaired groups, LSD's effects were not confined to input from a particular sensory modality.

Although the paired-versus-unpaired comparison indicated that LSD's effects acted through the contiguity component of the learning process, further assessment was required to determine whether the action of LSD would be

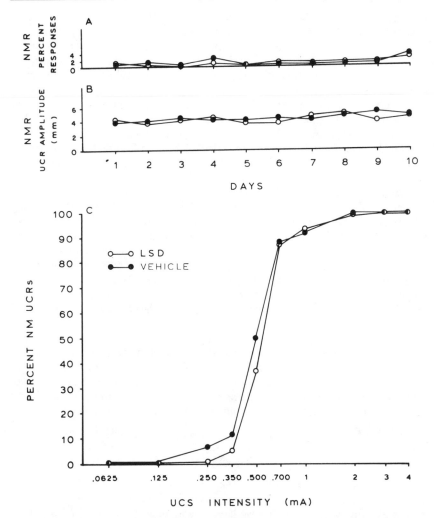

Figure 1.4. The frequency of baseline responding (*A*) and UCR amplitude (*B*) under LSD (30 nmol/kg) or vehicle for groups receiving unpaired CS and UCS trials, and the frequency of UCRs under LSD or vehicle as a function of UCS intensity (*C*).

characterized as altering the sensory processing of the stimuli, directly altering the effectiveness of the presumed associative connection, and/or facilitating motor output. However, an examination of the effects of LSD on the UCS–UCR relationship in the unpaired control groups on the 3-mA shock UCS-alone trials revealed no evidence of any effect of LSD, at any dosage level, upon UCR amplitude or frequency (see parts *A* and *B* of Figure 1.4). Moreover, in another experiment conducted to determine the effects of LSD (30 nmol/kg) on UCR amplitude for shock intensities ranging from 0.0625 mA to 4 mA,

LSD failed to lower the threshold for UCR evocation (see part *C* of Figure 1.4) or to alter UCR amplitude (Gormezano and Harvey 1980, experiment 2). Accordingly, in terms of the three-component model, the examination of the UCS–UCR function suggests that LSD did not affect sensory processing of the UCS or motor functioning of the UCR to provide facilitated conditioning. Consequently, an assessment of the effects of LSD on the sensory processing of the CS in relative isolation from the learning process was conducted by examining CR likelihood when CS intensity is varied after CR acquisition has occurred (Gormezano and Harvey 1980, experiment 3). Postacquisition manipulation of the CS intensity revealed, as seen in Figure 1.5, that the 30-nmol dose raised the likelihood of a CR by approximately 10 percentage points across the range of CS intensities, thus indicating a lowering of the sensory threshold for evoking a CR to a tone CS.

Since LSD administration augmented CR acquisition and CS intensity effects on CR performance but not measures of the UCS–UCR, it has been possible to conclude that LSD has effects functionally equivalent to increasing the physical intensity or "salience" of the CS. In more theoretical terms, LSD may be said to have a dynamogenic effect that (1) augments the increment in associative strength on each CS–UCS trial and (2) potentiates current responding. Thus, an alternation in the sensory processing of the CS ramifies into the associative process. Recently, we have similarly assessed the effects of haloperidol, and we have found converging evidence that it impairs CR acquisition by depressing CS processing (Harvey and Gormezano 1981). Consequently, our procedures are sensitive to both incremental and decremental effects of drugs. However, additional questions may be raised regarding the nature of the apparent alterations in CS processing by these drugs. Accordingly, we are currently in the process of further elucidating their modes of action by the systematic employment of other conditioning procedures (e.g., conditional discriminations).

CONCLUSION

Our original motive for developing the rabbit preparations was to obtain classical conditioning data of a quality suitable for studying associative learning processes. However, I hope that we have also demonstrated that our basic CS–CR preparations, and the NMR preparation in particular, are powerful tools for investigating a wide range of biological and psychological questions. In particular, Pavlov's intentions of using classical conditioning to study the neural pathways and processes of behavioral adaptation would appear to be on the way to being fulfilled. Frankly, it has been gratifying to see neurobehavioral investigators (e.g., Thompson 1976) adopting the rabbit NMR preparation as a model preparation, on the basis of its well-documented and robust parametric effects. Clearly, the use of the NMR preparation where the behavioral methods and laws are relatively well delineated leaves the neurobiological researcher free to discover the underlying neural processes with confidence that the behavioral phenomena are robust. For similar reasons, the use of the NMR preparation in psychopharmacology would appear to be promising.

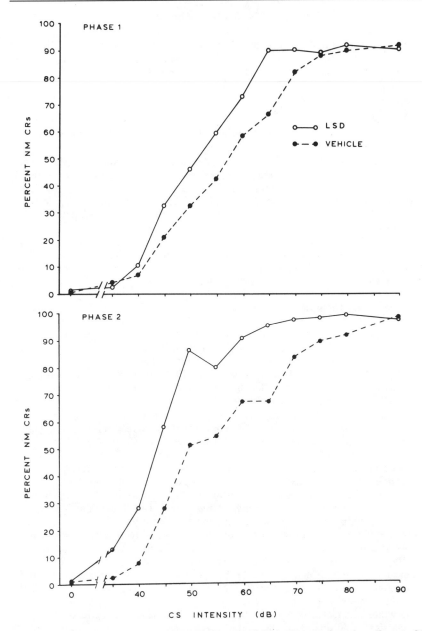

Figure 1.5. The mean percentage of CRs in 15-trial blocks as a function of tone–CS intensity under LSD or vehicle subsequent to 10 days of acquisition training (phase 1) and under drug reversal conditions (phase 2) initiated 6 days later.

ACKNOWLEDGMENTS

The development of our rabbit conditioning preparations has been supported by grants from the National Science Foundation and the research reported by grants from the National Science Foundation and NIMH Grant MH16841-15.

REFERENCES

Anderson, N. H. 1959. Response emission in time with applications to eyelid conditioning. In *Studies in Mathematical Learning Theory.* (R. R. Bush & W. K. Estes, eds.), pp. 125–134. Stanford: Stanford University Press.

Ayres, J. J. B., Benedict, J. O., and Witcher, E. S. 1975. Systematic manipulation of individual events in a truly random control in rats. *J. Comp. Physiol. Psychol.* *88*:97–103.

Benedict, J. O., and Ayres, J. J. B. 1972. Factors affecting conditioning in the truly random control procedure in the rat. *J. Comp. Physiol. Psychol. 78*:323–330.

Bernstein, A. L. 1934. Temporal factors in the formation of conditioned eyelid reactions in human subjects. *J. Gen. Psychol. 10*:173–197.

Brown, J. S. 1961. *The Motivation of Behavior.* New York: McGraw-Hill.

Cegavske, C. F., Thompson, R. F., Patterson, M. M., and Gormezano, I. 1976. Mechanisms of efferent neuronal control of the reflex nictitating membrane response in the rabbit. *J. Comp. Physiol. Psychol. 90*:411–423.

Deaux, E. G., and Gormezano, I. 1963. Eyeball retraction: Classical conditioning and extinction in the albino rabbit. *Science 141* (3581):630–631.

Gimpl, M. P., Gormezano, I., and Harvey, J. A. 1979. Effects of LSD on learning as measured by classical conditioning of the rabbit nictitating membrane response. *J. Pharmacol. Exp. Ther. 208*:330–334.

Gormezano, I. 1965. Yoked comparisons of classical and instrumental conditioning of the eyelid response; and an addendum on "voluntary responders." In *Classical Conditioning: A Symposium* (W. F. Prokasy, ed.), pp. 48–70. New York: Appleton-Century-Crofts.

    1966. Classical conditioning. In *Experimental Methods and Instrumentation in Psychology.* (J. B. Sidowski, ed.), pp. 385–420. New York: McGraw-Hill.

    1972. Investigations of defense and reward conditioning in the rabbit. In *Classical Conditioning.* vol. 2, *Current Research and Theory.* (A. H. Black and W. F. Prokasy, eds.), pp. 151–181. New York: Appleton-Century-Crofts.

    1980. Pavlovian mechanisms of goal-directed behavior. In *Neural Mechanisms of Goal-Directed Behavior and Learning.* (R. F. Thompson, L. H. Hicks, and V. B. Shvyrkov, eds.), pp. 39–56. New York: Academic Press.

Gormezano, I., and Coleman, S. R. 1975. Effects of partial reinforcement on conditioning, conditional probabilities, asymptotic performance, and extinction of the rabbit's nictitating membrane response. *Pavlov. J. Biol. Sci. 10*:80–89.

Gormezano, I., and Harvey, J. A. 1980. Sensory and associative effects of LSD in classical conditioning of the rabbit *(Oryctolagus Cuniculus)* nictitating membrane response. *J. Comp. Physiol. Psychol. 94*:641–649.

Gormezano, I., and Kehoe, E. J. 1975. Classical conditioning: Some methodological-conceptual issues. In *Handbook of Learning and Cognitive Processes,* vol. 2 (W. K. Estes, ed.), pp. 143–179. Hillsdale, N.J.: Erlbaum.

    1981. Classical conditioning and the law of contiguity. In *Advances in Analysis of*

Behavior. vol. 2, *Predictability, Correlation, and Contiguity.* (P. Harzem and M. D. Zeiler eds.), pp. 1–45. New York: Wiley.

Gormezano, I., Kehoe, E. J., and Marshall B. S. 1982. Twenty years of classical conditioning research with the rabbit. In *Progress in Psychobiology and Physiological Psychology.* (J. M. Sprague and A. N. Epstein, eds.), New York: Academic Press.

Gormezano, I., and Moore, J. W. 1969. Classical conditioning. In *Learning: Processes.* (M. H. Marx, ed.). New York: Macmillan.

Gormezano, I., Schneiderman, N., Deaux, E. G., and Fuentes, I. 1962. Nictitating membrane: Classical conditioning and extinction in the albino rabbit. *Science 138*:33–34.

Gray. T. S., McMaster, S. E., Harvey, J. A., and Gormezano, I. 1981. Localization of retractor bulbi: Motoneurons in the rabbit. *Brain Res. 226*:93–106.

Guthrie, E. R. 1959. Association by contiguity. In *Psychology: A Study of a Science,* vol. 2 (S. Koch, ed.). New York: McGraw-Hill.

Hall, J. F. 1976. *Classical Conditioning and Instrumental Learning: A Contemporary Approach.* Philadelphia: Lippincott.

Harvey, J. A., and Gormezano, I. 1981. Effects of haloperidol and pimozide on classical conditioning of the rabbit nictitating membrane response. *J. Pharmacol. Exp. Ther. 218*:712–719.

Hilgard, E. R. 1936a. The nature of the conditioned response. I. The case for and against stimulus substitution. *Psychol. Rev. 43*:366–385.

1936b. The nature of the conditioned response. II. Alternatives to stimulus substitution. *Psychol. Rev. 43*:547–564.

Holland, P. C., and Rescorla, R. A. 1975. Second-order conditioning with food unconditioned stimulus. *J. Comp. Physiol. Psychol. 88*:459–467.

Hull, C. L. 1943. *Principles of Behavior.* New York: Appelton-Century-Crofts.

1952. *A Behavior System.* New Haven: Yale University Press.

Ison, J. R., and Leonard, D. W. 1971. Effects of auditory stimuli on the amplitude of the nictitating membrane reflex of the rabbit *(Oryctolagus cuniculus). J. Comp. Physiol. Psychol. 75*:157–164.

Kehoe, E. J. 1979. The role of CS–US contiguity in classical conditioning of the rabbit's nictitating membrane response to serial stimuli. *Learn. Motiv. 10*:23–28.

Kehoe, E. J., Gibbs, C. M., Garcia, E., and Gormezano, I. 1979. Associative transfer and stimulus selection in classical conditioning of the rabbit's nictitating membrane response to serial compound CSs. *J. Exp. Psychol.: Anim. Behav. Proc. 5*:1–18.

Kimble, G. A. 1947. Conditioning as a function of the time between conditioned and unconditioned stimuli. *J. Exp. Psychol. 37*:1–15.

Kremer, E. F., and Kamin, L. J. 1971. The truly random control procedure: Associative or nonassociative effects in rats. *J. Comp. Physiol. Psychol. 74*:203–210.

Lashley, K. S. 1916. The human salivary reflex and its use in psychology. *Psychol. Rev. 23*:446–464.

Longo, N., Klempay, S., and Bitterman, M. E. 1964. Classical appetitive conditioning in the pigeon. *Psychonom. Sci. 1*:19–20.

McAllister, W. R. 1953, Eyelid conditioning as a function of the CS–UCS interval. *J. Exp. Psychol. 45*:417–422.

Patterson, M. M. 1970. Classical conditioning of the rabbit's *(Oryctolagus cuniculus)* nictitating membrane response with fluctuating ISI and intracranial CS. *J. Comp. Physiol. Psychol. 72*:193–202.

Pavlov, I. P. 1927. *Conditioned Reflexes* (G. V. Anrep, trans.). London: Oxford University Press.

Prokasy, W. F. 1965. Classical eyelid conditioning: Experimenter operations, task demands, and response shaping. In *Classical Conditioning* (W. F. Prokasy ed.), pp. 48–70. New York: Appelton-Century-Crofts.

Quinsey, V. L. 1971. Conditioned suppression with no CS–US contingency in the rat. *Can. J. Psychol./Rev. Can. Psychol. 25*:69–82.

Rescorla, R. A. 1967. Pavlovian conditioning and its proper control procedures. *Psychol. Rev. 74*:71–80.

1973. Second-order conditioning: Implications for theories of learning. In *Contemporary Approaches to Conditioning and Learning.* (F. J. McGuigan and D. B. Lumsden, eds.), Washington, D.C.: Winston.

Rescorla, R. A., and Cunningham, C. L. 1978. Within-compound flavor associations. *J. Exp. Psychol.: Anim. Behav. Proc. 4*:267–275.

Scavio, M. J., and Gormezano, I. 1974. CS intensity effects upon rabbit nictitating membrane conditioning, extinction, and generalization. *Pavlov. J. Biol. Sci. 9*:25–34.

Schneiderman, N. 1966. Interstimulus interval function of the nictitating membrane response in the rabbit under delay versus trace conditioning. *J. Comp. Physiol. Psychol. 62*:397–402.

1972. Response system divergences in aversive classical conditioning. In *Classical Conditioning,* vol. 2, *Current Theory and Research.* (A. H. Black and W. F. Prokasy, eds.), New York: Appelton-Century-Crofts.

Schneiderman, N., Fuentes, I., and Gormezano, I. 1962. Acquisition and extinction of the classically conditioned eyelid response in the albino rabbit. *Science 136*:650–652.

Schneiderman, N., and Gormezano, I. 1964. Conditioning of the nictitating membrane of the rabbit as a function of CS–US interval. *J. Comp. Physiol. Psychol. 57*:188–195.

Schneiderman, N., Smith, M. C., Smith, A. C., and Gormezano, I. 1966. Heart rate classical conditioning in rabbits. *Psychonom. Sci. 6*:39–40.

Schreurs, B. G., and Gormezano, I. 1980. Classical conditioning of the rabbit's nictitating membrane response to compound and component stimuli as a function of component pretraining. Paper presented at the 52nd Annual meeting of the Midwestern Psychological Association, St. Louis.

Skinner, B. F. 1938. *The Behavior of Organisms: An Experimental Analysis.* New York: Appelton-Century-Crofts.

Smith, M. C. 1968. CS–US interval and US intensity in classical conditioning of the rabbit's nictitating membrane response. *J. Comp. Physiol. Psychol. 66*:679–687.

Smith, M. C., Coleman, S. R., and Gormezano, I. 1969. Classical conditioning of the rabbit's nictitating membrane response at backward, simultaneous, and forward CS–US intervals. *J. Comp. Physiol. Psychol. 69*:226–231.

Smith. M. C., DiLollo, V., and Gormezano, I. 1966. Conditioned jaw movement in the rabbit. *J. Comp. Physiol. Psychol. 62*:479–483.

Thompson, R. F. 1976. The search for the engram. *Am. Psychol. 31*:209–227.

Woodruff, G., and Williams, D. R. 1976. The associative relation underlying autoshaping in the pigeon. *J. Exp. Anal. Behav. 26*:1–13.

2 · Comments on three Pavlovian paradigms

ROBERT A. RESCORLA

INTRODUCTION

This chapter discusses three somewhat complex Pavlovian conditioning paradigms: second-order conditioning, sensory preconditioning, and conditioned inhibition. None of these paradigms is new; in fact, each was originally studied by Pavlov. But their behavioral results do fit with modern concepts of Pavlovian conditioning. Many thinkers view Pavlovian conditioning as a way in which the organism represents relations among important events in its environment. Each of these paradigms illustrates the sophistication with which the organism can form such representations.

My reason for discussing these three paradigms here is twofold. First, it seems to me that their analytic advantages for the investigation of associative learning have not been fully appreciated. In particular, they have not been adequately exploited in the neurobiological study of learning. Second, some of the behavioral results that are already available from these paradigms should interest neurobiologists concerned with learning processes. Consequently, in what follows I describe each paradigm, then discuss some sample behavioral data resulting from its use in our laboratory, and finally mention some issues that each raises about the nature of associative learning.

SECOND-ORDER CONDITIONING

Early in his studies of conditioning, Pavlov (1927) discovered that a stimulus that signals an unconditioned stimulus (US) takes on an ability in addition to that of evoking new responses. Such a stimulus also develops the power to itself serve instead of the US to establish responding to other stimuli that it follows. Thus, an S_1 paired with a US could be then be paired with an S_2 with the result that a conditioned response (CR) would be established to S_2. Pavlov viewed the response to S_2 as second-order conditioning because S_2 was never itself directly paired with the US.

Although early conditioning results seemed to suggest that second-order conditioning was weak and temporary, more recent data give quite a different picture (see Rescorla 1980a). Substantial second-order conditioning has now been observed in a wide variety of paradigms, ranging from fear conditioning in rats (e.g., Rizley and Rescorla 1972), to "autoshaped" key pecking based on food USs in pigeons (e.g., Rashotte et al. 1977), to eyeblink conditioning in rabbits (Kehoe et al. 1981). There can remain little doubt that second-order conditioning is a powerful phenomenon readily observable over a broad range of conditions.

The simple occurrence of the phenomenon is of interest, because it suggests an ability of the organism to build on its prior knowledge, using associations to convert neutral stimuli into reinforcers. Moreover, it enables the psychologist to extend the theories originally developed for events of natural importance to a range of new, initially weak events. But much of the modern interest in the paradigm stems not from its occurrence but rather from its usefulness as a tool in the analysis of learning. Rescorla (1980a) has suggested ways in which second-order conditioning can be a powerful aid in analysis of three basic problems in the study of learning: what circumstances produce learning, what is the nature of the learning, and how that learning becomes evident in performance. This chapter will not rehearse these points in any detail. But it is worth briefly illustrating some of these uses both to demonstrate the power of second-order conditioning and to describe some samples of the particular information that they give about certain instances of associative learning.

Conditions producing learning

Let me begin with a sample of how second-order conditioning assists in analyzing the circumstances that produce learning. The special advantage in using second-order conditioning is the flexibility it permits in the selection of the things to be associated. This advantage is well illustrated by one program of research that has examined the possibility that perceptual relations among stimuli influence their associability. The best-studied relation is that of similarity, for which the question was originally raised by the Gestalt psychologist: Is it easier to associate two events if they are similar to each other? With simple Pavlovian associative preparations that is a difficult question to investigate because there are only a limited number of unconditioned stimuli available in any organism. With so few USs, one is correspondingly limited in the perceptual relations between CS and US that can be studied. However, second-order conditioning greatly expands the number of potent stimuli. In fact, one can select any two stimuli to which the organism is sensitive that have any perceptual relation one desires. One can then convert one of those stimuli into a reinforcer by pairing it with a US and study the subsequent ease of forming associations between the originally chosen stimuli. This greatly facilitates the study of perceptual relations in conditioning.

Using this logic, Rescorla and Furrow (1977) and Rescorla and Gillan (1980) found strong evidence that similarity promotes the formation of associations. Figure 2.1 shows the outcome of one such experiment carried out in pigeon subjects with an "autoshaping" procedure. Autoshaping is a Pavlovian preparation using a standard operant chamber in which the CS is the illumination of a keylight and the US is the delivery of food. Repeated pairings of those two events typically result in a conditioned response of a partcular form: pecking the illuminated key. Moreover, if one keylight (S_1) signals food, then

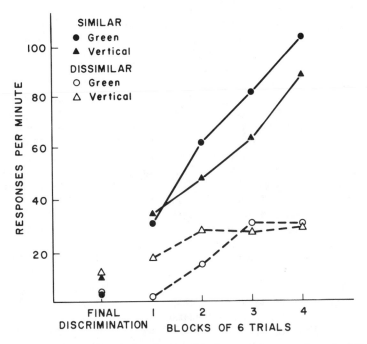

Figure 2.1 Acquisition of second-order conditioning to a green and vertical S_2 as a function of whether the reinforcing S_1 was similar or dissimilar to S_2. (From Rescorla and Furrow 1977)

preceding S_1 by another keylight (S_2) results in second-order conditioning of pecking to S_2. The results shown in Figure 2.1 are for responding to such an S_2 as it was repeatedly paired with an S_1. The substantial differences among the groups are due to variations in the degree of similarity of S_1 and S_2. When those stimuli were similar (color–color and orientation–orientation pairs), conditioning of S_2 was a good bit more rapid than when they were dissimilar (color–orientation and orientation–color pairs). Importantly, various details of the design (see Rescorla and Furrow 1977) help identify this superiority as due to differences in the formation of associations, rather than in such alternative processes as pseudoconditioning or stimulus generalization. Apparently, similars are more easily associated than are stimuli lacking a similarity relation.

This particular condition for the formation of associations is of interest because it suggests a preexperimental bias in the associations between certain kinds of events (homogeneous) rather than between others (heterogeneous). Modern behavioral theories of conditioning have largely ignored such variables. In this regard, the finding is like that popularized under the heading of "cue-to-consequence" (e.g., Garcia et al. 1966). However, the use of second-order conditioning has provided two advantages in the investigation of this possibility. First, as has been emphasized elsewhere (e.g., Bitterman 1975; Rescorla and Holland 1976; Schwartz 1974), it is actually exceedingly tricky to show that a variable of this sort affects the formation of associations rather than some other determinant of performance. One problem results from the fact that naturally occurring instances of "relatedness" between stimuli often have confounded features of the individual stimuli such as potency, duration, and the ability to produce pseudoconditioning. The contributions of these can be difficult to evaluate and separate from an impact of relatedness on associative learning. Such problems of analysis are more readily dealt with in the context of second-order conditioning because of the flexibility it permits in choosing stimuli. Second, by using second-order conditioning one can more easily carry out further analysis of the manner in which similarity has its impact.

Two recent observations in our laboratory will illustrate the latter point. First, we have recently found that the similarity that governs conditioning is defined not simply by the physical relation of the stimuli to be associated but rather by their similarity in comparison with the similarity of each to other events that the organism is experiencing. Two similar stimuli, S_2 and S_1, can be made more or less associable depending upon the similarity of each to other stimuli in the session. When those other events are unlike S_2 and S_1, the association proceeds rapidly; when those other events are like S_2 and S_1, the rate of associating S_2 with S_1 is depressed. That is, as suggested by the Gestalt psychologists, it is relative, rather than absolute, similarity that matters. It was the access that second-order conditioning provides to large numbers of stimuli that can bear arbitrary similarity relations to each other that made possible this finding. Second, we have also used second-order conditioning to ask whether similarity promotes the learning of a range of relations among events. A good illustration is the finding that similarity promotes learning not only when one event signals the occurrence of the other but also when one event signals the failure of the other to occur. That is, by using second-order conditioning one can find evidence that similarity promotes both excitatory and inhibitory learning (Rescorla 1980a). That suggests the possibility that similarity does not have its effect because of preexisting connections of a particular sort between the events. Rather it may serve as a kind of facilitator for the forming of *any* sort of connections. The two alternatives would have quite different implications at the biological level. They certainly have different consequences for those who would attempt to integrate such findings into general theories of learning.

These few examples illustrate, for one recently explored variable, the claim that second-order conditioning can be a useful tool in the study of the circumstances that produce learning. Other instances have been described elsewhere (Rescorla 1980a).

Content of learning

One of the most extensive modern uses of second-order conditioning is in the analysis of what is learned. Because it involves stimuli whose value is easily changed, second-order conditioning permits an investigation of what actually gets associated when two events are paired. To illustrate the logic, consider a situation in which we have paired two events, a potent S_1 and a more neutral S_2, and we now observe a Pavlovian conditioned response to S_2. Each of those

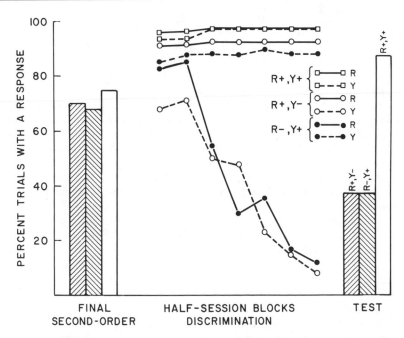

Figure 2.2 Responding in three groups of pigeons to a second-order stimulus (H) conditioned by a red – yellow (RY) compound first-order stimulus. Performance during H is shown at the end of second-order conditioning and then again after differential treatments of R and Y. Behavior during those differential treatments is displayed in the middle panel. (From Rescorla 1979a)

events may have multiple components, and it is important to determine which of those components are actually associated. For S_2 the question is easily addressed: Simply present the separate components of S_2 alone and infer which is associated with S_1 by which evokes the response. But determination of which features of the reinforcing S_1 are learned requires other steps; S_1 typically evokes no new response after its pairing with S_2, so there is little to be gained by simply presenting its components. However, if S_1 is a stimulus that has acquired its reinforcing power by previous pairing with a US, then a procedure is available. One may change the current value of the various components of S_1 and ask which change induces a change in the response to S_2. To the degree that a particular S_1 component has been associated with S_2, a change in its power might well induce a change in the response to S_2.

This kind of logic has been extensively applied in a variety of conditioning preparations. Figure 2.2 shows one example. That figure displays the results of a second-order experiment using an autoshaping preparation in pigeon subjects (Rescorla 1979a). In this experiment two colored lights, red (R) and yellow (Y), had separately been paired with food such that each evoked key pecking; then their joint side-by-side presentation was used as a compound S_1 to second-order condition key pecking to a set of black horizontal (H) lines on a white background. The left-hand portion of the figure shows that second-order conditioning rapidly occurred, resulting in substantial responding to H. But one may then ask about its basis; to what degree were each of the components of S_1 responsible for the performance? In order to investigate this, the animals were divided into three groups and given different patterns of reinforcement and nonreinforcement following red and yellow: $R+Y-$, $R-Y+$, or $R+Y+$. Behavior during these treatments is shown in the middle panel of Figure 2.2; nonreinforced colors showed extinction, whereas reinforced colors continued to evoke responding. Then the impact of these manipulations upon responding to H was tested. The right-hand portion of the figure shows the not especially surprising result: Extinction of either red or yellow partially, but incompletely, undermined performance to H. This suggests that both R and Y were associated with H and each was partially responsible for some of the performance to it.

This illustration of the technique is obviously just a starting point for analysis. It means we now have a procedure that will allow us to detail not only what is learned but also the conditions that influence what is learned. In the last few years we have spent a considerable amount of time using the technique for just such purposes, and a good deal is now known about the determinants of selection of S_1 aspects. I will only mention two of our most recent observations.

Both stem from an attempt to understand a puzzling early result obtained using this technique: Under some circumstances the organism appears to condition S_2 without learning about *any* of the stimulus features of S_1. Consider, for instance a fear conditioning experiment with rats that is analogous to the autoshaping experiment just described for pigeons. One can pair a tone with shock so that the rat comes to fear the tone; subsequent pairings of a light with the tone result in second-order fear conditioning of the light. But if one now repeatedly presents the tone without the shock until it no longer evokes fear, that manipulation has little effect on the animal's fear of the light. Figure 2.3 shows an illustration from Rizley and Rescorla (1972). The left-hand side of the figure shows the development of profound second-order fear conditioning of a light S_2. That fear is indexed in terms of the ability of the light to interrupt an ongoing operant performance. The dependent variable is a ratio of the form $A/(A + B)$, where A is the response rate during the CS and B is that in a comparable period prior to CS onset. After second-order conditioning but before a subsequent test of the light S_2, half the animals had extensive extinction presentation of the tone S_1 and half did not. The results of the test presentation of the light are shown in the second panel, separated according to whether or not the tone had been extinguished; clearly the treatment of the tone S_1 did not matter for the fear of the light. The final panel shows the results of a subsequent test with the tone – which verifies that differential extinction treatments indeed produced different fear of that stimulus in the two groups. But apparently, once the second-order fear conditioning was established to the light, it became independent of the level of fear evoked by the tone reinforcer.

There are various ways in which one can describe such results. But we chose a classical language that contrasted S–S and S–R alternatives. We suggested that just as changes in the response to S_2 induced by extinguishing S_1 indicate

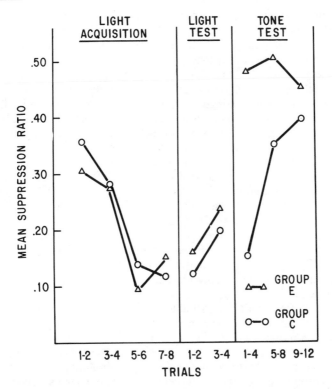

Figure 2.3. Conditioned suppression in rat subjects during the course of a second-order fear conditioning experiment. The left-hand panel shows suppression during trials in which a light S_2 was followed by a tone S_1 (which had previously been paired with shock). The middle panel shows test responding to the light after the tone had been extinguished (group E) or not (group C). The right-hand panel shows test responding to the tone. (From Rizley and Rescorla 1972)

the presence of learning about S_1, so the absence of changes indicates the absence of that learning. We argued that instead the organism had associated S_2 with a representation of the response evoked by S_1 during the S_2–S_1 pairings. That is, S_2 might be directly associated with the CR, bypassing S_1. If S_1 was not itself involved in the association, then changing its value would have no impact upon the response to S_2. For students of learning it becomes of primary importance to understand the circumstances under which one observes either S–S or S–R learning. Second-order conditioning turns out to be useful for such an analysis. The two recent observations I want to describe identify two conditions that matter.

The first variable is familiar from our previous discussion: stimulus similarity. One of the many differences between the pigeon autoshaping experiment described in Figure 2.2 and the rodent fear conditioning experiment described in Figure 2.3 is the degree to which S_2 and S_1 are similar. Two keylights are highly similar, whereas a light and a tone are quite dissimilar. Partly on the basis of this observation, we decided to explore the importance of the similarity of S_2 and S_1. For instance, Nairne and Rescorla (1981) introduced dissimilarity into the autoshaping preparation by using a tone S_1 signaled by a keylight S_2. Pigeons do not peck at a tone paired with food (a point to which we will return later); however, such a tone can be used as an S_1 to second-order conditioned pecking to a keylight S_2. But the interesting observation is that when the tone is subsequently extinguished, it leaves responding to the keylight fully intact. Figure 2.4 shows some sample data on this point. That figure displays second-order conditioning to a keylight S_2 that had been paired with either an auditory or a visual S_1. The data are displayed separately for the end of second-order conditioning, and after S_1 had either been extinguished ($-$) or had received further training with food ($+$). The differential training of visual S_1s resulted in differential subsequent responding to their S_2s; that result is consistent with the data shown in Figure 2.2. However, for auditory S_1s no such differential performance resulted. Responding to S_2s paired with auditory S_1s was indifferent to subsequent discrimination training with those auditory S_1s. That result is more in line with the rodent fear conditioning data shown in Figure 2.3. This pattern shows that within the pigeon autoshaping preparation similarity of S_1 to S_2 is an important determinant of whether learning will be S–S or S–R in nature.

Conversely, Rescorla (1980a) induced similarity into the fear conditioning situation by using two tones as S_2 and S_1. Under those circumstances, he found evidence for a reduction in the fear of S_2 when S_1 was extinguished; moreover, he was able to show that this reduction was not a simple consequence of stimulus generalization of extinction. Consequently, similarity appears not only to promote learning (as we argued earlier) but also to change its character. When similars are paired the stimulus properties of the potent stimulus are encoded; when dissimilars are paired those properties are not as well learned and instead the conditioning appears to be more S–R in nature.

The second variable recently found to affect the relative weights of S–S and S–R learning is also a commonly studied one in Pavlovian conditioning: the temporal relation between S_2 and S_1. It turns out that when two events occur simultaneously, S–S learning is especially encouraged; however, when they occur sequentially, the organism is more likely to learn about the response properties of the second event. Figure 2.5 shows illustrative data from a recent fear conditioning experiment with rats. That figure shows the level of fear (measured by low values on a suppression ratio index) in five groups of animals that received various second-order fear conditioning treatments. All animals initially received first-order fear conditioning involving presentations of a 30-sec tone S_1 ending in shock. Then two groups received second-order conditioning consisting of sequential pairings of a 30-sec light S_2 with that tone S_1; two groups received second-order conditioning consisting of simultaneous presentation of the light and tone; and the fifth group received tone and light in an unpaired fashion. Before testing with the light S_2, one of the sequential and one of the simultaneous groups received extinction presentations of the tone S_1; the other group was spared this extinction. The groups lacking the extinction treatment (labeled C) showed substantial fear of the light compared to the

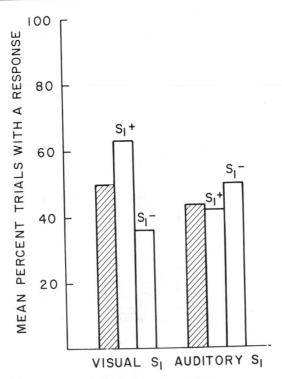

Figure 2.4. Responding in pigeon subjects to a keylight S_2 as a result of its pairing with either a visual or an auditory S_1 reinforcer. The crosshatched bar shows responding at the end of conditioning; the other bars show performance to S_2 after its S_1 had either received subsequent reinforcement $(+)$ or nonreinforcement $(-)$. (From Nairne and Rescorla 1981)

unpaired control. Moreover, the sequential group for which the tone was extinguished (E) continuted to fear the light S_2. That result replicates the findings of Figure 2.2. The finding of most interest is that after simultaneous light–tone presentations, the extinction of the tone S_1 had a detrimental effect on the fear of the light S_2. That outcome suggests that simultaneity can induce S–S learning in a paradigm that normally shows S–R learning for sequentially presented stimuli.

These experiments identify two important variables, stimulus similarity and temporal relationship, that bias the organism in its choice of S–S or S–R learning. Together with earlier findings, they are the beginnings of a specification of the rules determining what is learned. But equally important, they illustrate how one can use second-order conditioning to analyze questions of fundamental importance about the content of associative learning. With that technique we can hope to address the historically important question of what is learned and delineate the circumstances under which the question has different answers.

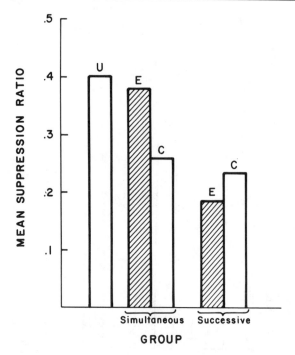

Figure 2.5. Conditioned suppression in rat subjects to a light S_2 that had been either unpaired (U) or presented in simultaneous or successive relation to a tone S_1. After that pairing the tone had either been extinguished (E) or not (C). (From Rescorla 1980b)

Performance

The final point I want to make about second-order conditioning has to do with performance. We are accustomed to using a particular measure of learning in Pavlovian conditioning: the change in the response to the CS. But in many instances that may not be the sole or even the best index of the formation of an association. The previously mentioned instance of pairing a tone with food in pigeon subjects is a case in point. In the context of an autoshaping procedure in which the response to keylight CSs is pecking, the absence of any direct pecking response to a tone CS might well lead to the suspicion that the organism has not learned the tone–food association. Of course, if one takes the kind of multiple response measurement that would result from direct observation of the organism's behavior, he would readily reject such a hypothesis. Pigeons become quite active during tones that signal food. But second-order conditioning also provides an excellent index of the tone–food learning. A tone paired with food, although evoking no pecking response of its own, will nevertheless condition pecking to a keylight S_2 as a result of S_2–S_1 pairings. Rescorla and Holland (1976) have argued that there may be many instances in which such

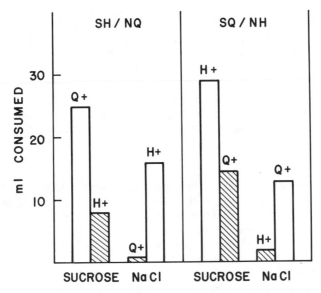

Figure 2.6. Consumption of sucrose (S) and salt (N) after their joint presentation with either quinine (Q) or hydrochloric acid (H) and after either H or Q had been poisoned ($+$). Consumption of flavors that had been presented with subsequently poisoned flavors is indicated by crosshatching. (From Rescorla and Cunningham 1978)

ability of an S_1 to accomplish second-order conditioning of an S_2 is an especially sensitive index of the first-order conditioning that that S_1 has undergone.

These examples point to the general utility of second-order conditioning in the analysis of associative learning. Its value extends to the elucidation of many of the basic questions about learning processes and their exhibition in behavior.

SENSORY PRECONDITIONING

If one simply exchanges the first two stages of a second-order conditioning paradigm, first pairing S_2 with S_1 and then pairing S_1 with a US, the result is a paradigm termed sensory preconditioning (Brogden 1939). To the degree that a final test of S_2 yields a response dependent upon those prior two stages, one infers the occurrence of an S_2–S_1 association. Like second-order conditioning, early instances of such learning appeared weak and short-lived. But in the last few years quite dramatic cases of such learning have been reported.

The results of one particularly strong demonstration are shown in Figure 2.6. That figure shows consumption data for rats who were given the choice between sucrose (S) and salt (N). Those data represent the final test after the animals had been exposed to a somewhat complex variation of the two-phase sensory preconditioning procedure. During the first phase, each animal was exposed to two solutions, one containing S and one containing N; those solutions also contained either quinine (Q) or HCl (H). Thus some animals received SQ and NH whereas others received SH and NQ. Then, each animal received pairings of either H or Q with an illness-inducing toxin (LiCl). The question was whether the animal would exhibit knowledge of the particular pairings he received during the first phase by rejecting S or N dependent upon whether or not its H or Q partner had been poisoned in the second phase of the experiment. Examination of Figure 2.6 reveals that there was a strong relative rejection of such solutions (shaded bars). For instance, animals that received SH and NQ rejected S when H was poisoned but rejected N when Q was poisoned. Similarly animals receiving SQ and NH rejected S when Q was poisoned but N when H was poisoned. This strong rejection occurred after only a few 10-min exposures to the compounds.

Clearly associations occur when flavors are paired. Moreover, recent evidence suggests that they occur between lights and tones presented to rats and between portions of a keylight presented to birds (e.g., Rescorla 1980b, 1981). The phenomenon of sensory preconditioning now seems well established.

Many of the virtues claimed above for second-order conditioning also apply to using sensory preconditioning in the study of associative learning. Before passing to the main comments I want to make about sensory preconditioning, I will briefly illustrate this by indicating the usefulness of sensory preconditioning for the case of studying the circumstances that produce learning. Two recent experiments from our laboratory used sensory preconditioning to address two long-standing questions about associative learning.

The first question is whether valuable stimuli enter into associations more readily than do less valuable stimuli. Is the current reinforcing value of the stimuli to be associated important in determining the strength of their association? One recent experiment addressed that question by measuring the strength of an association between salt (N) and quinine (Q) as a function of the attractiveness of N at the time of its joint occurrence with Q. Attractiveness was varied by administration of formalin, a substance that induces sodium deficit and consequently greatly increases the organism's intake of N. After restoration of the sodium balance, the strength of the $Q–N$ association was assessed by pairing N with a poison to make it aversive and then measuring the degree to which Q was rejected. Despite widely different values of N at the time of $Q–N$ pairing, subsequent poisoning of N and testing of Q revealed similar strength $Q–N$ associations. That is, in this situation the strength of the association formed by a pairing was independent of the value of the stimuli being associated. The virtue of sensory preconditioning in the study of this sort of question is that it permits the study of associations among stimuli whose values are readily subjected to experimental change so as to permit one to separate the value of the stimuli at the time of their pairing from that at the time of test.

A second classic question concerns what temporal relation between stimuli best results in their becoming associated. There are, of course, many experiments on this question using standard CSs and USs. However, we have recently argued that those experiments are necessarily biased in favor of a sequential temporal relation (Rescorla 1980b). One source of bias, for instance, is that a potent US may perceptually mask CSs with which it is simultaneously presented. It may be that the simultaneous relation is actually an excellent one for forming associations except that potent stimuli reduce perception of other events. For that reason, one might wish to use a sensory preconditioning procedure in which relatively neutral stimuli are paired. Using that paradigm, we have found some circumstances under which simultaneous presentation of events produces excellent learning, even better than does their sequential presentation (Rescorla 1980b). Such findings address fundamental issues in associative learning. Moreover, they point to ways in which sensory preconditioning can make a unique contribution to the study of associative processes.

But instead of elaborating on such uses I want to make two interpretative comments about sensory preconditioning. First, it seems natural to think of sensory preconditioning in terms of logical inference. The organism responds to S_2 because it is associated with S_1 that in turn is associated with the US. The organism seems to put two premises together (S_2 goes with S_1, and S_1 goes with US) to draw the conclusion (S_2 goes with the US) and therefore make a response. But some authors (e.g., Coppock 1958; Rescorla and Freberg 1978) have noted that the organism could use its associations in other ways that do not demand such an ability to integrate two associations at the time of the test. One alternative possibility is that the organism uses the S_2–S_1 association during the phase that pairs S_1 with the US. The presentation of S_1 in that phase may result in the activation of the representation of S_2 (via the S_2–S_1 association) with the consequence that it is active at the time of the US. As a result, the organism may form a direct S_2–US association that is responsible for the behavior to S_2 in the test session. Both accounts envision the organism as using the S_2–S_1 association to allow presentation of one of the stimuli to activate the representation of the other. But they view the organism as acting on reactivated representations in quite different ways. According to the logical inference account, that activated representation must in turn activate its associates to produce a response. According to the alternative, it must be available for developing associations with concurrently presented stimuli.

In fact, there is some evidence that reactivated representations serve both functions in sensory preconditioning paradigms. A recent experiment by Holland (1981) points to their participation in learning. Holland took advantage of the apparent special associability of flavors with toxins to provide a demonstration. He first paired a tone with a flavor and then followed the tone with a toxin. Subsequent testing revealed no evidence that the animal found the tone aversive, while simultaneously showing substantial aversion to the flavor. The most natural interpretation is that during the tone–toxin pairings the tone activated the representation of the flavor, thus permitting functional flavor–toxin pairings. That suggests that an activated representation can become associated with events occurring in the world. Conversely, a procedure like sensory preconditioning recently explored by Fudim (1979) suggests the involvement of a reactivated representation in an inferencelike performance process. Fudim exposed animals to compound flavors, one element of which was salt. He then induced a salt need and monitored consumption of the flavors. He found evidence for increased intake of substances that had previously co-occurred with salt. Notice that with such a procedure the change in value of S_1 does not involve its presentation in a paired relation to the US and so there is no opportunity for a reactivated representation of the salt-paired flavor to be paired with a US. Consequently, in order to increase intake of that flavor, the organism must integrate the S_2–S_1 association with the changed value of the salt, S_1.

The second comment I want to make about sensory preconditioning concerns its potential value as a procedure even in instances where the simple pairing of S_2 and S_1 is sufficient to yield a response. Many of us have adopted a kind of operational position for the detection of associations. For instance, Rescorla and Holland (1976) argued that whenever one finds differences in responding to an S_2 as a function of its bearing different relations to another event he has grounds for inferring the presence of an association. That position is especially valuable for its emphasis upon the acceptance of a broad range of responses as evidence for learning. But that position entails the assumption that the formation of an association is the only consequence of arranging a relation between two events. If that assumption should prove false, then differences in responding to S_2 as a function of different temporal relations might not be clear evidence for the presence of an association.

Pfautz et al. (1978) have suggested, however, one instance where the assumption may prove wrong. We know of many cases in which the simple exposure to a stimulus changes the response it evokes (e.g., habituation). Moreover, there is some evidence that the organism needs time after a stimulus exposure to process the experience in order to most effectively produce this change. Finally, there is evidence that if a second event occurs during that time, it will interfere with the processing and slow the course of the change. Consequently, if we were to arrange for some animals to have paired presentations of two events and others to have unpaired presentations, they might well develop different levels of responding to the first event even in the absence of any associative learning. That could result simply because the second event interferes with learning about the first. Yet such a difference is the hallmark of most demonstrations of associative learning.

There are several ways around the problems of interpretation that this raises. One that is relevant here involves treating the pairing as the first stage in a sensory preconditioning experiment. That is, after the S_2–S_1 pairing one may attempt to induce a change in the response to S_2 by pairing S_1 with a new event. If the change in response to S_2 is specific to the treatment of its previous paired S_1, that constitutes good evidence that an S_2–S_1 association has been formed. Indeed, this may be the best evidence for the presence of an association that one can obtain. Consequently, although the sensory preconditioning paradigm may not entail an ability to make logical inferences, it may indicate a rich ability to manipulate event representations and it may be an important tool for the firm identification of the presence of an association.

CONDITIONED INHIBITION

In recent years American psychologists have rediscovered what was obvious to Pavlov, that organisms can learn not only excitatory but also inhibitory relations among stimuli. There are several paradigms for the investigation of such learning, but the most popular is the conditioned inhibition paradigm. In that procedure, the organism is exposed to two kinds of trials involving two stimuli

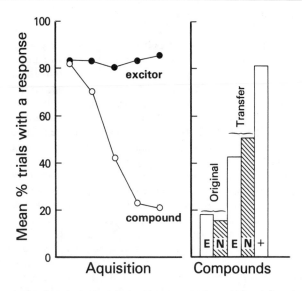

Figure 2.7. Conditioned inhibition in pigeon subjects. The left-hand panel shows acquisition of differential responding to excitatory line orientations reinforced when presented separately and nonreinforced when presented in compound colors. The right-hand panel shows test responding to excitors presented separately (+) and in compound with colors previously nonreinforced in the presence of the original excitor or some other (transfer) excitor. Some colors had been presented alone repeatedly (E) whereas others had not (N).

and a reinforcer. On one trial, type one stimulus, *A,* is presented alone and followed by the reinforcer. On intermixed trials, another stimulus, *B,* accompanies *A* and the joint *AB* presentation is nonreinforced. The result of this *A+/AB−* procedure is that *A* becomes excitatory and *B* becomes inhibitory with respect to the reinforcer. The evidence for these conclusions is that *A* comes to evoke a response but *B* is able to reduce that response when presented in conjunction with *A* (Rescorla 1969). Some sample data are shown in Figure 2.7, which displays responding from pigeons in an autoshaping procedure. These birds had been exposed to two different line orientations (*A* stimuli), each paired with food, and to two different nonreinforced compounds of colors (*B* stimuli) and line orientations. The figure shows substantial responding to the excitatory *A* stimuli and lower responding to the nonreinforced compounds. In addition it displays test responding during the two novel compounds consisting of trained excitors and inhibitors that had never previously co-occurred. It is clear that the colors successfully, but only partially, transferred their inhibition to the other excitor. Below we return to the failure of transfer to be complete; but for present purposes it is sufficient to note that the occurrence of some transfer indicates that the *B* stimuli have inhibitory power of their own even in novel contexts. This sort of observation, together with the failure of *B* to inhibit excitors based on other USs, has been taken as evidence that conditioned inhibition involves an association between the inhibitor and the US (e.g., Konorski 1948, 1967).

The precise nature of the relation between this inhibitory association and excitatory associations has been a matter of unresolved controversy. On the one hand, they seem to share many properties. Most behavioral theories assume that a common mechanism underlies the formation of both sorts of association, that the circumstances of their development can be described by a common theory. The most successful example of this is the quantitative theory of Rescorla and Wagner (1972). That theory attempts to capture an important feature of many Pavlovian preparations – the role of the information that a stimulus gives about the US in determining the level of conditioning. There is substantial evidence for both excitatory and inhibitory learning that simply arranging a contiguity between a CS and an event is often insufficient for learning. On the excitatory side the best example is the Kamin (1968, 1969) blocking effect. If a stimulus, *B*, is reinforced in conjunction with another stimulus, *A*, then *B* will acquire less conditioning if *A* has a history of already being paired with the US. The intuition is that since *A* already predicts the US, *B* is noninformative and therefore does not become conditioned to any substantial extent. On the inhibitory side the procedures that establish a conditioned inhibitor point to the involvement of a similar informational notion. Simply following *B* by nonreinforcement will not make it inhibitory; rather *B* must inform the animal that the nonreinforcement will occur. The simplest way of making *B* carry such information is to reinforce *A* when it is presented alone but not when it is accompanied by *B*. The Rescorla–Wagner model attempts to capture this symmetry.

However, even theories that have assumed comparable conditions for the formation of the relevant associations have differed on the issue of whether one or two types of associations are needed. Some authors (e.g., Konorski 1948; Rescorla and Wagner 1972) appear to assume that excitation and inhibition involve two different types of association with the same US representation. Others (e.g., Konorski 1967; Dickinson 1980) suggest instead that they are the same type of association but with different US representations. For instance, Konorski (1967) argued that conditioned inhibition is an excitatory association with a US representation having an inherently antagonistic relation to the original US (the so-called no-US center). These quite different concepts have generated surprisingly few differential predictions at the behavioral level.

However, two observations of dissimilarities between excitors and inhibitors may be relevant. First, separately presented Pavlovian conditioned inhibitors are notorious for their failure to generate easily observable responses. Historically, many psychologists were skeptical of the notion of inhibition for just this reason – simple presentation of an inhibitor typically does not evoke responses that can be measured. Of course, combining an inhibitor with an excitor, as in Figure 2.7, shows it to have the ability to reduce excitation; but the display of conditioned excitation needs no such assistance from the presence of inhibition. Second, there may be an asymmetry in the conditions that remove excitation and inhibition. An excitor can be extinguished by removing the condition that made it excitatory. If an excitor is presented alone in the absence of the reinforcer, it loses its excitation. However, the available evidence suggests that this way of removing the relation needed to establish an inhibitor does not remove its inhibition. To establish an inhibitor one embeds B in an $A+/AB-$ paradigm. But separately presenting B outside that paradigm does not remove that inhibition (Zimmer-Hart and Rescorla 1974). That is, inhibition does not extinguish in this manner. One example of this is shown in Figure 2.7. That figure displays separately the responding during compounds that contain inhibitors to which the extinction procedure has been applied (E) and during compounds that contain inhibitors lacking that procedure (N). It is clear that whether the inhibitor was tested against its original excitor or against another, repeated separate presentation did not undermine its inhibitory power. This observation turns out to be especially embarrassing for the Rescorla–Wagner model but is troubling for any theory that points to a symmetry in the processes.

One possible way of understanding both of these asymmetries was originally suggested by Konorski (1948) only to be abandoned by him in 1967 but advocated more recently by Rescorla (1979b). It seems possible that the inhibitory association between a CS and US does not act on the US representation in a manner opposite to that of an excitatory association. Instead, an inhibitory association may simply raise the threshold for the activation of the US center by excitors. In that case, one would not expect any observable response to a separately presented inhibitor nor would one anticipate its extinction (see Rescorla 1979b for a more detailed discussion). The distinction between stimuli that elicit and stimuli that modulate the ability of others to elicit has been common in the discussion of neural processes, but less popular among behavioral psychologists. It may have some value here.

Finally, I should mention something about the relation between conditioned inhibition and the other paradigms I have discussed. It will not have escaped attention that the conditioned inhibition paradigm has embedded within it procedures of the other paradigms. For instance, when A and B are jointly presented without reinforcement, B has the opportunity to develop not only an inhibitory association with the US but also an excitatory association with A in a manner that could be measured by sensory preconditioning or second-order conditioning. Although until recently that possibility has received little attention, it turns out to be important in at least some paradigms. By using the procedures of sensory preconditioning and manipulating the current value of A one can readily show that A and B *do* become associated in some conditioned inhibition paradigms. In fact, our recent work suggests that just about any conditioning paradigm that involves the joint presentation of multiple stimuli in relation to a US will result in the organism's learning the relation among the stimuli as well as their individual relations to the US (Rescorla and Durlach 1981). Although perhaps not surprising, this is nevertheless an important observation. Available theories describe conditioning situations as though all of the responsibility for responding to a stimulus falls upon its association with the US. But associations among the stimuli themselves can play an important (and sometimes theoretically confounding) role.

One example of the importance of such associations arises in conjunction with an observation about conditioned inhibition mentioned earlier–that its transfer to a new excitor is often incomplete. One possible interpretation has

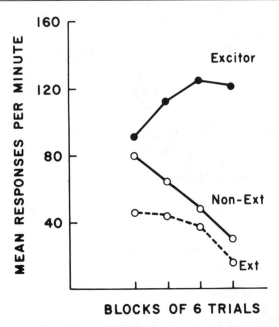

Figure 2.8. Transfer of conditioned inhibition to a new excitor in pigeon subjects. Responding is shown during a separately presented (and reinforced) excitor and during that excitor in compound with an inhibitor that had been subjected to an extinction treatment or not.

been that this reflects the incompleteness of the inhibitory association between the stimulus and the US. But it is also possible that there is nothing wrong with that inhibitory association between *B* and the US but rather the problem lies with the fact that the inhibitor also has an excitatory association with the *A* stimulus. When *B* is then presented in conjunction with a new excitor, it transfers both its inhibition with the US and its excitation with *A;* the two partially cancel, giving the impression of incomplete transfer of the inhibition. One way to assess this possibility is to follow the suggestion of sensory preconditioning paradigms: Extingush the excitatory *A* prior to testing the inhibitory *B* against the new excitor. If the incompleteness of *B*'s transfer lies in its borrowing excitation from *A,* such an operation should improve transfer.

Figure 2.8 shows the results of a recent autoshaping experiment that did just that. Birds were given two concurrent conditioned inhibition treatments, resulting in the establishment of two excitors and two inhibitors. The stimuli were treated equivalently, but each inhibitor was consistently nonreinforced in the presence of only one of the excitors. Then both inhibitors were tested by being trained in a conditioned inhibition paradigm with a third excitor; but prior to that test one of the original excitors was extinguished. Figure 2.8 shows performance during the new excitor and during compounds containing it and each of the inhibitors. It is clear from the lower response levels to the compounds than to the excitor alone that both inhibitors showed transfer to the new excitor; but that transfer was more immediate and stronger in the case of an inhibitor whose excitor had been extinguished. That suggests that the presence of an association between the inhibitor and its own excitor contributed to performance in such a way as to undermine our observation of transfer of inhibition.

This observation is by no means unique to the conditioned inhibition paradigm. We have recently identified the intrusion of such within-compound associations upon performance in a wide variety of standard Pavlovian preparations. Associations between *A* and *B* can affect performance to the individual stimuli whenever *A* and *B* have jointly appeared, whether that compound is reinforced (as in the case of blocking and overshadowing paradigms) or not (as in the case of sensory preconditioning and conditioned inhibition paradigms). In many instances, these associations are substantial enough that the performance they control has been mistaken for evidence of association between the elements and the US. Thus the interaction among the present paradigms is clearly important in governing behavior.

CONCLUSION

This chapter has discussed three "higher-order" Pavlovian paradigms. Even though in some absolute sense these are simple paradigms, involving only three total stimuli, one can readily see the complexity that they sometimes involve. One must be careful not to oversimplify conceptually what may be learned in such paradigms. But I am less interested in emphasizing the fact that simple organisms can develop complex associative structures than I am in pointing to the usefulness of these tools in understanding associative learning. Second-order conditioning, sensory preconditioning, and conditioned inhibition all provide extremely powerful means of analyzing Pavlovian associative processes. My view is that neurobiologists studying learning could profit considerably by including these paradigms in their armory of procedures.

ACKNOWLEDGMENT

This research was generously supported by grants from the National Science Foundation.

REFERENCES

Bitterman, M. E. 1975. The comparative analysis of learning: Are the laws of learning the same in all animals? *Science 188*:699–709.
Brogden, W. J. 1939. Sensory pre-conditioning. *J. Exp. Psychol. 25*:323–332.
Coppock, W. G. 1958. Pre-extinction in sensory preconditioning. *J. Exp. Psychol. 55*:213–219.
Dickinson, A. 1980. *Contemporary Animal Learning Theory*. New York: Cambridge University Press.

Fudim, O. K. 1978. Sensory preconditioning of flavors with a formalin-produced sodium need. *J. Exp. Psychol.: Anim. Behav. Proc.* 4:276–285.

Garcia, J., and Koelling, R. A. 1966. Relation of cue to consequence in avoidance learning. *Psychonom. Sci.* 4:123–124.

Holland, P. C. 1981. Acquisition of representation-mediated conditioned food aversions. *Learn. Motiv.* 12:1–18.

Kamin, L. J. 1968. Attention-like processes in classical conditioning. In *Miami Symposium on the Prediction of Behavior: Aversive Stimulation* (M. R. Jones ed.), pp. 9–33. Miami: University of Miami Press.

1969. Predictability, surprise, attention, and conditioning. In *Punishment and Aversive Behavior* (B. A. Campbell and R. M. Church, eds.). New York: Appleton-Century-Crofts.

Kehoe, E. J., Feyer, A. M., and Moses, J. L. 1981. Second-order conditioning of the rabbit's nictitating membrane response as a function of the CS2–CS1 and CS1–US intervals. *Anim. Learn. Behav.* 9:304–315.

Konorski, J. 1948. *Conditioned Reflexes and Neuron Organization.* Cambridge: Cambridge University Press.

1967. *Integrative Activity of the Brain.* Chicago: University of Chicago Press.

Nairne, J. S., and Rescorla, R. A. 1981. Second-order conditioning with diffuse auditory reinforcers in the pigeon. *Learn. Motiv.* 12:65–91.

Pavlov, I. P. 1927. *Conditioned Reflexes.* London: Oxford University Press.

Pfautz, P. L., Donegan, N. H., and Wagner, A. R. 1978. Sensory preconditioning versus protection from habituation. *J. Exp. Psychol.: Anim. Behav. Proc.* 4:286–295.

Rashotte, M. E., Griffin, R. W., and Sisk, C. L. 1977. Second-order conditioning of the pigeon's key-peck. *Anim. Learn. Behav.* 5:25–38.

Rescorla, R. A. 1969. Pavlovian conditioned inhibition. *Psychol. Bull.* 72:77–94.

1979a. Aspects of the reinforcer learned in second-order Pavlovian conditioning. *J. Exp. Psychol.: Anim. Behav. Proc.* 5:79–95.

1979b. Conditioned inhibition and extinction. In *Mechanisms of Learning and Motivation: A Memorial Volume to Jerzy Konorski* (A. Dickinson and R. A. Boakes, eds.), pp. 83–110. Hillsdale, N.J.: Erlbaum.

1980a. *Pavlovian Second-Order Conditioning.* Hillsdale, N.J.: Erlbaum.

1980b. Simultaneous and successive associations in sensory preconditioning. *J. Exp. Psychol.: Anim. Behav. Proc.* 6:207–216.

1981. Within-signal learning in autoshaping. *Anim. Learn. Behav.* 9:245–252.

Rescorla, R. A., and Cunningham, C. L. 1978. Within-compound flavor associations. *J. Exp. Psychol.: Anim. Behav. Proc.* 4:267–275.

Rescorla, R. A., and Durlach, P. J. 1981. Within-event learning in Pavlovian conditioning. In *Information Processing in Animals: Memory Mechanisms* (N. S. Spear and R. R. Miller eds.). Hillsdale, N.J.: Erlbaum.

Rescorla, R. A., and Freberg, L. 1978. The extinction of within-compound flavor associations. *Learn. Motiv.* 9:411–427.

Rescorla, R. A., and Furrow, D. R. 1977. Stimulus similarity as a determinant of Pavlovian conditioning. *J. Exp. Psychol.: Anim. Behav. Proc.* 3:203–215.

Rescorla, R. A., and Gillan, D. J. 1980. An analysis of the facilitative effect of similarity on second-order conditioning. *J. Exp. Psychol.: Anim. Behav. Proc.* 6:339–351.

Rescorla, R. A., and Holland, P. C. 1976. Some behavioral approaches to the study of learning. In *Neural Mechanisms of Learning and Memory* (M. R. Rosenzweig and E. L. Bennett, eds.), pp. 165–92. Cambridge, Mass.: MIT Press.

Rescorla, R. A., and Wagner, A. R. 1972. A theory of Pavlovian conditioning: Variations in the effectiveness of reinforcement and nonreinforcement. In *Classical Conditioning,* vol. 2 (W. F. Prokasy and A. H. Black, eds.), pp. 64–99. New York: Appleton-Century-Crofts.

Rizley, R. C., and Rescorla, R. A. 1972. Associations in second-order conditioning and sensory preconditioning. *J. Comp. Physiol. Psychol. 8*:1–11.

Schwartz, B. 1974. On going back to nature: A review of Seligman and Hager's *Biological boundaries of learning. J. Exp. Anal. Behav. 21*:183–198.

Zimmer-Hart, C. L., and Rescorla, R. A. 1974. Extinction of Pavlovian conditioned inhibition. *J. Comp. Physiol. Psychol. 88*:837–845.

3 · Conditioned disgust and fear from mollusk to monkey

JOHN GARCIA, DEBRA FORTHMAN QUICK, and
BRENT WHITE

ASSOCIATIONISM AND DARWINISM

The senses are our first teachers in whose home our mind is enclosed.... Of those
things perceived together, if one occurs it usually represents the other along with itself.
[Juan Luis Vives 1539 (1915)]

Psychologists are obsessed with associations. Students of learning in particular
are still beguiled by the empirical and associationistic analysis of the psyche
expounded by philosophers such as Juan Luis Vives, born in Valencia in 1492.
Vives was called the Father of Modern Psychology by Foster Watson (1915)
because he was the first to apply empiricism and associationism to psychological problems in a systematic way as the quotation (above) taken from Watson
indicates. According to Watson, Vives attributed a disgust to a prior association, recalling how he ate cherries while suffering a fever during his childhood;
"For many years afterwards whenever I tasted the fruit I not only recalled the
fever but also seemed to experience it again." Conditioned disgust is an apt
label because "disgust," from Latin for taste *(gustus),* denotes a change in the
emotional reaction to the taste associated with illness. Vives also discussed conditioned fear in animals, noting that if dogs "are beaten after being called, they
are frightened by the memory of the blows on hearing the same call again."

More than two centuries later, associationism was expanded and elaborated
by John Locke, whose contributions to taste aversion learning have been extensively quoted elsewhere (Garcia 1981). Here we will summarize only two
points made by Locke. First, Locke proposed a hypothetical mind furnished
with experiential "sensations" filtered through biased sensory windows, but he
gave equal status to inherent "reflections," including the motives and emotions
emanating from a reactive nervous system. Second, he described taste–illness
aversions in remarkably modern psychobiological terms.

About 50 years later, Hume (1739/1969) laid down the foundation for a
"black-box" theory of mind in his section "Division of the Subject." Reflections, he said, were posterior to sensations, and "the examination of our sensations belongs more to anatomists and natural philosophers than to moral."
Counting himself a moral philosopher, he chose to deal with ideas, "faint
images of thinking and reasoning," divorced from perception and feeling. In
his well-known argument, he rejects causality as a feeling evoked by the idea

of contiguity. Obviously, to natural philosophers, perceived causality is no different from perceived contiguity or perceived similarity; all are dependent upon sensory events integrated by the brain. Like moral philosophers, modern learning theorists have turned over the examination of the nature of sensory processes and expressive behavioral patterns to naturalists and ethologists. It is now apparent that they have given away too much.

Fortunately, psychobiological concerns are again in vogue, and the adaptive properties of conditioned responses are acceptable explanations for associative learning. This is a true renaissance because associationism was a necessary component of the theory of evolution from its inception. In Darwin's *Origin of Species* (1859/1936) associationism is explicitly or implicitly used to explain relationships of animal to plant and animal to animal. For example, flowers attract insects by pairing specific signals with special reinforcers in a series of conditioning trials to establish precise habits so that a given insect will visit a series of similar flowers in succession, thus insuring efficient pollination. Darwin preferred to speak of inherited structures and reflexive behaviors, but now we recognize that the honeybee has the capacity to associate new and arbitrary signals with commodities vital to the survival of the hive. Joachim Erber (1981) describes the honeybee as an excellent subject for behavioral studies of learning and memory, equal to any vertebrate. Because insects were well established in the Carboniferous period, over 100 million years before mammals appeared in the evolutionary record, learning may be an ancient evolutionary process.

Darwinians are obsessed with eating. All organisms are inexorably bound together in a great feeding chain. Actually, "feeding web" is a better metaphor, as all organisms must eat and are ultimately eaten by other organisms; herbivores feed on plants, carnivores feed on animals, scavengers feed on dead animals, omnivores feed on plants and animals, and plants recycle the wastes of feeding. Because of the universal pressure of incessant feeding, all successful living organisms have evolved elaborate mechanisms to select food and equally complex mechanisms to defend against becoming food, at least until after they have passed their genes on to their offspring.

The dual integrative structures associated with feeding and defense are apparent in the mollusk. These hardy creatures have a simply structured nervous system tolerating an extraordinary degree of experimental surgery, so they are ideal subjects for research on the neural substrate of acquired associations. After a trial or two, in which carrot juice is followed by poison, the garden slug *(Limax maximus)* will display an acquired disgust for carrot juice. Analysis of the neural mechanisms of conditioned disgust in the garden slug has been initiated by Gelperin and his associates (Chang and Gelperin 1980; Gelperin and Reingold 1980). On the other hand, the external defense system of the sea slug *(Aplysia)* has been studied by Kandel and his associates (Walters et al. 1981). The skin receptors of *Aplysia,* which inhabits tide pools, are sensitive to disturbances in the water that often signal a predatory attack. *Aplysia* learns to associate mild stimulation of the skin with a more intense insult to its tail. Chemical substances released in the seawater also act as distal signals; thus *Aplysia* quickly acquires a chemical–shock association as well. When alarmed by the chemical signal after conditioning, *Aplysia* exhibits an increased readiness to defend its body with a number of reflexive strategies, that is, escape locomotion, release of ink, and siphon withdrawal. The mollusk acts as if the conditioned stimulus (CS) produces a central motive state that has been described as "conditioned fear" (Carew et al. 1981).

THE GREAT SEGREGATION OF VERTEBRATE FUNCTION

This segregation of all sensory nerve fibers, except those of vision and olfaction, into only two receptive centers is the only well-defined localization of the sensory functions present in the medulla oblongata. It corresponds with the fundamental difference in behavior between internal visceral activities and somatic sensorimotor activities which have an external reference.

[C. Judson Herrick 1961]

The anatomy of the salamander reveals the great structural division of feeding and external defense according to Herrick (1961). Selective learning studies demonstrate the corresponding functional division in visceral learning and somatic learning throughout the animal kingdom (Garcia et al. 1977). The associative mechanisms defending the gut against poison are qualitatively and quantitatively distinct from those saving the skin from predatory attack. The effectiveness of a given signal or a given reinforcer in a conditioning trial depends upon which of the two systems is engaged.

This functional dichotomy of defense is illustrated in a study by Garcia et al. (1968). Food size or food flavor served as conditioned stimuli. Large-sized food pellets were cut into quarters to make the small size. Quartered pellets were rolled in either flour or powdered sugar to provide different flavor cues. The cues were balanced in the usual way, and four basic discrimination groups were formed: size–shock, size–x-ray, flavor–shock, and flavor–x-ray. The punished food was presented on conditioning days, and the safe food was presented on intervening days. The unconditioned stimulus (US) was either a mild foot shock delivered immediately after each nibble on the forbidden pellet or x-ray treatment delivered after a meal on forbidden pellets. The x-ray treatment was calculated to produce nausea an hour or so after exposure. This study, and others like it, are sometimes criticized because the shock parameters do not match the x-ray parameters, but subsequent experimentation with odor has shown this criticism, based on association by similarity, to be irrelevant, as we shall see later.

Figure 3.1 illustrates the differences in consumption of punished and safe food offered in separate 1-hr tests without punishment (since difference scores are graphed, the taller the bar, the stronger the conditioning). For shocked rats, size proved to be the more effective CS and latency the better measure of the conditioned response (CR), because the fearful animals waited to seize the forbidden pellet until the CS–US interval used in conditioning had been exceeded in the test. Then they ate avidly of the punished pellets for the rest of the hour, indicating they had no disgust for pellets associated with foot shock. Conversely, flavor was the more effective CS for x-rayed rats and amount eaten the better CR measure, indicating they did not fear the forbidden pellet, but were simply disgusted with the taste. There is little evidence of a flavor–shock

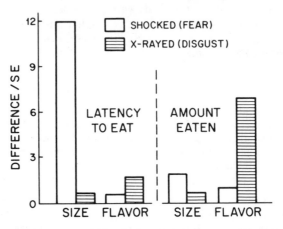

Figure 3.1. Selective learning within the internal and external domains of behavior. Mean differences between scores for punished and safe food are plotted after conditioning. See text for details.

association or a size–x-ray association. Perhaps with repeated trials, all four associations would be acquired by the rats and manifested in their quantitative scores; nevertheless, the functional dichotomy of internal–external defense would still be apparent in the expressive behavior of the rats.

Expressions of disgust and fear were observed in this experiment. In the beginning of conditioning, animals in the sugar–x-ray subgroup licked and nibbled the sweet powder off the pellets, but as the aversion developed, they attempted to shake and scrape the powder off and nibble at the unpowdered center of the pellet. This adaptive behavior nearly ruined our experiment by circumventing our designated flavor CS. When the quantitative tests were over, the rats of the big-size–shock subgroup were offered big and little pellets simultaneously in separate dishes. Occasionally an animal would snatch a large forbidden pellet and eat it sitting on the small safe pellets. Thus, adaptive behavior expressed more precisely than quantified behavior exactly what the animals had learned during conditioning: The x-rayed animal acquired a disgust for the powdery surface of the punished pellet whereas the shocked rats acquired a fear of the floor near the punished pellet.

The neural mechanisms underlying the two behavioral systems are also relatively discrete in mammals. This has been discussed elsewhere (Garcia and Ervin 1968; Garcia et al. 1974, 1982). The neural basis for selective taste–illness association has been outlined, if not clearly defined. Essentially, taste fibers and visceral fibers converge to the emetic mechanisms in the brain stem and midbrain that also receive the fibers from the area postrema where blood-borne toxins are monitored (rats cannot vomit effectively, but they gape and retch when poisoned). The mammalian pathways for auditory signals and cutaneous stimulation are more complex than in the salamander, but evidence for neural convergence of external information is still apparent.

WHAT ABOUT OLFACTION (AND VISION)?

Sense of smell. This sense is in close proximity to the organ of taste with which smell frequently cooperates; but we may consider it as placed at the entrance of the lungs to test the purity of the air we breathe.

[Alexander Bain 1868]

As Bain (1868) pointed out, olfaction is not an integral part of the feeding system. Olfaction, like vision, is explicitly excluded from the great segregation of sensory fibers in the medulla oblongata of the tiger salamander by Herrick (1961). Obviously these two sensory systems serve both feeding and external defense; however, the way they perform their dual role is surprising. The case has been clearly worked out for olfaction, but it is true for vision as well.

Odor plays one role in the feeding system and another completely different role in the defensive system. Only one trial, or at most several trials, were used in the studies that follow. This is essential because if a rat is given extended training, it is apt to learn many complex strategies, presumably employing redundant neural systems that obscure the more elementary associations revealed by one-trial learning. Under these minimal training conditions a recent study revealed a curious interaction between odor and taste during compound conditioning in the naive laboratory rat (Rusiniak et al. 1982). Thirsty animals were habituated to drinking from a water spout protruding from a nose cone. On the conditioning trial, novel almond food flavor on filter paper in the nose cone was used as an odor CS for *O* groups and a solution of saccharin and almond flavor in the water served as the odor–taste CS for *OT* groups. Some *O* groups and *OT* groups received a moderately high dose of lithium chloride after the single CS trial; other *O* groups and *OT* groups received three conditioning sessions with immediate mild foot shock repeated with variable intensity designed to prevent drinking of the water with the CS. (Additional groups received modified procedures: odor in the nose cone and taste in the water, odor alone in the water. The entire shock experiment was repeated using a single drinking session terminated by a single intense shock.)

Only selected groups are illustrated in Figure 3.2; the other groups exhibited the same pattern and were omitted for brevity and clarity (since suppression scores are plotted, the shorter the bar, the greater the conditioning). First, note that the *OT* groups drank less than the *O* groups before application of poison or shock (left panel), indicating that odor and taste summate to produce greater neophobia, as might be expected. Second, note that odor alone is tested in the center and right panels in the absence of further poison or shock. Odor alone proved to be a good signal for shock but a poor signal for delayed illness, supporting prior reports by Hankins et al. (1973, 1976) (*O* groups, center panel). In other words, odor alone acts as an external signal, similar to pellet size in the previous experiment. However, when odor is compounded with taste in acquisition, odor tested alone proves to be a potent cue for illness, as previously discussed by Garcia and Rusiniak (1980). Furthermore, under this same arrangement, odor is a poor signal for shock (*OT* groups, center panel). In other words, the presence of taste in acquisition reversed the role of odor in the test; odor now acts like taste, an internal cue.

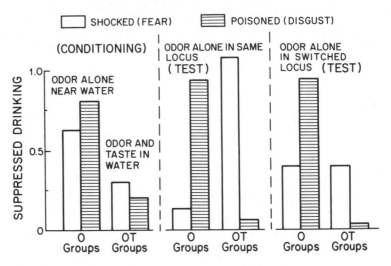

Figure 3.2. Potentiation of odor–illness and blocking of odor–shock associations by taste. Mean drinking scores are plotted as a proportion of water baseline. See text for details.

Finally, note that the locus of the odor was switched from the nose cone to the water for the *O* groups and from the water to the nose cone for the *OT* groups. This maneuver had no effect upon the performance of the poisoned groups, indicating that location of the disgusting odor is relatively trivial. But switching disrupted the performance of the shocked groups, indicating that the location of a feared odor is very important (right panel).

These results limit the generality of the classic principle of overshadowing or blocking in conditioning theory. When a compound CS is made up of a reliable component and an unreliable or redundant component, the reliable component will overshadow the unreliable component and block conditioning to the redundant component. Overshadowing makes good sense: If a clear signal is available, why bother with a weak one? The principle is backed by an impressive data base, but the data base is limited to the external sphere (see MacKintosh 1974). In the internal sphere of feeding, the exact opposite occurs. In Figure 3.2, taste is the reliable signal for poison, but it potentiates the unreliable odor component. Furthermore, taste is the unreliable signal for shock, but it blocks conditioning to the reliable odor component.

This experiment places a heavy burden on traditional learning theory in general because it is not so vulnerable to the criticism that the selective effects illustrated in Figure 3.1 may be due to the parametric differences between the shock US and the x-ray US. In Figure 3.2, with the odor CS and the shock US held constant, taste blocks the odor–shock association. And with the odor CS and the illness US held constant, taste potentiates the odor–illness association. Due to their theoretical significance, these effects are sure to be challenged experimentally. For example, Mikulka et al. (1982) reported negative results in their attempt to obtain odor potentiation by taste. However, they used hypertonic lithium injections that cause immediate pain, and the pain of foot-shock did not produce potentiation in Figure 3.2. It produced blocking.

Again a word of caution is in order: If many repeated trials are used, the rat is apt to learn too much and thus ruin the experimenter's plan. For example, when given repeated extinction trials with aversive odor in the nose cone and plain water in the spout, rats have been observed to claw out the filter paper containing the offensive odor from the nose cone and push the paper through slots in the floor at the back of the compartment, expressing their disgust. Then they return to drink from the water spout, running up their drinking scores and belying their odor aversions.

Visual cues, like odors, can be potentiated by taste. This was clearly demonstrated in hawks habituated to feeding on white mice (Brett et al. 1976). Novel black mice were followed by lithium poisoning, but the hawks did not utilize the visual cue to discriminate the poisoned mouse: After repeated poisoning the birds rejected both white and black mice. However, when the black mice were marked with a novel mildly bitter taste, the hawks frantically fluttered away when offered a second black mouse. Similar facilitation of visual cues has been demonstrated by Galef and Osborne (1978) testing rats with white and dark capsules. The rats made no use of the colored capsules unless differential colors were attended by differential flavors.

The most extensive work in taste potentiation of color is offered by Bow Tong Lett (in press). She finds an interesting pattern of results across avian species. Ducks, like hawks, do not readily form color–illness associations without taste potentiation when feeding. Geese can form direct color–illness associations, but such associations are potentiated significantly by adding taste to the colored food. All these birds have variable diets. In contrast, quail, pigeons, and chickens feed on relatively tasteless seeds to a much greater extent. And they make such excellent color–illness associations while feeding that taste potentiation is difficult to demonstrate, perhaps due to floor effects. However, taste potentiation can be demonstrated in the latter species while drinking by adding taste to colored water. After a review of most if not all of the odor and visual potentiation literature to date, Lett rejects traditional learning explanations in favor of the adaptive explanation offered by Garcia and Hankins (1977). The capacity to acquire taste–poison associations is a phylogenetically old mechanism common to most vertebrates. The ability to make direct color–poison associations is a more recent feeding specialization in species that subsist on visually distinctive food with little taste; it adds to, but does not replace, the taste–illness mechanism.

SENSORY GATES FOR OLFACTION (AND VISION)

The point to be emphasized here, just as in heightened attention, is not only the excitation of one centre and simultaneous inhibition of the other centres but, doubtlessly, also increased activity of the particular centre as a result of incidental stimulations.
[V. M. Bechterev 1928 (1973)]

The results displayed in Figure 3.2 are reminiscent of the notion of dominant focus in Russian reflexology espoused by Bechterev (1973), who was, indubitably, a Darwinian conditioning theorist. The dominant focus of excitatory activity in the central nervous system inhibits the activity of other systems. Such a mechanism might account for the differential role of odor.

We offer an explanation similar to dominant focus, illustrating it with a simple block diagram in Figure 3.3. The external defense system is depicted across the top of the figure; noise followed by a cutaneous shock is an ideal CS–US combination, mimicking the footsteps and bite of a predator. The animal employing inherent defensive strategies quickly learns to avoid pain on cue in a given place. The internal defense system is depicted across the bottom of the diagram; taste followed by internal malaise is the ideal CS–US combination, mimicking the consequences of eating and absorbing poisonous plant foods in the natural world. The animal employs inherent feeding strategies and quickly acquires an aversion for the taste no matter where it is found. Odor, depicted in the center, passes through a sensory gate operated by the presence or absence of taste. If taste is absent, odor is shunted to the external defense system where it exhibits the temporal parameters displayed by distal signals for external events, and where it may be subject to interference from other external signals. When taste is present, odor is gated into the long-term memory stores associated with feeding, and there it is protected from external interference. Once in the feeding system, odor exhibits tastelike parameters; the effectiveness of a constant odor component is a function of the logarithm of the varying taste concentrations used to potentiate the odor (Rusiniak et al. 1979). Furthermore, while odor alone yields a steep delay of reinforcement gradient when the CS–US interval is varied, odor potentiated by taste exhibits the same long shallow delay-of-illness gradient as taste. Finally, the interaction between taste and odor is asymmetrical; the taste component exerts a huge influence on the odor component, but the odor has very little if any effect on taste (Palmerino et al. 1980). This is appropriate because the taste receptors stand guard at the entry into the gut, which processes the food.

The dominant focus has another interesting property in common with potentiation. The focus can be made ready to react by very weak preliminary influences from the appropirate receptive field (Rusinov 1973). In feeding, taste is the appropriate field. So profound is the potentiating effect of taste on odor that a taste too weak to acquire any significant conditioning itself is sufficient to potentiate an odor (Rusiniak et al. 1979). Even when gustatory neocortical lesions are used to disrupt taste–poison associations, taste potentiates odor powerfully in the absence of any evidence of a taste aversion (Kiefer et al. 1982). Apparently the taste signal is split; one part proceeds to the gustatory neocortex to facilitate taste discrimination, and the other probably goes to the limbic system to potentiate odors and other distal feeding cues (Garcia et al. 1982).

The dominant focus exhibits yet another property in common with potentiation. According to Bechterev (1928), the dominant activity is increased by incidental stimulation. Lasiter and Braun (1981) discovered potentiating interaction between peripheral pain and internal nausea. It goes as follows. When a sweet fluid is followed by nausea, a rat will drink less thereafter so the effect is negative. Normally, a rat has a difficult time associating a taste with shock, especially if the shock is delayed for a minute or two. However, if a rat is returned to a situation in which it has previously been shocked, it is apt to drink

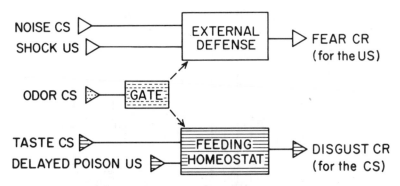

Figure 3.3. The gating of odor cues into either external or internal defense. Odor cues are sequestered internally by the presence of taste and externally by its absence. See text for details.

more sweet water. In other words, the effect of incidental (nonassociative) shock is nil or positive. Lasiter and Braun demonstrated that when sweet water was followed by nausea plus foot shock, shock produces an associative effect upon water intake, but now the effect is negative. That is, nausea plus shock produces a greater decrement than nausea alone. Paradoxically, a positive effect added to a negative effect results in increased negativity. It is as if nausea gated the punishing effect of shock into the dominant feeding focus, where it enhanced the aversive effect of nausea.

MAMMALIAN EXPRESSIONS OF DISGUST AND FEAR

The state of agitation, in either appetite or aversion, is exhibited externally by increased muscular tension; by static and phasic contractions of many skeletal and dermal muscles, giving rise to bodily attitudes and gestures which are easily recognized signs or "expressions" of appetite or of aversion.

[Wallace Craig 1917]

Behavioral expressions of fear and disgust are readily distinguishable in the rat, and these distinctive patterns support the notion of sensory gating of odor by taste. We presented two groups of rats with water flavored with the taste of saccharin and the odor of almond. One group was shocked to the feet and the other group was poisoned with lithium chloride. Later the rats were tested with odor alone, taste alone, and with the odor–taste compound. Expressive behavior and amount of water consumed in each test was recorded; the mean results are illustrated in Figure 3.4. Behavioral patterns were recorded as mean Hansen frequencies for a 30-min period. (Essentially the period is divided into 1-min intervals, and one or more responses observed in any 1-min interval is scored as "one.")

The frequency of sniffing is high under all conditions, indicating that olfaction is a busy information channel as the rat constantly samples many scents (upper left, Figure 3.4). The shocked rats drank more than the poisoned rats under all conditions (upper right). The *OT* compound produced the greatest

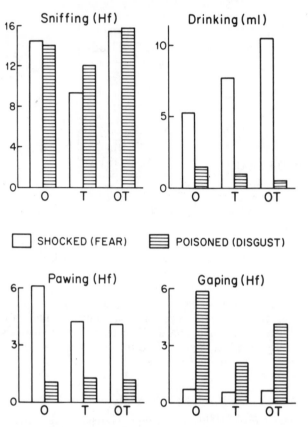

Figure 3.4. Expressions of disgust for poisoned cues and agonistic behavior for shocked cues. Mean Hansen frequencies observed after conditioning are plotted. See text for details.

and O the smallest decrement in the poisoned rats. Converse effects were observed in the shocked rats, paralleling the results illustrated in Figure 3.2. The shocked animal also engaged in more "pawing" than the poisoned animal (lower left). This pawing or "boxing behavior," delivered from an upright position as the rat rears up into a defensive posture, is often observed in agonistic encounters (Sbordone and Garcia 1977) or when the rat is fending off a predator such as the ferret (Rusiniak et al. 1976). Note the reciprocal relationship between drinking and pawing in the shocked rats, indicating that the two behaviors are not compatible.

The poisoned rat does more gaping than the shocked rat (lower right). Gaping is part of the disgust pattern; the rat lowers its head, often rubbing its chin on the floor, and opens its mouth gagging and retching. Odor apparently elicits more gaping than taste because the rat can avoid the taste by refusing to lick the spot whereas the odor diffusing through the air continues to assault its nose.

Odor, sensed by the receptors in the respiratory passages, has now gained control over the final common pathway of food rejection, normally controlled by the taste receptors stationed at the oral entrance to the digestive tube.

Similar patterns of disgust in other species have been reviewed elsewhere (Garcia et al. 1977). Here we only point out that each species reacts with its species-specific disgust pattern so that observation of a single animal is often quite sufficient. For example, after one venison–poison trial a cougar went through an entire pan of venison, picking up each piece with its mouth and dropping it, indicating that though olfactory cues are useful, taste is the final arbiter for ingestion. Then the cougar turned away from the meat, shaking each paw in the feline sign of disgust. And after one mutton–poison trial, a coyote retched and rolled on the offensive meat, finally kicking dirt on it with its hind legs, expressing the canine signs of disgust. Several days afterward, the hungry coyote eagerly charged a living lamb, ears up and tail high, indicating that the visual system had not been affected by conditioning. On contacting the lamb with nose and mouth, the disgusted coyote turned away and retched with ears back and tail down, indicating that the taste and odor of lamb were as aversive as the mutton bait. The behavior of a frightened coyote is markedly different. Like many fearful mammals, it bares its teeth and snarls, and the hair along its back rises. Piloerection is a virtually universal sign of fear and rage among mammals, whereas most avian species will fluff their feathers and raise their crests when threatened.

Signs of disgust in the monkey are intuitively obvious to the human observer, who is, after all, a close relative. For example, when presented with a novel food, the monkey will eagerly seize it. Visual avoidance is rare. Disgust is manifested by excessive sniffing and manipulation often followed by breaking and squashing the food item, then wiping and brushing its hands. This may be attended by finicky sampling with retracted lips, using only the bared incisors and the tongue pressed over the lower incisors. If the food is mildly disgusting, the monkey may bite the food item and spit out the food repeatedly, sniffing constantly, finally letting the food drop from a limp hand while looking the other way. If the food is extremely disgusting, due to prior conditioning or to some inherent property, the animal is apt to fling it away and retreat from the food as if in fear. If tested in a social situation, urged on by the competitive challenges of the other monkeys, the monkey may immediately stuff a tainted banana into its cheek pouches. Later it will expel the offending fruit into its hands and exhibit the same signs of disgust described above (which is enough to elicit empathetic nausea in the human observer).

The behavior of the frightened monkey also contrasts sharply with that of the disgusted monkey. Vocalization and intention movements predominate as the animal wavers between attack and retreat. Initially, the threatened monkey may stare, open-mouthed, at the intruder. This pattern may dissolve into squeals and screams interspersed with slapping and grabbing at its tormenter. In panicked retreat, the tail is held high and straight up, and the animal urinates and defecates. If retreat is blocked, the frightened monkey will bite and scratch the intruder. If it is unable to escape from a fearful place, it may huddle motionless or even solicit aid from the human observer.

CONVERGENT AND DIVERGENT EVOLUTION OF LEARNING

It is incredible that the descendents of two organisms, which had originally differed in a marked manner, should ever afterwards converge so closely as to lead to a near identity throughout their whole organization.

[Charles Darwin 1859 (1936)]

Darwin might have been incredulous if he had witnessed how closely the garden slug's learning converged upon that of the rat, the two descendants of organisms that diverged in pre-Cambrian times. To train a slug, one must know a slug, at least as well as we know the rat. Accordingly, Sahley et al. (1981) began with a Darwinian analysis, characterizing the garden slug as an opportunistic herbivore for whom vegetables and plant poisons provided ideal stimuli for conditional pairing. Employing the associationism inherent in adaptation, she pointed out that vegetable odors signaled food in the slug's natural niche. Once the appropriate stimuli to guide and motivate were selected, the rest was easy. The slug performed like a rat in the associationistic paradigms considered de rigueur to prove learning; it mastered classical conditioning, higher-order conditioning, and blocking with remarkable ease. But the slug is not a rat.

The learning paradigms could not distinguish slug from rat because they are based upon physical, not biological, interpretations of two of the three laws of association, contiguity and similarity. Thus interpreted, the two laws only describe some pervasive ecological forces to which both slug and rat must conform during evolution. It is adaptive for any animal to interpret discrete contiguous events as if they are actually connected. It is also adaptive for any animal to categorize and act to similar events as if they were the same event. Given this interpretation, it seemed that general laws could be formulated without examining the details of the sensory system and the integrative circuitry of any specific organism. Any organism would do, so long as the physical inputs from the environment and the motor outputs to the environment were recorded. Most thoughtful learning theorists knew that contiguity and similarity might look different to different beasts, but they assumed that this biological adjustment could be taken care of by a little empirical spadework (e.g., Estes 1959). But very little comparative analysis has ever been carried out with the necessary qualification of species differences. Quite naturally, the resultant laws of learning tell us a little about the global ecology of learning but nothing about the biological learning mechanisms.

The third law, association by causality, cannot be interpreted in terms of physics; it leads inevitably to the examination of intuitions, emotions, expressions, and ultimately to biological mechanisms. For this very reason, Hume rejected causality and the modern learning theorists followed suit. But causality remains a necessary part of a Darwinian analysis. Causality evolves out of contiguity and similarity, as over epochs of time pervasive environmental correlations are impressed upon the integrative systems of organisms. Through specialization, different sequential correlations, such as the chemical stimuli impinging upon first the foregut and then the hindgut when food is engulfed, are handled by one integrative system. Other correlations, such as the vibrations in substrate preceding an attack on the peripheral surface, are handled by another system. Ultimately contiguity and similarity are reinterpreted by each system according to its likes, thereby limiting the usefulness of the general formulations. Such selective perception and specialized integration give rise to the causal perceptions: Taste causes illness, noises cause pain. This inherent perception of causality promotes survival in the natural niche and selective learning in the laboratory. Through adaptive radiation, different organisms encounter unique correlations in their respective niches and reinterpret causality in specialized ways. And the psychologist must become a naturalist if he is to interpret individual behavior of such animals.

As we make detailed studies of specific organisms, as well as detailed analyses of the sensory systems that carry the CS and the US, important functional differences capture our attention, and the unifying associationistic philosophy binding learning to biology fades from view. The gross differences in the neural organization of slug and rat point to convergent evolution of associative learning. Unity must be sought elsewhere. At the molecular level, there may exist equally important similarities in membrane potentials and synaptic transmission pointing to divergent evolution of more basic learning mechanisms; and appropriately so, for the unifying notion that all life diverged from a single source has a molecular basis.

ACKNOWLEDGMENTS

This work was supported by National Institutes of Health Research Grant NS 11618, Program Project Grants HD 05958 and AA 03513, and D. Forthman Quick's Fulbright Award 5048211.

REFERENCES

Bain, A. 1868. *The Senses and the Intellect.* London: Longman, Green.
Bechterev, V. M. 1973. *General Principles of Human Reflexology.* New York: Arno Press. (Originally published, 1928.)
Brett, L. P., Hankins, W. G., and Garcia, J. 1976. Prey-lithium aversions. III. Buteo hawks. *Behavioral Biology 17*:87–98.
Chang, J. J., and Gelperin, A. 1980. Rapid taste-aversion learning by an isolated molluscan central nervous system. *Proc. Nat. Acad. Sci. 77*:6204–6206.
Craig, W. 1917. Appetites and aversions as constituents of instincts. *Biol. Bull. Marine Biol. Lab. 33*:91–107.
Darwin, C. 1936. *The Origin of Species by Means of Natural Selection: Or, the Preservation of Favoured Races in the Struggle for Life and the Descent of Man and Selection in Relation to Sex.* New York: Modern Library. (Originally published, 1859, 1871.)
Erber, J. 1981. Neural correlates of learning in the honeybee. Paper presented at the Primary Neural Substrates of Learning and Behavioral Change Conference, Princeton University, Princeton, N.J., October 2–4.
Estes, W. K. 1959. The statistical approach to learning theory. In *Psychology: A Study of a Science* (S. Koch, ed.), p. 455. New York: McGraw-Hill.
Galef, B. G., and Osborne, B. 1978. Novel taste facilitation of the association of visual cues with toxicosis in rats. *J. Comp. Physiol. Psychol. 92*:907–916.
Garcia, J. 1981. Tilting at the paper mills of academe. *Am. Psychol. 36*:149–158.

Garcia, J., and Brett, L. P. 1977. Conditioned responses to food odor and taste in rats and wild predators. In *The Chemical Senses and Nutrition*, (M. Kare, ed.), pp. 277–289. New York: Academic Press.

Garcia, J., and Ervin, F. R. 1968. Gustatory–visceral and telereceptor–cutaneous conditioning: Adaptation in the internal and external milieus. *Comm. Behav. Biol. 1*:389–415.

Garcia, J., and Hankins, W. G. 1977. On the origin of food aversion paradigms. In *Learning Mechanisms in Food Selection* (L. M. Barker, H. M. Best, and M. Domjan, eds), pp. 3–19. Waco, Texas: Baylor University Press.

Garcia, J., Hankins, W. G., and Rusiniak, K. W. 1974. Behavioral regulation of the milieu interne in man and rat. *Science 185*:824–831.

Garcia, J., McGowan, B. K., Ervin, F. R., and Koelling, R. A. 1968. Cues: Their relative effectiveness as a function of the reinforcer. *Science 160*:794–795.

Garcia, J., and Rusiniak, K. W. 1980. What the nose learns from the mouth. In *Chemical Signals* (D. Muller-Schwarze and R. M. Silverstein, eds.), New York: Plenum Press.

Garcia, J., Rusiniak, K. W., and Brett, L. P. 1977. Conditioning food-illness aversions in wild animals: *Caveant canonici*. In *Operant–Pavlovian Interactions* (H. Davis and H. Hurwitz, eds.), pp. 273–316. Hillsdale, N.J.: Erlbaum.

Garcia, J., Rusiniak, K. W., Kiefer, S. W. and Bermudez-Rattoni, F. 1982. The neural integration of feeding and drinking habits. *Conditioning: Representation of Involved Neural Function* (C. D. Woody, ed.), pp. 567–579. New York: Plenum Press.

Gelperin, A., and Reingold, S. C. 1980. Plasticity of feeding responses emitted by isolated brain of a terrestrial mollusc. *Adv. Physiol. Sci. 23*:249–266.

Hankins, W. G., Garcia, J., and Rusiniak, K. W. 1973. Dissociation of odor and taste in baitshyness. *Behav. Biol. 8*:407–419.

Hankins, W. G., Rusiniak, K. W., and Garcia, J. 1976. Dissociation of odor and taste in shock-avoidance learning. *Behav. Biol. 18*:345–358.

Herrick, C. J. 1961. *The Evolution of Human Nature*. New York: Harper.

Hume, D. 1969. A treatise on human nature. In *The Essential David Hume* (R. P. Wolff ed.). p. 35. New York: New American Library (Originally published, 1739.)

Kiefer, S. W., Rusiniak, K. W., and Garcia, J. 1982. Flavor–illness aversions: Potentiation of odor by taste in rats with gustatory neocortex ablations. *J. Comp. Physiol. Psychol. 96*:540–548.

Lasiter, P. S., and Braun, J. J. 1981. Shock facilitation of taste aversion learning. *Behav. Neural Biol. 32*:277–281.

Lett, B. T. In press. Taste potentiation in poison avoidance learning. In *Quantitative Analysis of Behavior: Acquisition*, vol. 3 (M. L. Commons, R. J. Herrnstein, and A. R. Wagner, eds.), Cambridge: Ballinger.

Mackintosh, N. 1974. *The Psychology of Animal Learning*. London: Academic Press.

Mikulka, P. J., Pitts, E., and Philput, C. 1982. Overshadowing not potentiation in taste aversion conditioning. *Bull. Psychonom. Soc. 20*:101–104.

Palmerino, C. C., Rusiniak, K. W., and Garcia, J. 1980. Flavor–illness aversions: The peculiar roles of odor and taste in memory for poison. *Science 208*:753–755.

Rusiniak, K. W., Gustavson, C. R., Hankins, W. G., and Garcia, J. 1976. Prey–lithium aversions. II. Rats and ferrets. *Behav. Biol. 17*:73–85.

Rusiniak, K. W., Hankins, W. G., Garcia, J., and Brett, L. P. 1979. Flavor–illness aversions: Potentiation of odor by taste in rats. *Behav. Neural Biol. 25*:1–17.

Rusiniak, K. W., Palmerino, C. C., Rice, A. G., Forthman, D. L., and Garcia, J. 1982. Flavor–illness aversions: Potentiation of odor by taste with toxin but not shock in rats. *J. Comp. Physiol. Psychol. 96*:527–539.

Rusinov, V. S. 1973. *The Dominant Focus: Electrophysiological Investigations.* New York: Consultants Bureau.

Sahley, C. L., Rudy, J. W., and Gelperin, A. 1981. An analysis of associative learning in a terrestrial mollusc: Higher-order conditioning, blocking and a transient US pre-exposure effect. *J. Comp. Physiol. 144*:1–8.

Sbordone, R., and Garcia, J. 1977. Untreated rats develop "pathological" aggression when paired with a mescaline-treated rat in a shock-elicited aggression situation. *Behav. Biol. 21*:451–461.

Vives, J. L. 1915. Quoted in F. Watson, The Father of Modern Psychology. *Psychol. Rev. 22*:333–353. (Originally published, 1539).

Walters, E. T., Carew, T. J., and Kandel, E. R. 1981. Associative learning in *Aplysia:* Evidence for conditioned fear in an invertebrate. *Science 211*:504–506.

Watson, F. 1915. The father of modern psychology. *Psychol. Rev. 22*:333–353.

PART II

MODEL SYSTEMS

Introduction to Part II

JOSEPH FARLEY

In Part II, a representative sample of the major model system preparations currently used to study learning is presented. The chapters can be grouped along a number of orthogonal dimensions, three of which warrant comment here. First, the various research efforts differ with regard to the degree to which they explicitly aspire to elucidate general principles of cellular function that underlie learning phenomena as they occur across phylogeny. Chapters 4 through 6, concerned with neural correlates of vertebrate classical conditioning, are thus directly relevant to the pursuit of cellular insights into a form of learning that both is ubiquitous and has direct parallels in human learning (Spence 1940; Prokasy 1972). Indeed, it has been argued that the essential behavioral and psychological characteristics of associative learning and memory in humans – exclusive of those features typically attributed to linguistic processes – are recapitulated in their entirety in the classically conditioned nictitating membrane response of the rabbit (Chapter 4).

Research efforts with molluscan preparations similarly aspire to uncover cellular mechanisms that underlie a family of associative learning processes presumed to share common features across phylogeny (see chapters 8, 10, 11, 12, and 15). Although patterned after studies of classical conditioning or related paradigms in vertebrates, the various examples of associative training paradigms nevertheless differ in some important respects. Perhaps foremost among these is the choice of invertebrate response systems that are known, or suspected, to be heavily influenced by nonassociative learning processes (Chapters 8, 11, and 12), which require minimal experience for full expression and which may resolve relations of temporal contiguity with much less fidelity than vertebrate preparations. Although an argument can be made that the cellular mechanisms believed to underlie habituation are somewhat general for vertebrates and invertebrates (Kandel 1976), any such claim for associative learning is simply premature at this point.

The complementary approach to a study of general-process learning mechanisms is that of the neuroethologist (Chapters 13 and 14), whose concern is more with the unique, species-specific characteristics of learning and memory in those preparations in which these capacities are both highly developed as well as narrowly circumscribed. In direct contrast to the majority of the vertebrate and invertebrate model system research represented here, Chapters 12,

13, and 14 have explicitly concerned themselves with determining the role that learning might play within the overall adapative framework of an animal's behavior. In so doing, they bring a necessary balance to the concern for general principles and mechanisms.

The concern for maximizing the generality of the learning processes and mechanisms studied also forms the basis for a second dichotomy represented here: the distinction between nonassociative and associative forms of learning and memory. It is generally recognized that the bricks of human knowledge structures are "associations" (Bower and Anderson 1973), though there is a great diversity of opinion as to how such metaphysical units of cognition are best conceptualized and whether they are sufficient for the construction of a data base that any reasonably complex information-processing system might have recourse to (Wilson 1980). Nevertheless, because of the paramount importance typically attached to associative learning and memorial processes, much of the recent research with invertebrate preparations is explicitly oriented toward robust and reliable demonstrations of associative learning. This is not to say that nonassociative processes such as habituation and sensitization (Chapter 8) play no role in vertebrate, especially human, learning; however, there are quite obvious logical limitations in the extent to which such processes can be invoked in the control of human mentation and behavior.

Nonetheless, nervous systems often avail themselves of both nonassociative and associative mechanisms in the production of learned behavior, and it is important to clarify their interactions in many cases (Chapter 8). Indeed, one of the interesting possibilities raised by the associative training results with the gill withdrawal and siphon withdrawal responses in *Aplysia* (Chapter 8), and perhaps for *Limax* as well (Chapters 11 and 12), is that the *apparently* associative behavioral changes, which are being studied here, may in fact represent an intermediate form of nonassociative behavioral change in which temporal contiguity between events nonetheless plays a critical role.

It is important to recognize, however, that quite different views exist, even among specialists, as to the relationship of conditioning to "higher cognitive processes" in human learning and memory. On the one hand, there are those who believe that what one studies in conditioning experiments are the mechanism and processes of conditioning per se, which may be circumscribed in their generality both within and across species. A far more influential view, especially among those interested in general process theories of learning and memory, is that conditioning refers to a collection of processes, most clearly revealed in situations that strictly conform to the operational definition of classical conditioning perhaps, but that nonetheless operate in many if not all forms of learning (Hull 1943; Pavlov, 1927; Konorski 1967; Spence 1951, 1956; Bolles 1975; Estes 1973; Mackintosh 1975).

Even here one encounters at least two distinct conceptions as to the relation between conditioning and higher mental processes. On the one hand, there is the view that associative conditioning processes may account for all, or nearly all, of the mental life of less evolved organisms and may in a sense represent the least common denominator of shared cognitive processes. With progressively more evolved animals, the degree to which apparently intelligent and adaptive behavior may be interpretable within an associative conditioning framework diminishes, until in humans the relatively simple processes of associative conditioning are superseded by, indeed are subordinate to, the more complex processes of perception, selective attention, verbal and linguistic systems, and so forth. To be sure, classical conditioning can be observed in relatively pure form within certain restricted domains – such as visceral and autonomic nervous system functioning, certain reflexes, and emotional behavior. But even here, it requires considerable experimental care to excise the influence of the more complex processes.

A second view, however, is that human learning and the conditioning of animals with simpler nervous systems share a common core of associative learning processes and mechanisms, which are evidenced by a number of striking functional similarities in their learned behavior – for example, the commonly observed findings that argue for phenomenally distinct "stages" of information storage, the differential effects upon retention of massed versus distributed training trials, and the like. However, even in cases where close operational correspondences exist between two different examples of associative learning for different species, the issue as to similarity of mechanisms can only be answered through experiment. Such a comparison presumes, of course, that at minimum neural correlates of learning phenomena have been identified.

Finally, there is the obvious division of model system preparations into vertebrate and invertebrate categories. One of the major advantages of the use of invertebrate model systems for the cellular analysis of learning and memory that is often given is the relatively small number of neurons in central ganglia of invertebrates (Kandel 1976). The implications of this statement are often assumed to be twofold: (1) It should, in principle, be relatively easier to locate neural correlates of a learning-induced behavioral change in these preparations; and (2) it should, in principle, be relatively easy to establish more or less complete wiring diagrams of the behavior in question that undergoes robust learning.

In our opinion, there is little validity for the former expectation. A host of neural correlates for a wide variety of associative training paradigms, both classical and operant conditioning, as well as those corresponding to phenomenally more complex examples of learning and memory, have been described and have been documented repeatedly. Indeed, a simple comparison of the number of distinct neuronal correlates that have been reported for associative learning paradigms for vertebrates versus invertebrates would no doubt reveal greater numbers for the vertebrates by a factor of at least 3 or 4. Thus, it is certainly incorrect to assert that neural correlates of learning are more readily found within invertebrate nervous systems.

Furthermore, there are the numerous examples of relatively long-term changes in the electrophysiological properties of various regions of the mammalian brain (such as hippocampal long-term potentiation), which, although not directly correlated with any specific example of a learning-induced behavioral change in a model system, occur in structures that have been indirectly implicated in various learning phenomena (Swanson et al. 1982). Thus, the belief that basic physiological mechanisms of neuronal and synaptic change are necessarily more easily addressable in invertebrate preparations is also a gross oversimplification.

The potential advantages of at least some of the invertebrate preparations lie in three areas. First, there is the ability to precisely describe the connectivity of the neural systems responsible for producing the change (i.e., the initial conditions for learning). Second, it is possible to describe how these changes are propagated throughout the neural networks subserving behavioral changes (rules relating learning to performance). Finally, in those cases in which the neural changes responsible for learning can be identified as occurring in a small, physically accessible number of identified sites (somata, synapses, dendritic arborizations, etc.), it may be possible to propose and directly assess subcellular biophysical and biochemical mechanisms of learning.

At least two critical issues must be confronted when evaluating the potential causal relevance for a given neural correlate of a learning-induced behavioral change. First, the degree to which a neural network, in which changes occur, provides a complete account of the behavior in question must be ascertained. Secondly, the degree to which a given neural correlate of learning provides a complete explanation of output changes in the neural circuit must also be determined.

Ideally, a proposed neural circuit for behavior should be both necessary and sufficient for the expression of the salient, criterial features of behavioral change as it occurs in the intact animal. A corollary to this is that the major functional components of such a network must be identified cells (or populations) whose physiological and morphological characteristics can be repeatedly examined and demonstrated using conventional electrophysiological, anatomical, and histological methods. It is also imperative that the various ways in which components can interact with one another be fully characterized in such networks. In other words, it is in the functioning of the system in its entirety in which a complete description and account of learning and behavior is to be found. In many of the invertebrate preparations (see Chapters 8 and 15) this task of specifying the exchange and flow of information through a network is facilitated by the ability to simultaneously record from pre- and postsynaptic neurons. In vertebrate preparations, such direct means are generally impractical at best, and so additional strategies must be employed. One common and powerful technique, exemplified by the contributions of Woody and Cohen (Chapters 5 and 6), is to utilize information concerning the sequence of neural events, in conjunction with neuroanatomically established patterns of connectivity, to formulate a tentative flowchart for the transmission of information throughout the nervous system. In any event, a common goal in both cases is a full and complete mathematical model of the actual nervous system, a goal yet to be realized for any preparation.

With a tentative model of such a network in hand, it is then possible to proceed with more acute means of establishing the precise contribution of individual components to behavior. Crucial to evaluations of whether or not a given class of neurons are necessary for some behavioral (or learning) function are demonstrations that the behavior, as it occurs in the intact animals, depends upon the functional integrity of the neural structures in question. In those rare, relatively simple response systems among the vertebrates, lesion, ablation, or extirpation strategies often suffice to demonstrate the mediation of behavior by a given component of the nervous system (see Chapters 4 and 7). In invertebrates, this strategy can be carried one step further. Precise and reversible microlesions of single neurons can be used to establish the contribution of individual cells to the control of behavior, without the interpretive problems presented by the phenomena of sprouting and synaptogenesis that often accompany the more acute lesions in vertebrate central nervous systems. For example, it is now quite clear from research using such techniques in a host of invertebrate preparations that single "command" neurons often play a key and appreciable role in this initiation, patterning, and modulation of complex coordinated patterns of movement. In a somewhat similar vein, I have found in my own work (Farley and Alkon in press) that a single optic ganglion cell is largely responsible for a synaptic facilitation of the light response of photoreceptors, and is both necessary and sufficient for production of the short-term neural changes in photoreceptors that represent the mechanism for acquisition of learning-induced long-term neural changes.

It is nevertheless important to bear in mind that even in the simplest examples of stimulus-evoked behavior in animals, sensory divergence and parallel processing are the physiological facts. Thus, for example, in the mammalian brain visual information may be diverted to a number of distinct anatomical sites for feature extraction and analysis before being resynthesized for the purposes of object recognition, perception, the retrieval of acquired hedonic value, and so on.

Similarly, in the case of those neural correlates of learning that have been identified, numerous training-correlated differences have been reported even within the same preparation (see Chapters 5, 6, 8, 10 and 15). The task then becomes one of distinguishing primary neural changes from those secondary or tertiary consequences of learning. At this point, the exclusive reliance upon a circuit diagram can be as misleading as it is informative. It may be misleading primarily because it is incomplete. For example, from the wiring diagrams generated for the neural control of gill withdrawal and siphon withdrawal responses in *Aplysia* and the control of phototaxic behavior in *Hermissenda,* localization of a neural correlate of learning at relatively peripheral levels of the neural circuit (the sensorimotor–neuron synapse and the type B photoreceptor) seems to imply that changes occurring at these initial nodes within the network *must* be expressed – and indeed are sufficient to explain – changes recorded elsewhere to which these sites project. How then is one to interpret recent reports that in *Aplysia,* the neural control of gill and siphon withdrawal in the freely behaving animal appears quite different from that of the more reduced preparation (Kanz et al. 1979)? How can one be absolutely certain that in *Hermissenda* the changes in the type B photoreceptors are sufficient to explain the other differences that have been observed? Again, which changes are essential and which are accidental? In addition to the obvious strategy of further circuit *analysis,* I have adopted a complementary tactic of behavioral *synthesis* in my own work. Thus, in *Hermissenda* we have successfully demonstrated that experimental induction of the same membrane changes in type B photoreceptors of untrained, intact animals, which normally occur in these cells as a result of the visual and vestibular neural systems' transformation of the information provided by pairings of light and rotation, is both necessary and sufficient to produce long-lasting changes in phototaxis (Farley et al. 1983).

As can be seen from the selections represented here, establishing the precise causal relevance of those correlates of learning that have been identified in many of the model system preparations is one of the most pressing and formidable problems facing the field.

REFERENCES

Bolles, R. C. Learning, motivation, and cognition. 1975. In *Handbook of Learning and Cognitive Processes,* vol. 1 (W. K. Estes, ed.), Hillsdale, N.J.: Erlbaum.

Bower, G., and Anderson, J. A. 1973. *Human Associative Memory.* Washington, D.C.: Winston.

Estes, W. K. 1973. Memory and conditioning. In *Contemporary Approaches to Conditioning and Learning* (F. J. McGuigan and P. B. Lumsden, eds.). Washington, D.C.: Winston.

Farley, J., and Alkon, D. L. In press. In vitro associative conditioning of *Hermissenda* Cumulative depolarization of type B photoreceptors and short-term associative behavioral changes. *J. Neurophysiol.*

Farley, J., Richards, W., Ling, L., Liman, E., and Alkon, D. L. 1983. Membrane changes in a single photoreceptor cause associative learning in *Hermissenda. Science, 221:*1201–3.

Hull, C. L. 1943. *Principles of Behavior.* New York: Appleton-Century-Crofts.

Kandel, E. R. 1976. *Cellular Basis of Behavior.* San Francisco: Freeman.

Kanz, J. E., Eberly, L. B., Cobbs, J. S., and Pinsker, H. M. 1979. Neuronal correlates of siphon withdrawal in freely behaving Aplysia. *J. Neurophysiol. 42:*1538.

Konorski, J. 1967. *Integrative Activity of the Brain.* Chicago: University of Chicago Press.

Mackintosh, N. J. 1974. *The Psychology of Animal Learning.* New York: Academic Press.

Pavlov, I. P. 1927. *Conditioned Reflexes.* Oxford: Oxford University Press.

Prokasy, W. F. 1972. Developments with the two-phase model applied to human eyelid conditioning. In *Classical Conditioning, II: Current Research and Theory* (A. H. Black and W. F. Prokasy, eds.), pp. 119–147. New York: Appleton-Century-Crofts.

Spence, K. W. 1951. Theoretical interpretations of learning. In *Comparative Psychology* (C. P. Stone, ed.). Englewood Cliffs, N.J.: Prentice-Hall.

Spence, K. W. 1956. *Behavior Theory and Conditioning.* New Haven: Yale University Press.

Swanson. L. W., Teyler, T. J., and Thompson, R. F. 1982. Hippocampal long-term potentiation: Mechanisms and implications for memory. *Neurosci. Res. Prog. Bull. 20:*613.

Wilson, K. V. 1980. From associations to structure. In *Advances in Psychology,* vol. 6 (G. E. Stelmach and P. A. Vroon, eds.). New York: North-Holland.

4 · Neuronal substrates of associative learning in the mammalian brain

RICHARD F. THOMPSON, JACK D. BARCHAS,
GREGORY A. CLARK, NELSON DONEGAN, RONALD
E. KETTNER, DAVID G. LAVOND, JOHN MADDEN IV,
MICHAEL D. MAUK, and DAVID A. McCORMICK

The nature of the memory trace has proved to be among the most baffling questions in science. At present, analysis of brain mechanisms of learning and memory faces problems that are both empirical and conceptual. In order to analyze mechanisms of information storage and retrieval it is first necessary to identify and localize the brain systems, structures, and regions that are critically involved. With a very few exceptions, such information has not yet been obtained. It would seem necessary to know where such processes occur before they can be analyzed. This problem of localization has been perhaps the greatest barrier to progress in the field.

There are conceptual issues attendent upon any particular experimental approach to brain mechanisms of learning and memory. Most workers would agree that learning is not a unitary phenomenon. There is much less agreement on how many "kinds" of learning exist and whether they reflect one, two, or several types of basic processes. In terms of underlying brain processes, this issue can be dealt with empirically. Having demonstrated that a brain process is critically involved in a given learning paradigm, it can then be examined in other paradigms. The same approach can be used to answer objections that a given brain process might be limited to one particular response system or one species. For the most part, this issue of generality has not been resolved in studies of brain substrates of learning, making general conclusions and comparison across laboratories difficult. It is essential to explore the degree of generality of brain substrates of learning.

In recent years the "model system" approach to analysis of the neuronal substrates of learning and memory has been valuable and productive. The basic notion is to utilize a preparation showing a clear form of learning or behavioral plasticity in which neuronal analysis is possible. Habituation proved to be a particularly good example – it exhibits similar behavioral properties and, to the extent that it has been analyzed, similar neuronal mechanisms in a range of animals from mollusks to mammals (Castellucci and Kandel 1976; Thompson and Glanzman 1976).

Each approach and model preparation has particular advantages. Mollusk preparations (e.g., Alkon 1979; Walters et al. 1979; Davis and Gillette 1978; and Chapter 15, this volume) appear particularly useful in terms of the feasibility of cellular analysis. Classical leg flexion conditioning of the acute spinal cat (Durkovic 1975; Patterson 1976; Patterson et al. 1973) appears to offer similar potential advantages. The difficulties of cellular analysis of learning in the intact mammal are formidable. However, the other side of the coin is that they provide potential models for a basic understanding of information processing and learning in higher animals and ultimately in humans. In behavioral terms it is clear that higher vertebrates have developed increasing capacities for learning and have made use of these capacities in adaptive behavior. It would seem that the evolution of the mammalian brain has resulted in systems specially adapted for information processing, learning, and memory.

We have adopted a particularly clear-cut and robust form of learning in the intact mammal as a model system: classical conditioning of the rabbit nictitating membrane (NM) response, first developed for behavioral analysis of learning by Gormezano (see Chapter 1 and Gormezano et al. 1962). This simple form of learning is extremely well characterized behaviorally, thanks largely to the extensive studies of Gormezano and his associates, and seems particularly well-suited for neurobiological analysis (see Thompson et al. 1973, 1976). Eyelid conditioning exhibits the same basic laws of learning in a wide range of mammalian species, including humans, and is prototypic of classical conditioning of striated muscle responses.

A word is in order about the nature of the conditioned response. Investigators typically record either extension of the NM, which is a largely passive consequence of eyeball retraction (Cegavske et al. 1976) or closure of the external eyelid. However, with standard procedures for NM conditioning, both become conditioned simultaneously and synchronously, together with some degree of contraction of the periorbital facial musculature (McCormick et al. 1982b). We will discuss the motor control in more detail below. The major components are NM extension (eyeball retraction) and eyelid closure. In recent work we have measured and/or observed both. When we refer to the conditioned response (CR) below we mean both the NM and eyelid. All effects reported here occur equally for both. This fact maps very nicely into the large animal and human behavioral literature on eyelid conditioning.

Several laboratories have adopted the "model systems" approach to analysis of associative learning in the intact vertebrate – Cohen and associates use classical conditioning of heart rate in the pigeon (see Cohen 1980), as do Kapp and associates (1979) in rabbit and Smith and associates (1980) in baboon. Woody uses classical conditioning of the very short latency click evoked "alpha" eye blink response with a glabellar tap unconditioned stimulus (UCS) in cat (see Brons and Woody 1980). Gabriel uses instrumental avoidance learning in the rabbit (see Gabriel et al. 1980). Weinberger (1980) uses classical conditioning of the pupillary response in the paralyzed cat. Olds and associates (e.g., 1972) and Segal (Segal and Olds, 1973) have used a combined classical and instrumental task in rat with food reward.

In the discussion that follows we will use classical conditioning of striated muscle responses (particularly the rabbit eyelid and NM responses) as the basic paradigm. In this paradigm, the essential condition for associative learning involves effective "pairing" of the conditioned stimulus (CS) and UCS. The essential condition for associative learning in all paradigms, including instrumental learning, involves some form of concatenation among stimuli, responses, and reinforcement–some processes of "pairing" analogous to the pairing of the CS and UCS. When we refer to learning we will generally use classical conditioning as the prototype but assume that the reader will grant the generalization to other instances of at least simpler forms of learning as well. In all cases the critical requirement is some degree of "contiguity." Taste aversion learning appears to stand at one extreme on the contiguity continuum, but even here there are temporal limits.

The first issue that must be addressed is the identification and localization of neuronal structures and systems that appear to be involved in learning. We approached this question using electrophysiological recording of neural unit activity as the initial method of identification. More detail will be given later. It is sufficient to note here that in terms of engagement of neuronal activity during learning, at least the following structures are involved: regions of the brain stem and midbrain, the cerebellum, the hippocampus, and related structures and portions of the neocortex. Fortunately, a number of other structures do not appear to be involved, for example, basal ganglia and most nuclei of the amygdala.

It seems a reasonable assumption that the various brain structures and systems that, judged from their electrophysiological activity, do become involved in learning and memory retrieval play roles in learning and memory. In particular, higher brain regions appear under normal circumstances to play such roles. The conditioned response is selectively and reversibly abolished by spreading depression of the contralateral motor cortex (Papsdorf et al. 1965) – a clear effect of memory retrieval. The conditioned response cannot be learned if training is given only during periods of penicillin-induced hippocampal seizures (Thompson et al. 1980). Yet the conditioned response can be learned in the absence of the neocortex (Oakley and Russell 1972) or hippocampus (Solomon and Moore 1975). All these data hold for the standard-delay conditioned response (in which the UCS overlaps the termination of the CS). In fact, higher brain structures assume essential roles if more demands are placed on the learning and memory system. Thus, animals with prior bilateral ablation of the hippocampus are unable to learn a trace conditioned response (CR), in which a period of no stimulation intervenes between CS offset and UCS onset (Weisz et al. 1980). The same is true for a number of other phenomena of learning – for example, latent inhibition, blocking, discrimination reversal (see Orr and Berger 1981; Solomon 1980).

The hippocampus is of particular interest. Even in the short-delay paradigm, activity of pyramidal neurons always grows rapidly over training to form a temporal model of the behavioral conditioned response. These neurons develop a very clear model of the behavioral response with correlations between the pattern of unit increase and the behavioral NM response as high as .90 (see Figure 4.1). This model of the behavioral response generated by pyramidal neurons is not present in the first few trials of training, develops rapidly, initially in the UCS period and then in the CS period as the learned behavioral response develops, and grows considerably in amplitude. The initial growth in unit activity in the UCS period is the earliest sign of learning over the trials of training that we have seen in the brain. Note that the increased unit activity models the entire behavioral response, both the learned CS period response and

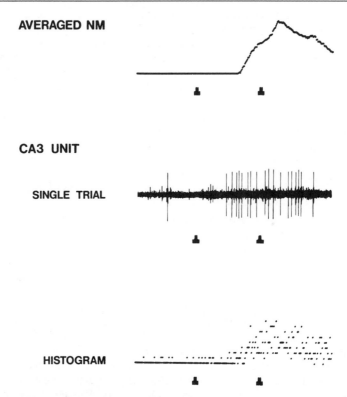

Figure 4.1. Examples of the discharge pattern of an identified hippocampal pyramidal neuron responding during trial periods in a rabbit well trained in the conditioned NM/ eyelid response. Upper trace: NM response averaged over a number of trials. Center trace: single trial example of the discharge pattern of the pyramidal neuron. Lower trace: histogram (3-msec time bins) of the cell discharge over the same number of trials as for the NM response shown above. First cursor: Tone CS onset. Second cursor: Airpuff UCS onset. Total trace duration 750 msec. Note that the pattern of increased frequency of cell discharge closely models the amplitude–time course of the behavioral NM response. (From Berger and Thompson 1978b)

the response in the UCS period (which probably reflects both learned and reflexive behavioral components). This learning-induced increase in pyramidal neuron activity does not develop in unpaired control animals.

We have characterized the learning-induced hippocampal unit response over a wide range of conditions, including acquisition and extinction (Berger and Thompson 1978a,b, 1981) variation of CS–UCS interval (Hoehler and Thompson 1980), temporal alternation (Hoehler and Thompson 1979), and in another learning paradigm – leg flexion conditioning (Thompson et al. 1980). We have reviewed this work in recent chapters and need not repeat it here (e.g., Thompson et al. 1980; Berger et al. 1980b). In brief, over a wide range of conditions that impair or alter acquisition, maintenance, or extinction of the learned response, *the learning-induced increase in hippocampal unit activity invariably precedes and accurately predicts subsequent behavioral learning performance*. The hippocampal response has all the properties of a direct measure of the inferred processes of learning and memory in the brain. In terms of mechanisms, the neuronal plasticity is exhibited by identified pyramidal neurons of fields CA3–CA1 but not in general by other types of hippocampal neurons and has many formal similarities to the process of long-term potentiation (LTP) (Swanson et al. 1982).

Rabbits can learn the standard-delay NM extension CR following ablation of all brain tissue above the level of the thalamus (Enser 1976), and cats with high decerebration can learn the delay conditioned eye blink response (Norman et al. 1977). Several possible inferences can be made from these results. Perhaps the most reasonable is that a "primary memory trace" circuit exists below the level of the thalamus for the standard-delay conditioned response. We have adopted this as a working assumption. This is not to say that higher brain structures do not normally play important roles and, in fact, develop substantial learning-induced neuronal plasticity. Indeed, the hippocampus does so.

Many workers still place their greatest faith in the lesion approach to localization of the memory trace. Certainly if a specific brain lesion causes selective and permanent abolition of a learned response in a given learning task it is strong presumptive evidence that the region, structure, or pathway plays an essential role in memory. However, this might involve necessary sensory pathways or motor nuclei. If the CS is auditory, bilateral ablation of the cochlear nuclei will abolish the learned response. Similarly, destruction of the motor neurons necessary for performance of the task will abolish the learned response. If these possibilities can be excluded, then the lesion may well have destroyed a significant part of the circuitry containing the essential neuronal plasticity for the learned response.

The lesion technique is considerably less helpful with negative results. The absence of a lesion deficit in no way implies that the structure so damaged or destroyed is not normally involved in learning. Assume that three partially separate systems in the brain all code the engram. Destruction of any one or even two might have no effect on the learned response. Just such a situation was found by Cohen (1980) in terms of the necessary visual pathways for the CS in classical conditioning of the heart rate response in the pigeon. Any one of three separate visual pathways can support the response. It remains an article of faith but, at this point in time, a fairly reasonable faith, that if two or more neuronal systems can each function to code the learned response, then destruction of all of them will eliminate the learned response. Again, the lesion approach per se will not necessarily tell us how these systems function in learning and memory. Electrophysiological recording of neuronal activity seems at present to be the technique best suited to provide such information.

THE CS CHANNEL: THE AUDITORY SYSTEM

In studies involving simple learning paradigms where an acoustic CS has been used, training-related changes in unit activity have been reported at virtually all levels of the auditory system in at least some studies: cochlear nucleus (Oleson et al. 1975), inferior colliculus (Disterhoft and Stuart 1976), medial genic-ulate body (Buchwald et al. 1966; Disterhoft and Stuart 1977; Ryugo and Weinberger 1978; Gabriel et al. 1975), auditory cortex (Disterhoft and Stuart 1976; Kitzes et al. 1978), and association cortex (Woody et al. 1976; Disterhoft et al. 1982). Negative results were also reported in several of these studies for certain areas, including the inferior colliculus, medial geniculate body, and auditory cortex. In our own earlier work on the rabbit NM response, we did not see consistent changes over training in auditory unit activity in the cochlear nucleus and inferior colliculus (Lonsbury-Martin et al. 1976). Weinberger and associates (Ryugo and Weinberger 1978; Weinberger 1980) found training-induced changes in the medial but not the ventral division of the medial genic-ulate body. Similarly, Gabriel et al. (1975) found such changes more medially than ventrally. Note that the negative result occurs in the "mainline" auditory specific portion of the medial geniculate body, the ventral division.

The observation that tone-CS-evoked activity of units in an auditory relay nucleus shows an increase as a result of training per se is simply a statement of a correlation, with its attendant problems of interpretation. Olds attempted to deal with the problem by specifying that the increase must be the shortest latency event within a trial in order to be a critical aspect of the neuronal sub-strate of learning (Olds et al. 1972). Gabriel (1976) pointed out that a change in "bias" – the influence of another structure or system on the auditory relay nuclei – could result in short latency increases. Such a bias increase need not play an essential role in learning. It might, for example, reflect a process of conditioned "arousal" that itself might not be a necessary part of the neuronal changes that form the essential substrate of the behavioral learning. We sug-gested that any neuronal changes that form the essential substrate for learning must develop earlier than (or certainly no later than) the appearance of behav-ioral signs of learning over the trials of training (Thompson et al. 1976). How-ever, we did not argue that this is a sufficient condition, only a necessary one. In any event, studies that report increases in unit activity in auditory structures with training do not necessarily find any clear relationship between such change in neural activity and the development of behavioral learning over the trials of training.

In extensive recent studies we have examined neuronal unit activity over the course of initial training of the rabbit NM response to an acoustic CS (Kettner and Thompson 1982) in the anteroventral cochlear nucleus and the central nucleus of the inferior colliculus. There were absolutely no changes in back-ground or CS-evoked levels of neuronal activity for any recording in either structure over the course of training.

These data are strongly negative but still at the level of simple correlation, in this case the absence of a correlation. A much more powerful test of the possible role of the auditory relay nuclei in learning would be to manipulate the dependent behavioral variable – the occurrence of the learned NM response – while holding the independent variable – the acoustic stimulus – constant. We have developed just such a paradigm–signal detection (Kettner et al. 1980). In brief, the animal is trained, overtrained, and taken to threshold using a staircase procedure so that the animal responds 50 percent of the time to the same-intensity acoustic CS. A white noise CS is used to avoid problems related to tonotopic representation. At constant CS threshold a 25 percent reinforcement schedule is used with data collected only on the 75 percent of white-noise-alone trials. The behavioral NM response at threshold is extremely reliable and well behaved. It is in fact dichotomous, being clearly present on detection trials and completely absent on nondetection trials.

Learning and memory processes are fundamental to signal detection. In order to indicate behaviorally the detection of a threshold level stimulus the organism must make use of a learned response. When an organism is detecting a constant-intensity acoustic stimulus at threshold, the learned response is activated on 50 percent of the trials. The neuronal circuitry that plays an essential role in this memory retrieval must be activated on behavioral detection trials and either not activated or not sufficiently activated on nondetection trials. Consequently, any neuronal regions or systems in the brain that do show dichotomous, or at least differential, activation by detected and nondetected stimuli are candidate substrates for the learning–memory circuitry (note the hippocampal detection response in Figure 4.2). The only alternative is that such structures are a part of the "motor" system generating the learned behavioral response.

Conversely, neurons and circuits activated identically, by detected and nondetected stimuli are not a part of the neuronal substrate of the learning–memory circuitry. We have now completed an extensive analysis of the anteroventral cochlear nucleus, the central nucleus of the inferior colliculus, and the ventral division of the medial geniculate body, using multiple-unit recording and a single-unit analysis of the central nucleus of the inferior colliculus (87 cells). The central nucleus of the inferior colliculus is particularly important because it is an obligatory relay nucleus for all fibers of the primary auditory system ascending from the brain stem. Results are completely consistent and clear (see Figure 4.2). For every recording electrode in every primary auditory relay nucleus there is a substantial white noise–evoked unit response at threshold that is *identical* on detection and nondetection trials. In our judgment this provides very strong evidence that the principle relay nuclei of the auditory system are not a part of the neuronal plasticity that codes learning.

In sum, the primary auditory relay nuclei from the cochlear nucleus through the medial geniculate body are not a part of the memory trace–the neuronal plasticity that codes learning and memory. On the other hand, they obviously must transmit the information to other brain structures that the acoustic CS is occurring.

THE "ALPHA RESPONSE" PATHWAY

The relatively long latency of the standard NM/eyelid CR – typically at least 80 msec (see Figure 4.3) – and the similarly long CS–UCS interval necessary for learning have implications for the brain systems and pathways that underlie learning. For an auditory CS, there is a relatively direct pathway from the auditory system to the cranial motor nuclei, particularly the seventh nucleus, as indicated by the short latency of the startle or alpha response to a loud or sudden sound. Woody reports a latency of about 20 msec for the alpha eyelid response in the cat to a click stimulus (Woody and Brozek 1969). Woody's

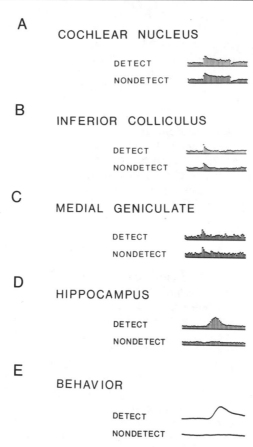

Figure 4.2. Comparison of multiple-unit responses from auditory relay nuclei and hippocampus during detect vs. nondetect trials. *A* through *D:* Average poststimulus histograms (15-msec time bins) created by averaging from 200 to 300 trials (obtained from several testing sessions) for cochlear nucleus (*A*), inferior colliculus (*B*), medial geniculate (*C*), and hippocampus (*D*) on detect (upper histogram) vs. nondetect (lower histogram) trials. *E:* Average nictitating membrane response for detect (upper trace) vs. nondetect (lower trace) trials. Note the large difference between hippocampal responses during detect vs. nondetect trials in comparison with no differences in responses to the auditory relay nuclei. (From Thompson et al. 1980)

learning paradigm, incidentally, is an interesting model of associative plasticity – pairing a click and glabellar tap leads to an increase in the "alpha" response that does not occur with unpaired stimulus presentations. The properties of this very-short-latency alpha eyelid response are quite different from those of the standard eyelid (and NM) conditioned response. In any event, if the essential neuronal plasticity coding the standard-delay learned NM/eyelid response develops within the alpha response pathway, the learned response would have a latency of about 20 msec. This is much too short. Consequently, the alpha response system would seem not to be the locus of the memory trace.

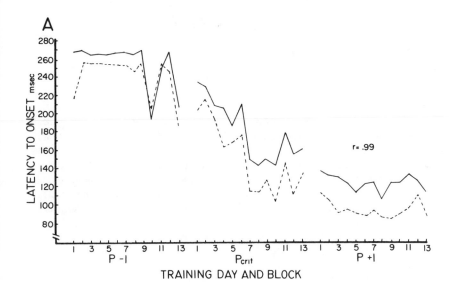

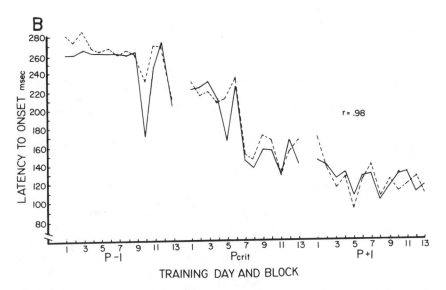

Figure 4.3. Latency to onset of the behavioral responses from the onset of the tone in milliseconds. Graph *A* shows the latencies found for the left eyelid and left NM of all eight animals. The solid line represents the latencies for the left NM while the dashed line represents the latencies for the left eyelid. Graph *B* represents the latencies for the right eyelid and left NM for six animals which showed bilateral conditioning. The solid line represents the latencies of the NM while the dashed line represents the latencies of the right eyelid. P_{crit} is the day on which the animals reached criterion performance of eight conditioned responses on any nine consecutive trials. $P - 1$ is the training day prior to P_{crit}, and $P + 1$ is the training day after P_{crit}. (From McCormick et al. 1982a)

THE UNCONDITIONED REFLEX RESPONSE PATHWAYS AND MOTOR NEURONS

The afferent limb of the reflex pathway for the NM response is limited to fibers of the trigeminal nerve. The efferent limb involves several cranial nerve nuclei. The primary response that produces NM extension is retraction of the eyeball (Cegavske et al. 1976). A major muscle action producing this response is contraction of the retractor bulbus, which in rabbit is innervated by motoneurons of the abducens and accessory abducens nuclei (Gray et al. 1981). However, it appears that most of the extrinsic eye muscles may contract synchronously with the retractor bulbus (Berthier and Moore 1980). In addition, the external eyelid (which is normally held open in NM conditioning) extends, under control of motoneurons in the seventh nucleus. Finally, there is a variable degree of contraction of facial muscles in the vicinity of the eye. In sum, the total response is a coordinated defense of the eye involving primarily eyeball retraction (NM extension) and eyelid closure with some contraction of periorbital facial musculature (see McCormick et al. 1982b).

Simultaneous recordings from one NM and both eyelids during conditioning of the left eye show essentially perfect correlations in both amplitude and latency of the conditioned responses as they develop over the course of training (see Figure 4.3). Other evidence indicates that the relevant motor nuclei (left accessory/abducens, left seventh and right seventh) are not tightly coupled (McCormick et al. 1982b). If the essential neuronal plasticity develops at the cranial motor nuclei, it would have to do so independently at each. If this were the case, it is very difficult to imagine how the several motor nuclei could generate perfectly correlated conditioned responses. Consequently, there would seem to be a common central system at some point that acts synchronously on all the cranial motor nuclei engaged in generation of the conditioned response.

Perhaps the simplest common central system would be some components of the reflex pathways. If it were possible to develop experimental manipulations that could independently vary the CR and the UCR, it would provide very strong evidence that the unconditioned reflex pathways – trigeminal afferents, interneurons, motoneurons – are not a critical part of the essential neuronal plasticity coding the learned response. Recently, we have succeeded in doing this for the conditioned response using morphine (Mauk et al. 1982b). In brief, animals are trained to criterion (8 CRs in 9 successive trials), given two 9-trial blocks of additional training, injected with morphine (5 to 10 mg/kg i.v.), run for five blocks, and then injected with naloxone (0.1 mg/kg). Results are striking (see Figure 4.4). There is an immediate and complete abolition of the CR but no effect at all on the UCR. In unpaired control animals, morphine has no effect at all on the UCR, the reflex response to corneal airpuff. Similarly, CRs can be selectively abolished by spreading depression of the cerebral cortex (Papsdorf et al. 1965; Megirian and Bures 1970) and by cerebellar lesions (McCormick et al. 1981). This provides strong evidence against essential participation of the unconditioned reflex response pathways and motor neurons in the memory trace.

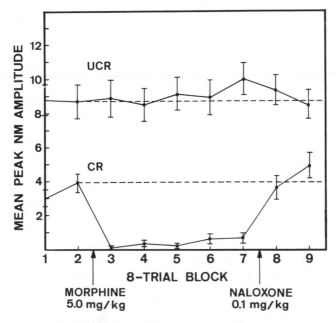

Figure 4.4. The effect of administration of morphine on conditioned (CR) and unconditioned (UCR) NM responses. Scores for the CR and UCR were determined by the peak amplitude of the NM extension during the CS and UCS periods, respectively. Dotted lines indicate baseline response amplitudes prior to morphine injection. (From Mauk et al. 1982b)

TWO PROCESSES OF LEARNING IN THE BRAIN?

The highly specific naloxone-reversible action of morphine on the conditioned response appears to provide a powerful tool for study of the learning circuitry in the brain. It must somehow inactivate some portion of the essential neuronal plasticity coding the learned response. Any brain structure or system showing learning-induced neuronal plasticity that is strongly and reversibly influenced by morphine becomes a candidate for the memory trace. One possibility is that conditioned aversiveness is an essential part of the memory trace in eyelid conditioning – an aversive component of the corneal airpuff becomes conditioned to the tone CS – a form of conditioned fear. If such is the case, then morphine might act on this aspect of the associative network. A large body of literature suggests that both morphine and the endogenous opioids act more on learned fear or anxiety than on pain per se (see, e.g., Jaffe and Martin 1980; Julien 1981; Martinez et al. 1981; Wikler, 1958).

Behavioral analyses of aversive learning suggest that it may occur as two processes or phases, the first involving classical conditioning of a central state (e.g., "conditioned fear") and the second concerned with learned performance of discrete, adaptive motor responses (Brush 1971; Konorski 1967; Miller 1948; Mowrer 1947; Prokasy 1972; Rescorla and Solomon 1967). The first process presumably involves primarily brain stem – hypothalamic mechanisms and hormonal actions, and the second would involve whatever mechanisms underlie the learning of the specific adaptive motor responses, in this case the NM and eyelid responses. Insofar as aversive learning is concerned, learning theorists have not necessarily specified that the process of "conditioned fear" occurs first and is essential for subsequent development of the second process – learning of the discrete adaptive response. In general agreement with Konorski (1967), we suggest that this may be the case. The first process is the basic association between a neutral stimulus and an aversive stimulus and/or its consequences. The second process involves learning of the set of adaptive motor responses best to deal with the situation.

Weinberger (1982) recently surveyed aversive classical conditioning in infrahuman animals in terms of rate of learning and noted two clearly distinct categories. "Nonspecific" responses, indices of conditioned "fear," are acquired in 5 to 15 trials, but specific skeletal muscle responses require many more trials, from 50 to several hundred. Nonspecific responses are so defined because they are not specific to the UCS; they do not permit the animal to avoid the UCS, whereas specific responses do. Nonspecific responses are mostly autonomic – heart rate, blood pressure, pupil diameter, galvanic skin response – but include nonspecific skeletal motor activity as well. Under normal conditions of aversive training both kinds of responses are learned. Evidence summarized below suggests that the neuronal substrates for these two aspects of aversive learning may differ, at least in part.

In this context it would seem reasonable to suggest that the selective morphine depression of the CR might be due to an action on the conditioned fear aspect of learning in the NM/eyelid paradigm, as noted above. Studies in which both the NM/eyelid and heart rate are recorded during conditioning (using a periorbital shock rather than corneal airpuff UCS) show that conditioned slowing of the heart rate develops in a few trials and then fades away as the discrete NM/eyelid response is learned (Powell et al. 1974; Schneiderman et al. 1969). Conditioned heart rate slowing seems a very good candidate for an autonomic expression of an initial process of conditioned fear. As the animal "masters" the situation and learns to make the discrete, adaptive NM/eyelid response, conditioned fear decreases. Interestingly, in current work we have found that the learned NM/eyelid response becomes relatively impervious to the action of morphine if animals are given overtraining.

Systemically administered morphine could potentially act in a variety of ways to abolish a conditioned response. The effect could be mediated by nonspecific action or by specific activation of one or more classes of opiate receptors. Moreover, activation of opiate receptors at both central (Kapp et al. 1979) and peripheral (Martinez et al. 1981) sites has been implicated in the mediation of opiate effects on aversive conditioning. Thus, as a first step in the localization and characterization of the opiate sensitive processes involved, we conducted a series of experiments examining the relative effects of both systemic and central administration of several opiate agonists and antagonists on the expression of NM/eyelid CRs.

In our initial studies we demonstrated that microinfusion of morphiceptin (Tyr-Pro-Phe-Pro-NH$_2$), a highly selective mu (opiate) receptor agonist

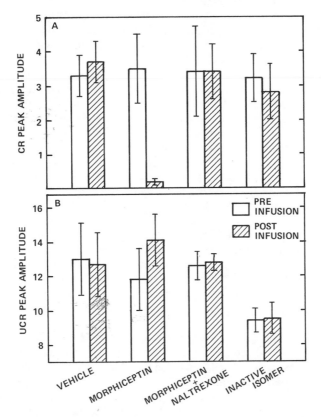

Figure 4.5. Mean conditioned response (*A*) and unconditioned response (*B*) peak amplitudes for the two training blocks prior to i.v. infusion (open bars) and for the three blocks following infusion (shaded bars). (From Mauk et al. 1982a)

(Chang et al. 1981) produces complete and selective abolition of CRs. This effect is blocked by concomitant administration of the opiate antagonist naltrexone (Figure 4.5). Administration of the (D-Pro^2) analog of morphiceptin, which has been previously shown to be void of opiate agonist activity (Chang et al. 1981), was also ineffective in abolishing CRs. The fact that this isomer was ineffective supports the stereospecific nature of the morphiceptin opiate-sensitive site interaction under investigation. In a subsequent study we found that central administration of as little as 12 nmol of (N-Me-Phe^3-D-Pro^4) morphiceptin, a potent, long-lasting, and highly mu receptor–specific analog (Chang et al. 1981), similarly produced a marked abolition of CRs. This effect was dose-dependent and completely reversed by subsequent administration of the opiate antagonist naloxone. Collectively, the effects of these substances on the expression of the CR are consistent with the known pharmacological properties that opiate agonists display through activation of opiate receptors, that is, high affinity, reversibility, stereospecificity, and the blockage and/or reversibility by opiate antagonists. Thus, we have accumulated strong evidence that opiate-induced abolition of conditioned responding is produced by direct, receptor-specific binding to opiate receptors. Further, highly specific binding to the mu receptor is sufficient to produce the effect. Whether selective binding to the mu receptor is necessary to produce the effect or whether binding to other opiate receptors (e.g., delta receptors) will also abolish CRs are important questions currently under investigation.

Having established that opiate effects on aversive classical conditioning are mediated by action at opiate receptors, the next question is the primary site of action. The experiments mentioned above demonstrated that central administration of very small doses of morphiceptin (200 nmol) and (N-Me-Phe^3-D-Pro^4) morphiceptin (12 nmol) were quite effective in abolishing conditioned responses. This alone suggests that the primary site of action is within the central nervous system. However, one may argue that these centrally administered peptides are leaving the brain in sufficient quantities to produce effects through peripheral action. To the extent that this is true, systemically administered morphiceptin should produce effects on CRs at doses equal to or even less than those effective centrally. We have tested systemically administered morphiceptin and (N-Me-Phe^3-D-Pro^4) morphiceptin in doses ranging from 0.1 to 10 times those administered centrally. A central site of action is supported by the fact that there was no effect on conditioned responding at these doses (Figure 4.6).

Evidence that this opiate effect on learned responses is mediated by central action is supported by an additional experiment that compared the effects of naloxone and its quaternary (Q-naloxone) analog, which does not cross the blood–brain barrier (Valentino et al. 1980, 1981), on reversing opiate-induced abolition of CRs. As noted above, low doses of naloxone rapidly and completely reverse the effects of opiates on CRs. In contrast, serial, systemic administration of Q-naloxone had no effect on morphine-induced abolition of the CR (Figure 4.7). The range of doses utilized were comparable in potency to the dose of naloxone subsequently used to completely reverse the effect of morphine.

In summary, we have demonstrated that opiates produce complete and selective abolition of aversively motivated conditioned responses. This effect is the result of specific binding to opiate receptors; in fact, highly selective activation of mu receptors is sufficient to produce a complete effect. Further, the essential site of action is within the central nervous system. Preliminary analysis indicates that the periaqueductal gray–periventricular region of the fourth ventricle appears to be a particularly sensitive site for producing this opiate-receptor–mediated abolition of the CR, relative to other sites studied. Microinfusion of comparable amounts of morphiceptin into the lateral ventricles produced an abolition of the CR; however, this effect typically had a delayed onset and a considerably shorter and more variable action. Moreover, bilateral microadministration of this peptide into either the medial septal region or the amygdaloid complex had no effect on the CR.

Finally, in current work we have found that infusion of the same small amount of (N-Me-Phe^2-D-Pro^4) morphiceptin (12 nmol) into the fourth ven-

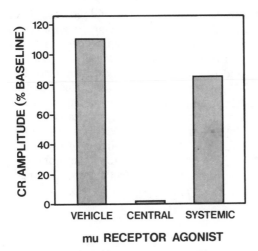

Figure 4.6. Central administration (fourth ventricle) of (N-Me-Phe³-D-Pro⁴) morphiceptin abolishes the conditioned NM/eyelid response (but has no effect on the unconditioned response and is blocked by naltrexone). One group of animals received central infusion of the vehicle (left box), another group received central infusion of the drug (center box). Peripheral (i.v.) injection of the drug in other animals (right box) had no effect at doses from 0.1 to 10 times the central dose. (Adapted from Mauk et al. 1982a)

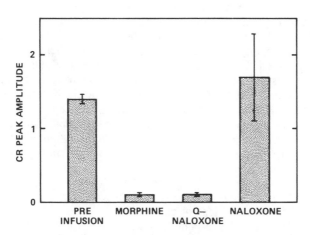

Figure 4.7. Mean conditioned response peak amplitudes prior to any injection, after i.v. administration of morphine, following several doses of Q-naloxone, and finally after injection of naloxone. (From Mauk et. al. 1982a)

tricle that abolishes the just-learned NM/eyelid response also naloxone-reversibly abolishes the conditioned heart-rate-slowing response (Lavond et al. 1982). It is tempting to suggest that one and the same neuronal system in the vicinity of the fourth ventricle codes the learned fear that is indexed by conditioned heart rate slowing and that is the necessary driving force for learning of the discrete, adaptive NM/eyelid response. However, this is at present only a speculation. The amygdala and a portion of the hypothalamus are also essential for conditioned heart rate response in several species (Cohen and MacDonald 1974; Kapp et al. 1979; Pribram et al. 1979; Smith et al. 1980). Interestingly, administration of opiates to the central nucleus of the amygdala markedly impairs the conditioned heart rate response in the rabbit (Kapp et al. 1982; Gallagher et al. 1981) but has no effect on the just-learned NM/eyelid response (Mauk et al. 1982b).

THE CEREBELLUM: LOCUS OF THE MEMORY TRACE FOR DISCRETE LEARNED RESPONSES?

If the selective opiate action on learned responses is in fact primarily on conditioned fear, we are still left with the question of the possible neuronal substrates of the second process – learning of discrete adaptive motor responses. Current evidence from our laboratory suggests that the cerebellum is critically involved in this phase or process of learning.

In the course of our signal detection studies we obtained recordings from 10 electrode placements mostly in the cerebellar cortex just dorsal to the cochlear nucleus (Kettner 1981; Kettner and Thompson 1982). These recordings showed a clear dichotomous response – a substantial increase in unit firing on detection trials that closely modeled the behavioral detection response and an absence of such a response on nondetection trials (see Figure 4.8). Interestingly, there was typically a short-latency noise-onset-evoked auditory response that was identical on both detection and nondetection trials.

As noted above, any brain structure that exhibits a dichotomous or differential response on detection and nondetection trials is a candidate for the decision–memory system – either that or a part of the motor system generating the behavioral response. At the level of the motor nuclei, the neuronal response always closely models the behavioral response, whether learned or reflexive, and would of course show a dichotomous response on detection versus nondetection trials. The cerebellum is commonly viewed as a "motor" structure and might be considered a part of the neuronal system generating the behavioral response in the detection paradigm. However, the neuronal detection response in the cerebellum precedes the behavioral detection response by about 50 msec (see Figure 4.8), too long a time period even to participate in the reflex response. (With the conditions of our studies, the mean onset latency of the unconditioned reflex eyelid response is 7.3 msec and the NM is 22.4 msec.)

Since 1980 we have been in the process of completing an extensive and detailed mapping of the midbrain–brain stem, recording neuronal activity in already trained animals (McCormick et al. 1983). For this purpose we developed a chronic micromanipulator system that permits mapping of unit activity

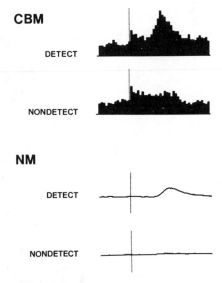

Figure 4.8. Histograms of neuronal activity recorded from cerebellar cortex (CBM) dorsal to cochlear nucleus on detection (above) and nondetection (below) trials to an identical threshold-level white noise stimulus (225 detection trials and 225 nondetection trials). (Note the identical short latency stimulus-onset-evoked response on both and the later "decision–memory" response that occurs only on detection trials.) Behavioral nictitating membrane (NM) detection and nondetection responses summed over the same trials are shown in the lower panel. Vertical line indicates white noise onset. (From Kettner and Thompson 1982)

in a substantial number of neural loci per animal. Data from the mapping study will not be given here in any detail, except to indicate that learning-related increases in unit activity – an increase in unit activity that forms a temporal model within a trial of the learned behavioral response – were prominent in certain regions of the cerebellum, both in cortex and deep nuclei, certain regions of the pontine nuclei, the red nucleus, the inferior olive, and the motor cortex. Such unit activity is also seen in certain regions of the reticular formation and of course in the cranial motor nuclei engaged in generation of the behavioral response – portions of the third, sixth, and accessory sixth and seventh nuclei. The results to date of the mapping study point to substantial engagement of the cerebellar system in the generation of the conditioned response.

Current studies in which we have recorded the neuronal unit activity from the deep cerebellar nuclei (dentate and interpositus nuclei) over the course of training have in some locations revealed a striking pattern of learning-related growth in activity (McCormick et al. 1982a). In the example shown in Figure 4.9, the animal did not learn on day 1 of training. Unit activity showed evoked responses to tone and airpuff onsets but no response in association with the reflex NM response, in marked contrast to unit recordings from the cranial

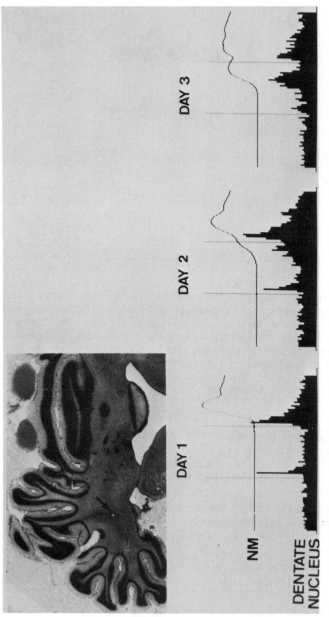

Figure 4.9. Histograms of unit cluster recordings obtained from the medial dentate nucleus during classical conditioning of NM/eyelid response. The recording site is indicated by the arrow. Each histogram bar is 9 msec in width and each histogram is summed over an entire day of training. The first vertical line represents the onset of the tone and the second vertical line represents the onset of the airpuff. The trace above each histogram represents the averaged movement of the animal's NM for the same day, with up being extension of the NM across the cornea. The total duration of each histogram and trace is 750 msec. The pattern of increased discharges of cerebellar neurons appears to develop a neuronal "model" of the amplitude–time course of the learned behavioral response. (From McCormick et al. 1982b)

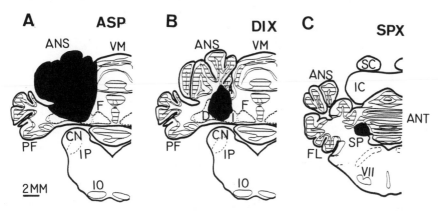

Figure 4.10. Reconstructions of cerebellar lesions effective in abolishing the ipsilateral conditioned NM/eyelid response. *A:* Typical unilateral aspiration of the lateral cerebellum and dentate/interpositus nuclei. *B:* Unilateral electrolytic lesion of the dentate/interpositus nuclei (DIX) in which the overlying cortex is spared. *C:* Localized unilateral lesion of the superior cerebellar peduncle (SPX). All reconstructions are through the broadest extent of each lesion. Abbreviations are as follows: ANS, ansiform lobe; CN, cochlear nucleus; D, dentate nucleus; F, fastigial nucleus; ANT, anterior lobe; FL, flocculus; I, interpositus nucleus; IC, inferior colliculus; IO, inferior olive; IP, inferior cerebellar peduncle; PF, paraflocculus; SC, superior colliculus; SP, superior cerebellar peduncle; VM, vermal lobes; VII, seventh nucleus.

motor nuclei. On day 2 the animal began showing CRs, and unit activity in the dentate nucleus developed a model of the conditioned response. On day 3 the learned behavioral response and the cerebellar model of the learned response are well developed, but there is still no clear model of the reflex behavioral response. The cerebellar unit model of the learned response precedes the behavioral response significantly in time. In this animal, a neuronal model of the *learned* behavioral response appears to develop de novo in the cerebellum. (Note the contrast between this response and the learning-induced response in the hippocampus that models the entire behavioral response – see Figure 4.1.)

In current work, we have found that lesions ipsilateral to the trained eye in several locations in the neocerebellum – large ablations of the lateral portion of the hemisphere, localized electrolytic lesions of the dentate/interpositus nuclei and surrounding fibers, and small, discrete lesions of the superior cerebellar peduncle – permanently abolish the CR but have no effect on the UCR and do not prevent learning by the contralateral eye (McCormick et al. 1981, 1982a, b; Thompson et al. 1982a, b; Clark et al. 1982) Figures 4.10 to 4.12). If training is given after unilateral cerebellar lesion, the ipsilateral eye cannot learn, but the contralateral eye subsequently learns as though the animal is normal and new to the situation (Lincoln et al. 1982) (see Figure 4.12). If training is given before unilateral cerebellar lesion, the contralateral eye learns rapidly, with significant savings (McCormick et al. 1981, 1982b). Lesions in several locations in the ipsilateral pontine brain stem produce a similar selective abolition of the CR (Desmond and Moore 1982; Lavond et al. 1981; Moore

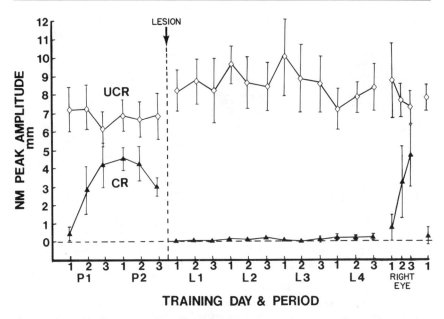

Figure 4.11. Effects of ablation of left lateral cerebellum on the learned NM/eyelid response (six animals). Solid triangles: Amplitude of conditioned response (CR). Open diamonds: Amplitude of unconditioned response (UCR). All training was to left eye (ipsilateral to lesion) except where labeled "Right Eye." The cerebellar lesion completely and permanently abolished the CR of the ipsilateral eye but had no effect on the UCR. P1 and P2 indicate initial learning on the two days prior to the lesion. L1–L4 are four days of postoperative training to the left eye. The right eye was then trained and learned rapidly. The left eye was again trained and showed no learning. Numbers on abscissa indicate 40-trial periods, except for "Right Eye," which are 24-trial periods. (From McCormick et al. 1982b)

et al. 1982; Thompson et al. 1982a). Although some uncertainty still exists, the learning-effective lesion sites in the pontine brain stem appear to track the course of the superior cerebellar peduncle.

Taken together, these results indicate that the cerebellum is an obligatory part of the learned response circuit for eyelid/NM conditioning. Since decerebrate animals can learn the response, this would seem to localize an essential component of the memory trace to the ipsilateral cerebellum and/or its major afferent/efferent systems. The fact that a neuronal unit "model" of the learned behavioral response develops in the cerebellar deep nuclei and may precede the behavioral response by as much as 50 msec would seem to localize the process to cerebellum or its afferents. This time period, interestingly, is not too much less than the minimum onset latency for CRs in well-trained animals (about 80 msec) and is very close to the minimum CS–UCS interval that can support learning.

The possibility that unilateral cerebellar lesions produce a modulatory disruption of a memory trace localized elsewhere in the brain seems unlikely. If

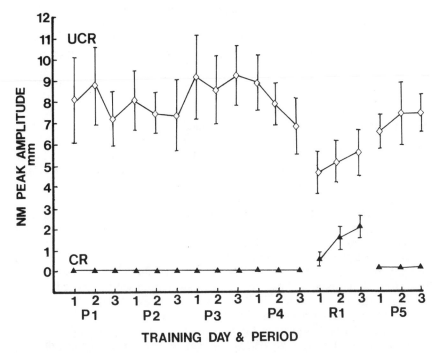

Figure 4.12. Effects of ablation of left lateral cerebellum on learning of the nictitating membrane (and eyelid) responses (six animals). Solid triangles: Amplitude of conditioned response (CR). Open diamonds: Amplitude of unconditioned response (UCR). All training was to left eye (ipsilateral to lesion) except where labeled R1. The cerebellar lesion prevented conditioning of the ipsilateral eye but had no effect on the UCR. P1–P4 indicate the four days of postlesion training to the left eye. The right eye was then trained (R1) and learned at a rate comparable to that of initial learning of nonlesioned animals. The left eye was again trained (P5) and showed no learning. Numbers on abscissa indicate 40-trial blocks. (From Lincoln et al. 1982)

so, it must be efferent from the cerebellum, since discrete lesions of the superior cerebellar peduncle abolish the behavioral learned response. Yet the neuronal model of the learned response is present within the ipsilateral cerebellum. In current studies we have trained animals with both eyes and made bilateral ablations of the lateral cerebellum. These lesions permanently abolish the conditioned NM/eyelid response on both sides. One such animal, which shows no obvious signs of motor dysfunction, has been trained repeatedly postoperatively over a period of 3 months and shows no relearning or recovery at all of the learned response in either eye.

Finally, in current work in progress we have obtained evidence that the cerebellum is also essential for classical conditioning of the hindlimb flexion reflex. Rabbits are initially trained with a 2.0-mA shock (60 Hz, 100 msec) to the left hind paw using the same tone CS as in NM/eyelid conditioned (350 msec, 1 kHz, 85 dB, interstimulus interval 250 msec). Animals are given 45 trials – 30 paired, 10 CS-alone, and 5 UCS alone – per day with a day with a 1.5-min intertrial interval. Electromyographic (EMG) activity is recorded from flexor muscles of both hindlimbs. Interestingly, in the rabbit both hindlimbs develop an equivalent conditioned flexor response, consistent with the fact that rabbits typically move their hindlimbs synchronously together when they locomote.

Ablation of the left lateral cerebellum permanently abolishes this conditioned flexor response in both hindlimbs when training is continued to the left hindlimb (see Figure 4.13). Training was then given to the right hindlimb, and both hindlimbs developed clear conditioned responses. Training was then shifted back to the left hindlimb and the conditioned response in both hindlimbs extinguished, even though reinforced training continued to the left hindlimb. These results demonstrate that the cerebellar lesion does not simply prevent the animal from making the conditioned response in the left hindlimb – normal CRs are given by the left hindlimb when training is given to the right hindlimb. This finding supports the view that the memory trace for classical conditioning of the hindlimb flexor response is established unilaterally in the cerebellar hemisphere ipsilateral to the leg being trained.

From these results, we suggest that the cerebellum may be the locus of the primary memory trace coding the learning of specific adaptive responses, at least for all instances of classical conditioning of discrete striated muscle responses with an aversive UCS. Since instrumental avoidance training is classical until the occurrence of the first avoidance response, we suggest that the cerebellum may be the locus of the memory trace for instrumental avoidance learning as well. Interestingly, there is an earlier Soviet report indicating that in two dogs well trained in leg flexion conditioning complete removal of the cerebellum apparently permanently abolished the ability of the animals to make a discrete leg flexion response but did not abolish conditioned generalized motor activity (see description by Karamian et al. 1969).

The cerebellum has been suggested by several authors as a possible locus for the coding of learned motor responses (Albus 1971; Eccles et al. 1967; Ito 1970; Marr 1969). Cerebellar lesions impair a variety of skilled movements in animals (Brooks 1979; Brooks et al. 1973). In addition, neuronal recordings from Purkinje cells of the cerebellar cortex have implicated these cells in the plasticity of various responses (Gilbert and Thach 1977; Dufosse et al. 1978).

It seems a very reasonable possibility that learning of classical and instrumental discrete, adaptive motor responses occurs in the cerebellum. Perhaps the most prominent feature of such learned responses is their precise timing. At least in aversive learning, the CR is under very strong control by the CS–UCS interval in terms of onset latency and temporal morphology and is always timed to be maximum at or shortly before the time when the onset of the UCS occurs. The cerebellum is very well designed to provide such precise timing (see, e.g., Eccles et al. 1967).

In terms of the relation between the initial process of "conditioned fear" and the subsequent development of the discrete adaptive striated muscle CR, we suggested above that conditioned fear is necessary for later development of the specific adaptive response. However, the latter may develop some degree of independence. Consistent with this is the fact that infusion of certain opioids into the fourth ventricle abolished both the conditioned heart rate response and

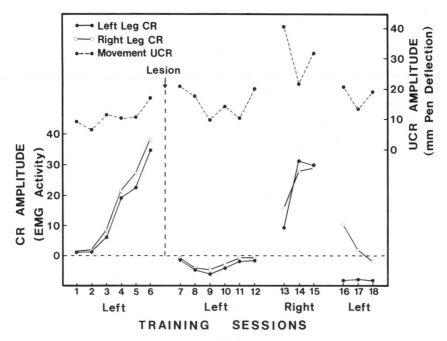

Figure 4.13. Example of the effects of lesioning the cerebellar deep nuclei ipsilateral to the side of training on leg flexion conditioned responses (CRs) and unconditioned responses (UCRs) in the rabbit. In each session 30 training trials were presented wherein a 350-msec auditory CS overlapped and terminated with a 100-msec paw shock UCS. During sessions 1–12 and 16–18 the shock UCS was delivered to the left paw (ipsilateral to the side of the lesion), and during sessions 13–15 the shock UCS was shifted to the right paw. The CR was measured as EMG activity recorded from flexor muscles in the left and right hind legs. The bilateral leg flexion UCRs to the paw shock UCS were measured by means of a stabilimeter device. As can be seen, initial learning to left paw shock is bilateral. Lesioning the left cerebellar deep nuclei resulted in abolition of CRs in both legs but had no effect on UCR amplitude. Switching the shock UCS to the paw contralateral to the lesion (right side) resulted in reacquisition of bilateral CRs. Switching the UCS back to the lesioned side resulted in a loss of the CR. (From Donegan et al. 1983)

the just-learned NM/eyelid response but that overtraining protects the conditioned NM/eyelid response against the effects of opiates. Such results are nicely consistent with our data, suggesting that the memory trace for discrete adaptive responses becomes established in the cerebellar system. Once established, the cerebellar memory trace can develop some degree of functional autonomy. The aversive UCS is still of course necessary to maintain the specific conditioned response. In its absence the CR will eventually extinguish.

This neuronal dual-trace hypothesis may also account for the fact that, following extinction of a CR, if the CS–UCS pairings are reintroduced, relearning proceeds much more rapidly than initial acquisition (see Scavio and

Thompson 1979 for an example of the savings effect seen in the rabbit NM preparation). Once established, the memory trace for the discrete response in the cerebellar system may remain relatively unchanged but the conditioned fear memory trace may not; that is, only the latter may extinguish. Because acquisition of conditioned fear requires only a few trials, reacquisition, for example, of the NM CR would also be rapid. Once reestablished, conditioned fear can again activate the cerebellar trace system.

A final implication of our neuronal dual-trace hypothesis is that the memory trace for the discrete, adaptive learned response in the cerebellar system would presumably not be essential for the development of the "conditioned fear" memory trace. In current work we have found that following bilateral neocerebellar lesions, rabbits can learn the conditioned heart rate slowing response but are of course unable to learn the conditioned NM/eyelid response with either eye (Lavond et al. in press).

ACKNOWLEDGMENTS

The work reported here was supported in part by research grants from the National Science Foundation (BNS 81-17115), the National Institutes of Health (NS23368), the National Institute of Mental Health (MH26530), and the McKnight Foundation.

REFERENCES

Albus, J. S. 1971. A theory of cerebellar function. *Math. Biosci. 10*:25–61.

Alkon, D. L. 1979. Voltage-dependent calcium and potassium ion conductances: A contingency mechanism for an associative learning model. *Science 205*:810–816.

Berger, T. W., Clark, G. A., and Thompson, R. F. 1980a. Learning-dependent neuronal responses recorded from limbic system brain structures during classical conditioning. *Physiol. Psychol. 8*(2):155–167.

Berger, T. W., Laham, R. I., and Thompson, R. F. 1980b. Hippocampal unit-behavior correlations during classical conditioning. *Brain Res. 193*:229–248.

Berger, T. W., and Thompson, R. F. 1978a. Identification of pyramidal cells as the critical elements in hippocampal neuronal plasticity during learning. *Proc. Nat. Acad. Sci. 75*(3):1572–1576.

1978b. Neuronal plasticity in the limbic system during classical conditioning of the rabbit nictitating membrane response. I. The hippocampus. *Brain Res. 145*(2):323–346.

Berthier, N. E., and Moore, J. W. 1980. Role of extraocular muscles in the rabbit *(Oryctolagus cuniculus)* nictitating membrane response. *Physiol. Behav. 2*:931–937.

Brons, J. F., and Woody, C. D. 1980. Long term changes in excitability of cortical neurons after Pavlovian conditioning and extinction. *J. Neurophysiol. 44*(3):605–615.

Brooks, V. B. 1979. Control of intended limb movements by the lateral and intermediate cerebellum. In *Integration in the Nervous System* (H. Asanuma and V. J. Wilson, eds.), pp. 321–356. New York: Igaku-Shoin Press.

Brooks, V. B., Kozlovskaya, I. B., Atkin, A., Horvath, F. E., and Uno, M. 1973.

Effects of cooling dentate nucleus on tracking-task performance in monkeys. *J. Neurophysiol. 36*:974–995.

Brush, F. R. (ed.) 1971. *Aversive Conditioning and Learning.* New York: Academic Press.

Buchwald, J. S., Halas, E. S., and Schramm, S. 1966. Changes in cortical and subcortical unit activity during behavior conditioning. *Physiol. Behav. 1*:11–22.

Castellucci, V., and Kandel, E. R. 1976. An invertebrate system for the cellular study of habituation and sensitization. In *Habituation: Perspectives from Child Development, Animal Behavior, and Neurophysiology* (T. J. Tighe and R. N. Leaton, eds.). New York: Wiley.

Cegavske, C. F., Thompson, R. F., Patterson, M. M., and Gormezano, I. 1976. Mechanisms of efferent neuronal control of the reflex nictitating membrane response . in the rabbit. *J. Comp. Physiol. Psychol. 90*:411–423.

Chang, K. J., Killian, A. Hazum, E., and Cuatrecasas, P. 1981. Morphiceptin (NH_4-TYR-PRO-PHE-PRO-$CONH_2$): A potent and specific agonist for morphine (M) receptors. *Science 212*:75–77.

Clark, G. A., McCormick, D. A., Lavond, D. G., Baxter, K., Gary, W. J., and Thompson, R. F. 1982. Effects of electrolytic lesions of cerebellar nuclei on conditioned behavioral and hippocampal neuronal responses. *Soc. Neurosci. Abstr. 8*:22.

Cohen, D. H. 1980. The functional neuroanatomy of a conditioned response. In *Neural Mechanisms of Goal-Directed Behavior and Learning.* (R. F. Thompson, L. H. Hicks, and B. V. Shryrkov, eds.), pp. 283–302. New York: Academic Press.

Cohen, D. H., and MacDonald, R. L. 1974. A selective review of central neural pathways involved in cardiovascular control. In *Cardiovascular Psychophysiology* (P. A. Obrist, A. H. Black, J. Brener, and L. V. DiCara, eds.), Chicago: Aldine.

Davis, W. J., and Gillette, R. 1978. Neural correlate of behavioral plasticity in command neurons of *Pleurobranchaea. Science, 199*:801–804.

Desmond, J. E., and Moore, J. W. 1982. Brain stem elements essential for classically conditioned but not unconditioned nictitating membrane response. *Physiol. Behav. 28*:1029–33.

Disterhoft, J. F., Shipley, M. T., and Krans, N. 1982. Analyzing the rabbit NM conditioned reflex arc. In *Conditioning: Representation of Involved Neural Functions* (C. D. Woody, ed.), pp. 433–49. New York: Plenum Press.

Disterhoft, J. F., and Stuart, D. K. 1976. The trial sequence of changed unit activity in auditory system of alert rat during conditioned response acquisition and extinction, *J. Neurophysiol. 39*:266–281.

1977. Differentiated short latency response increases after conditioning in inferior colliculus neurons of alert rat. *Brain Res. 130*:315–333.

Donegan, N. H., Lowry, R. W. and Thompson, R. F. 1983. Effects of lesioning cerebellar nuclei on conditioned leg-flexion responses. *Neuroscience Abstr. 9*:331 (Abstract No. 100.7).

Dufosse, M., Ito, M., Jastreboff, P. J., and Miyashita, Y. 1978. A neuronal correlate in rabbit's cerebellum to adaptive modification of the vestibuloocular reflex. *Brain Res. 195*:611–616.

Durkovic, R. G. 1975. Classical conditioning, sensitization, and habituation of the flexion reflex of the spinal cat. *Physiol. Behav. 14*:297–304.

Eccles, J. C., Ito, M., and Szentagothai, J. 1967. *The Cerebellum as a Neuronal Machine.* New York: Springer-Verlag.

Enser, D. 1976. Ph.D. thesis. University of Iowa.

Gabriel, M. 1976. Short-latency discriminative unit response: Engram or bias? *Physiol. Psychol., 4*(3):275–280.

Gabriel, M., Foster, K., and Orona, E. 1980. Unit activity in cingulate cortex and anteroventral thalamus during acquisition and overtraining of discriminative avoidance behavior in rabbits. In *Neural Mechanisms of Goal-Directed Behavior and Learning* (R. F. Thompson, L. H. Hicks, and V. B. Shvyrkov, eds.). New York: Academic Press.

Gabriel, M., Saltwich, S. L., and Miller, J. D. 1975. Conditioning and reversal of short-latency multiple-unit responses in the rabbit medial geniculate nucleus. *Science 189*:1108–1109.

Gilbert, P. F. C., and Thach, W. T. 1977. Purkinje cell activity during motor learning. *Brain Res. 128*:309–328.

Gormezano, I., Schneiderman, N., Deaux, E., and Fuentes, I. 1962. Nictitating membrane: Classical conditioning and extinction in the albino rabbit. *Science 138*:33–34.

Gray, T. S., McMaster, S. E., Harvey, J. A., and Gormezano, I. 1981. Localization of retractor bulbi motoneurons in the rabbit. *Brain Res., 226*:93–106.

Hoehler, F. K., and Thompson, R. F. 1979. The effect of temporal single-alternation on learned increases in hippocampal unit acitivty in classical conditioning of the rabbit nictitating membrane response. *Physiol. Psychol. 7*:345–351.

1980. Effect of the interstimulus (CS–UCS) interval on hippocampal unit activity during classical conditioning of the nictitating rabbit, *Oryctolagus cuniculus. J. Comp. Physiol. Psychol. 94*:201–215.

Ito, M. 1970. Neurophysiological aspects of the cerebellar motor control system. *Int. J. Neurol. 7*:162–176.

Jaffe, J. H., and Martin, W. R. 1980. Opioid analgesics and antagonists. In *The Pharmacological Basis of Therapeutics* (A. G. Goodman, L. S. Goodman, and A. Gilman, eds.). Macmillan, New York.

Julien, R. M. 1981. *A Primer of Drug Action.* San Francisco: Freeman.

Kapp, B. S., Frysinger, R. C., Gallagher, M., and Haselton, J. R. 1979. Amygdala central nucleus lesions: effect on heart rate conditioning in the rabbit. *Physiol. Behav. 23*:1109–1117.

Kapp, B. S., Gallagher, M., Applegate, C. D., and Frysinger, R. C. 1982. The amygdala central nucleus: Contributions to conditioned cardiovascular responding during aversive pavlovian conditioning in the rabbit. In *Neural Substrates of Conditioning* (C. D. Woody, ed.), pp. 581–99. New York: Plenum Press.

Karamian, A. I., Fanardijiam, V. V., and Kosareva, A. A. 1969. The functional and morphological evolution of the cerebellum and its role in behavior. In *Neurobiology of Cerebellar Evolution and Development, First International Symposium,* (R. Llinas, ed.). Chicago: American Medical Association.

Kettner, R. N. 1981. Ph.D. thesis. University of California, Irvine. ·

Kettner, R. N., Shannon, R. V., Nguyen, T. M., and Thompson, R. F. 1980. Simultaneous behavioral and neural (cochlear nucleus) measurement during signal detection in the rabbit. *Percep. Psychophys. 28*(6):504–513.

Kettner, R. E., and Thompson, R. F. 1982. Auditory signal detection and decision processes in the nervous system. *J. Comp. Physiol. Psychol. 96*:328–331.

Kitzes, L. M., Farley, G. R., and Starr A. 1978. Modulation of auditory cortex unit activity during the performance of a conditioned response. *Exp. Neurol. 62*:678–698.

Konorski, J. 1967. *Integrative Activity of the Brain.* Chicago: University of Chicago Press.

Lavond, D. G., Lincoln, J. S., McCormick, D. A. and Thompson, R. F. In press. Effect of bilateral lesions of the dentate and interpositus cerebellar nuclei on conditioning of heart-rate and nictitating membrane/eyelid responses in the rabbit. *Brain Res.*

Lavond, D. G., McCormick, D. A., Clark, G. A., Holmes, D. T., and Thompson, R. F. 1981. Effects of ipsilateral rostral pontine reticular lesions on retention of classically conditioned nictitating membrane and eyelid responses. *Physiol. Psychol. 9*:335–339.

Lavond, D. G., Mauk, M. D., Madden IV, J., Barchas, J. D., and Thompson, R. F. 1982. Central opiate effect on heart-rate conditioning. *Soc. Neurosci. Abstr., 8*:319.

Lincoln, J. S., McCormick, D. A., and Thompson, R. F. 1982. Ipsilateral cerebellar lesions prevent learning of the classically conditioned nictitating membrane/eyelid response. *Brain Res. 242*:190–193.

Lonsbury-Martin, B. L., Martin, G. K., Schwarts, S. M., and Thompson, R. F. 1976. Neural correlates of auditory plasticity during classical conditioning in the rabbit. *J. Acoust. Soc. Am. 60*:S82.

McCormick, D. A., Clark, G. A., Lavond, D. G., and Thompson, R. F. 1982a. Initial localization of the memory trace for a basic form of learning. *Proc. Nat. Acad. Sci. 79*:2731–2735.

McCormick, D. A., Lavond, D. G., Clark, G. A., Kettner, R. E., Rising, C. E., and Thompson, R. F. 1981. The engram found? Role of the cerebellum in classical conditioning of nictitating membrane and eyelid response. *Bull. Psychonom. Soc. 18*:103–105.

McCormick, D. A., Lavond, D. G., and Thompson, R. F. 1982b. Concommitant classical conditioning of the rabbit nictitating membrane and eyelid response: Correlations and implications. *Physiol. Behav. 28*:769–775.

McCormick, D. A., Lavond, D. G., and Thompson, R. F. 1983. Neuronal responses of the rabbit brainstem during performance of the classically conditioned nictitating membrane (NM)/eyelid response.

Marr, D. 1969. A theory of cerebellar cortex. *J. Physiol. 202*:437–470.

Martinez, Jr., J. L., Jensen, R. A., Mesing, R. B., Righter, H., and McGaugh, J. L. 1981. *Endogenous Peptides and Learning and Memory Processes.* San Francisco: Academic Press.

Mauk, M. D., Madden IV, J., Barchas, J. D., and Thompson, R. F. 1982a. Opiates and classical conditioning: Selective abolition of conditioned responses by activation of opiate receptors within the central nervous system. *Proc. Nat. Acad. Sci. 79:*7598–7602

Mauk, M. D., Warren, J. T., and Thompson, R. F. 1982. Selective, naloxone-reversible morphine depression of learned behavioral and hippocampal responses. *Science 216*:434–435.

Megirian, D., and Bures, J. 1970. Unilateral cortical spreading depression and conditioned eyeblink responses in the rabbit. *Exp. Neurol. 27*:34–45.

Miller, N. E. 1948. Studies of fear as an acquirable drive. I. Fear as motivation and fear-reduction as reinforcement in learning of new responses. *J. Exp. Psychol. 38*:89–101.

Moore, J. W., Desmond, J. E., and Berthier, N. E. 1982. The metencephalic basis of the conditioned nictitating membrane response. In *Conditioning: Representation of Involved Neural Functions* (C. D. Woody, ed.), pp. 459–482. New York: Plenum Press.

Mowrer, O. H. 1947. On the dual nature of learning: A reinterpretation of "conditioning" and "problem-solving." *Harvard Educ. Rev. 17*:102–148.

Norman, R. J., Buchwald, J. S., and Villablanca, J. R. 1977. Classical conditioning with auditory discrimination of the eyeblink in decerebrate cats. *Science 196*:551–553.

Oakley, D. A., and Russell, I. S. 1972. Neocortical lesions and classical conditioning. *Physiol. Behav. 8*:915–926.

Olds, J., Disterhoft, J. F., Segal, M., Hornblith, C. L., and Hirsch, R. 1972. Learning centers of rat brain mapped by measuring latencies of conditioned unit responses. *J. Neurophysiol. 35*:202–219.

Oleson, T. D., Ashe, J. H., and Weinberger, N. M. 1975. Modification of auditory and somatonsensory system activity during pupillary conditioning in the paralyzed cat. *J. Neurophysiol. 38*:1114–39.

Orr, W. B., and Berger, T. W. 1981. Hippocampal lesions disrupt discrimination reversal learning of the rabbit nictitating membrane response. *Neurosci. Abstr.* 7:648.

Papsdorf, J. D., Longman, D., and Gormezano, I. 1965. Spreading depression: Effects of applying KCl to the dura of the rabbit on the conditioned nictitating membrane response. *Psychonom. Sci.* 2:125–126.

Patterson, M. M. 1976. Mechanisms of classical conditioning of spinal reflexes. In *Advances in Psychobiology,* vol. 3 (A. H. Riesen and R. F. Thompson, eds.). New York: Wiley.

Patterson, M. M., Cegavske, C. F., and Thompson, R. F. 1973. Effects of a classical conditioning paradigm on hindlimb flexor nerve response in immobilized spinal cat. *J. Comp. Physiol. Psychol. 84:*88–97.

Powell, D. A., Lipkin, M., and Milligan, W. L. 1974. Concomitant changes in classically conditioned heart rate and corneoretinal potential discrimination in the rabbit *(Oryctolagus cuniculus)*. *Learn. Motiv.* 5:532–547.

Pribram, K. H., Reitz, S., McNeil, M., and Spevack, A. A. 1979. The effect of amygdalectomy on orienting and classical conditioning in monkeys. *Pavlov. J. Biol. Sci. 14*:203–217.

Prokasy, W. F. 1972. Developments with the two-phase model applied to human eyelid conditioning. vol. 2, In *Classical Conditioning, Current Research and Theory,* (A. H. Black and W. F. Prokasy, eds.), pp. 119–147. New York: Appleton-Century-Crofts.

Rescorla, R. A., and Solomon, R. L. 1967. Two-process learning theory: Relationships between Pavlovian conditioning and instrumental learning. *Psychol. Rev. 74*:151–182.

Ryugo, D. K., and Weinberger, N. M. 1978. Differential plasticity of morphologically distinct neuron populations in the medial geniculate body of the cat during classical conditioning. *Behav. Biol. 22*:275–310.

Scavio, M. J., and Thompson, R. F. 1979. Extinction and reacquisition performance alternations of the conditioned nictitating membrane response. *Bull. Psychonom. Soc. 13*(2):57–60.

Schneiderman, N., VanDercar, D. H., Yehle, A. L., Manning, A. A., Golden, T., and Schneiderman, E. 1969. Vagal compensatory adjustment: Relationship to heart rate classical conditioning in rabbits. *J. Comp. Physiol. Psychol. 68*:175–183.

Segal, M., and Olds, J. 1973. The activity of units in the hippocampal circuit to the rat during classical conditioning. *J. Comp. Physiol. Psychol. 82*:195–204.

Smith, O. A., Astley, C. A., DeVito, J. L., Stein, J. M., and Walsh, K. E. 1980. Functional analysis of hypothalamic control of the cardiovascular responses accompanying emotional behavior. *Fed. Proc. 39*(8):2487–2494.

Solomon, P. R. 1980. A time and place for everything? Temporal processing views of hippocampal function with special reference to attention. *Physiol. Psychol. 8*(2):254–261.

Solomon, P. R., and Moore, J. W. 1975. Latent inhibition and stimulus generalization for the classically conditioned nictitating membrane response in rabbits *(Oryctolagus cuniculus)* following dorsal hippocampal ablation. *J. Comp. Physiol. Psychol. 89*:1192–1203.

Swanson, L. W., Teyler, T. J., and Thompson, R. F., eds. 1982. *Hippocampal LTP: Mechanisms and Functional Implications.* Neurosciences Research Program Bulletin. Cambridge, Mass: MIT Press.

Thompson, R. F., Berger, T. W., Berry, S. D., Clark, G. A., Kettner, R. E., Lavond, D. G., Mauk, M. D., McCormick D. A., Solomon, P. R., and Weisz, D. J. 1982a. Neuronal substrates of learning and memory: Hippocampus and other structures. In *Conditioning: Representation of Involved Neural Functions* (C. D. Woody, ed.), pp. 115–29. New York: Plenum Press.

Thompson, R. F., Berger, T. W., Berry, S. D., Hoehler, F. K., Kettner, R. E., and Weisz, D. J. 1980. Hippocampal substrate of classical conditioning. *Physiol. Psychol. 8*(2):262–279.

Thompson, R. F., Berger, T. W., Cegavske, C. F., Patterson, M. M., Roemer, R. A., Teyler, T. J., and Young, R. A. 1976. The search for the engram. *Am. Psychol. 31*:290–227.

Thompson, R. F., Cegavske, C., and Patterson, M. M. 1973. Efferent control of the classically conditioned nictitating membrane response in the rabbit. *Psychonom. Soc. Abstr.,* pp. 15–16.

Thompson, R. F., and Glanzman, D. L. 1976. Neural and behavioral mechanisms of habituation and sensitization. In *Habituation: Prespectives from Child Development, Animal Behavior, and Neurophysiology* (T. J. Tighe and R. N. Leaton, eds.), pp. 49–93. Hillsdale, N.J.: Erlbaum.

Thompson, R. F., McCormick, D. A., Lavond, D. G., Clark, G. A. Kettner, R. E., and Mauk, M. D. 1982b. The engram found? Initial localization of the memory trace for a basic form of associative learning. In *Progress in Psychobiology and Physiological Psychology* (A. N. Epstein, ed.). New York: Academic Press.

Valentino, R. J., Herling, S., Woods, J. H., Medzihradsky, J., and Merz, H. 1980. Comparison of narcotic antagonists and their quaternary derivatives. *Fed. Proc. 39*(3):760.

Walters, E. T., Carew, T. J., and Kandel, E. R. 1979. Classical conditioning in *Aplysia californica. Proc. Nat. Acad. Sci. 76*:6675–6679.

Weinberger, N. M. 1980. Neurophysiological studies of learning in association with the pupillary dilation conditioned reflex. In *Neural Mechanisms of Goal-Directed Behavior and Learning* (R. F. Thompson, L. H. Hicks, and V. B. Shvyrkov, eds.), pp. 241–61. New York: Academic Press.

1982. Effects of conditioned arousal on the nervous system. In *The Neural Basis of Behavior* (A. L. Bechman, ed.), pp. 63–91. New York: Spectrum Publications.

Weisz, D. J., Solomon, P. R., and Thompson, R. F. 1980. The hippocampus appears necessary for trace conditioning. *Bull. Psychonom. Soc. Abstr. 19*:244.

Wikler, A. 1958. *Mechanisms of Action of Opiates and Opiate Antagonists: A Review of Their Mechanisms of Action in Relation to Clinical Problems.* Public Health Monograph No. 52. Washington, D.C.: U.S. Government Printing Office.

Woody, C. D., and Brozek, G. 1969. Changes in evoked responses from facial nucleus of cat with conditioning and extinction of an eye blink. *J. Neurophysiol. 32*:717–726.

Woody, C. D., Knispel, J. D., Crow, T. J., and Black-Cleworth, P. A. 1976. Activity and excitability to electrical current of cortical auditory receptive neurons of awake cats as affected by stimulus association. *J. Neurophysiol. 39*(5)1045–1061.

5 · The electrical excitability of nerve cells as an index of learned behavior

C. D. WOODY

INTRODUCTION

Neural excitability to locally injected current is a useful index of neural plasticity supporting changes in behavior. In addition, changes in the excitability of brain tissue measured by direct electrical stimulation can provide an indication of local neuronal change. This chapter concerns the results of applying measurements of neural excitability to studies of conditioning and related behavior.

Studies of the excitability of the motor cortex

In 1870, Fritsch and Hitzig found that stimulating the motor cortex with galvanic currents resulted in the production of contralateral limb movements. In years thereafter attention focused on correlating the movement produced by electrical stimulation with the area of the cortex stimulated (Grunbaum and Sherrington 1901; Penfield and Boldrey 1937; Garol 1942; Woolsey et al. 1952; Lilly 1958). Maps of the homuncular representation of motor function such as that shown on the left in Figure 5.1 (Penfield and Rasmussen 1957) were constructed and accepted. The strict representation of body form over the topography of the cortex was further supported by later studies of Asanuma and colleagues (Asanuma et al. 1968; Stoney et al. 1968; Asanuma 1973; Asanuma and Rosen 1973) in which microstimulation of a few cells at a time confirmed a columnar topographic organization consistent with the surface homunculus determined by gross stimulation.

There were, however, areas of uncertainty. For example, it was unclear whether movements (i.e., muscle group activation) or single-muscle activation was the principal effect of the more discrete forms of stimulation. In 1947, Chang et al. reported that single muscles could be activated by gross stimulation but noted that usually more than one muscle contracted when the same technique was applied from area to area. There were other experiments in which homuncular representation was found to be uncertain or redundant. For example, in cats, eye blinks were found to be represented at at least three different cortical areas: the classical anterior cruciate region, the area of Garol (1942), and that of Livingston and Phillips (1957). Subsequent investigations

(Woody and Yarowsky 1972) found that an electromyographically measured eye blink could be elicited by microampere stimulation of almost any area of the coronal pericruciate cortex of the awake cat.

These uncertainties and redundancies of motor representation could be explained, in part, by plastic changes in function occurring from moment to moment or from animal to animal. Or the uncertainties might be reflections of the underlying cell projections, and it might be more accurate to construct the homunculus of motor projection more like the "quantum" tiger shown on the right in Figure 5.1, the latter being a portrayal of uncertainties found by scientists in the field of chemistry. The "quantum" homunculus of motor function would then arise as a construct of the probability density function of the projections of cells in the area stimulated.

The possibility that each cell might have more than one motor projection has become more likely as techniques for measuring projection have become more sensitive. Many neurons of the motor cortex have now been shown to send their messages to more than one muscle group. Some evidence for this arises from the application of weak currents intracellularly while recording (and averaging) myographic responses from the peripheral musculature (Woody and Black-Cleworth 1973; Fetz and Finocchio 1975; Brons and Woody 1980). When but two facial muscles are sampled, about 25 percent of the cortical cells tested project to the one, 25 percent to the other, and 25 percent to both muscles (Woody and Black-Cleworth 1973). Other evidence arises from the anatomical finding of multiple-axon collaterals emanating from the axons of single

Figure 5.1. *Top:* "Motor homunculus. A cross-section of the left cerebral hemisphere is shown, and the parts of the right side of the body which move when the surface of the cortex is stimulated electrically are laid upon the surface, indicating the relative areas of cortex connected to various parts of the body." (After Penfield and Rasmussen 1957; from Bindman and Lippold 1981) *Bottom:* "At this moment a terrible roar filled the air and their elephant jerked so violently that Mr. Tomkins almost fell off. A large pack of tigers was attacking their elephant, jumping simultaneously from all sides. Sir Richard grabbed his rifle and pulled the trigger, aiming right between the eyes of the tiger nearest to him. The next moment Mr. Tomkins heard him mutter a strong expression common among hunters; he shot right through the tiger's head without causing any damage to the animal.

"'Shoot more!' shouted the professor. 'Scatter your fire all round and don't mind about precise aiming! There is only one tiger, but it is spread around our elephant and our only hope is to raise the Hamiltonian . . .'

"'Who is this Hamiltonian?' asked Mr. Tomkins . . . 'Is he some famous hunter you wanted to raise from the grave to help us?'

"'Oh!' said the professor . . . 'Hamiltonian is a mathematical expression describing the quantum interaction between two bodies . . . by shooting more quantum bullets we increase the probability of the interaction between the bullet and the body of the tiger. In the quantum world, you see, one cannot aim precisely and be sure of a hit. Owing to the spreading out of the bullet, and of the aim itself, there is always only a finite chance of hitting, never a certainty.' In our case we fired at least thirty bullets before we actually hit the tiger; and then the action of the bullet on the tiger was so violent that it hurled its body far away. The same things are happening in our world at home but on a much smaller scale." (From Gamow 1957)

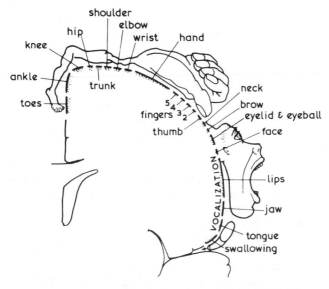

A large pack of tigers was attacking their elephant, jumping simultaneously from all sides

neurons and, for the case of cortical pyramidal cells, the finding, both electro-physiologically and anatomically, that such collaterals may project to different regions of the central nervous system (e.g., Sholl 1956; Allen et al. 1977).

Early studies of plasticity of the motor cortex

There remains the possibility that inconsistencies in motor responses elicited by electrical stimulation could arise from local neuronal adaptation. As early as 1875, Ferrier noted a continuing effect of electrical stimulation that per-turbed subsequent measurements (Ferrier 1875). Further studies by Ukhtom-sky (1926) and by Graham-Brown and Sherrington (1912; Graham-Brown 1914, 1915a, b, c, d) disclosed that repeated stimulation tended to facilitate the production of subsequent movements. Usually, though not invariably, the movements facilitated were those produced by the original stimulation. This phenomenon was termed "reflex dominance" by Ukhtomsky and "secondary facilitation" by Graham-Brown (1915c). It is termed "latent facilitation" herein because the manifestations await elicitation by the proper stimulus and sometimes result in an increased rate of acquisition of subsequently learned motor behavior (Matsumura and Woody 1982; Woody 1982b).

Further investigations confirmed that application of weak polarizing cur-rents to the motor cortex could produce a facilitation of motor response and increase the rate of spontaneous neuronal discharge (Burns 1957; Rusinov 1958, 1959; Lippold et al. 1961; Bindman et al. 1962a, b, 1971; Spehlmann and Kapp 1964; Voronin and Solntseva 1969). Some of these changes were found to be quite long-lasting (Burns et al. 1962; Burns and Smith 1962; Bind-man et al. 1964, 1976a, b, 1979). Recent studies by Bindman and colleagues (1979) have demonstrated that facilitation may occur after stimulation of the pyramidal tract (PT) and last for hours. Other studies have shown additional short-term effects of PT stimulation (Phillips 1959; Stefanis and Jasper 1964; Brooks and Asanuma 1965; Brooks et al. 1968; Oshima 1969; Renaud et al. 1974); in fact, a reduction in excitability is the predominant effect seen in cells that are activated antidromically by low-frequency stimulation (Tzebelikos and Woody 1979). In contrast, an increase in excitability is the predominant effect in cells activated orthodromically (transsynaptically) by PT stimulation (Tzebelikos and Woody 1979) or antidromically by intense, high-frequency stimulation (Bindman et al. 1979). The direction of excitability change and the duration of change are frequency-, intensity-, and neurotransmission-depen-dent (see Woody et al. 1978; Bindman et al. 1979; Tzebelikos and Woody 1979; Woody 1982b).

Electrical excitability and learned behavior

Many have attempted to link persistent changes in neural activity with some specific form of learned behavior. Initial studies dealt with changes in gross potentials (e.g., Yoshii 1957; John 1961; Galambos and Sheatz 1962; Woody 1970) or patterns of unit activity (e.g., Jasper et al. 1960; Woody et al. 1970; Woody and Engel 1972; O'Brien et al. 1973) in relation to conditioning. A review of those results is available elsewhere (Woody 1982a). Two further approaches have been used successfully to investigate neural substrates of conditioning, and it is the results of these studies that I now wish to discuss.

The first approach is that of measuring the excitability of single cells (or cell areas) as a function of conditioning (e.g., Woody 1970; Woody et al. 1970, 1976b; Woody and Black-Cleworth 1973; Brons and Woody 1980). The means by which this is done is discussed in the next section. The other approach involves use of electrical stimulation of brain tissue as a conditioned stimulus (CS). It is widely thought that changes in behavior elicited by delivery of an electrical stimulus into a local area of the brain may reflect a conditioned change in neural excitability *if* the stimulus is presented as one of a repeated *associative* pair. (The possibility that such changes can be related to *non*associative latent facilitation is discussed further below.) This approach was begun by Loucks (1933) and advanced by Giurgea (1953) and Doty and colleagues (1956). Their studies showed that a weak electrical stimulus in the microampere range was a satisfactory CS when applied to almost any region of the brain.[1] Generalization of the resulting conditioned response (CR) to stimulation of other areas of the brain was as specific as that to natural CSs.

ELECTRICAL STIMULATION OF SINGLE (OR A FEW) CELLS

Extracellular microampere technique

The threshold for eliciting an electromyographic (EMG) response by stimulating through metal electrodes may be defined as the lowest (microampere) current at which a response of fixed latency occurs for at least 50% of the stimulus presentations. Cathodal currents of 1–20 μA are passed through the same electrode used to record unit activity.

When loci within the motor cortex of awake cats were stimulated in this manner, gross facial movements could be elicited at slightly higher levels of stimulation than the threshold current for eliciting EMG activity (Woody et al. 1970; Woody and Engel 1972). In the studies cited, 50-msec trains of stimuli of 0.2 msec duration and 2.5 msec interval were used. EMG electrodes were placed in the orbicularis oculi and levator orii muscles. If cortical stimulation with currents of up to 20 μA produced repeatable EMG activity, the recorded cell was classified as being projective to that musculature. Units of each projective classification were usually encountered along each tract of electrode penetration. The latencies of the EMG responses (Figure 5.2) were equivalent to the conduction time from the coronal–pericruciate cortex to peripheral muscles – about 8 msec (Woody 1970; Woody and Engel 1972; Woody and Yarowsky 1972). A storage oscilloscope was used to superimpose successive EMG responses. In this way spontaneous EMG activity could be separated from that evoked by electrical stimulation. Although the classification of the projection of the recorded unit could have been in error for any single cell (neighboring cells might have instead responded to the stimulating current), the consistency of the results suggested that the classifications were accurate and repeatable.

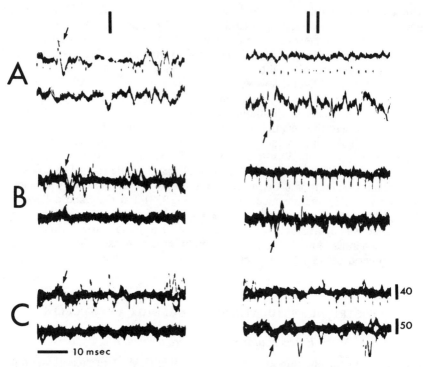

Figure 5.2. Responses of 8- to 10-msec latency following microstimulation of eye areas (I) and nose areas (II) of coronal pericruciate cortex. *A:* Single response to a 10-µA stimulus train. Upper trace recorded from electrode pair in eye muscle; lower trace recorded from electrode pair in nose muscle (*A–C*). *B:* Three superimposed responses at same areas and stimulating current as in *A*. *C:* Three superimposed responses to 10-µA stimulus train recorded from two different loci along one electrode tract with the position of EMG electrodes unchanged between recordings. Different animal than in *A–B*. Amplitude (µV) and time calibration for *A–C* as shown. Delivery of initial stimulus synchronous with start of sweep in all traces. Stimulus artifacts in II*A* were indistinct on face of storage oscilloscope, but appeared in II*B* with superposition of sweeps. (From Woody and Engel 1972)

Intracellular nanoampere technique

The threshold level of depolarizing current (pulses of 10 msec duration and 10 Hz frequency) required to evoke spike activity on at least 50 percent of the pulse presentations has been used as a measure of unit excitability to intracellular stimulation. In our studies cells of the motor cortex were classified as projective to eye muscles, nose muscles, both eye and nose muscles, or neither of these muscles depending on the type of EMG response that was evoked. Figure 5.3 illustrates the results of applying the constant-current pulses intracellularly, and Figure 5.4 shows the effects of these stimuli on the EMG activity of facial muscles. Although small variations in the latency of evoked spikes

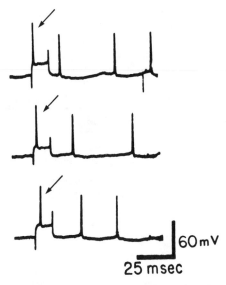

Figure 5.3. An example of intracellular nA stimulation (with 10-msec pulses at 10 Hz) through a bridge circuit. Threshold current (0.8 nA) elicited spikes on each current pulse with a latency of 1–3 msec. (From Brons and Woody 1980)

occurred due to random fluctuations of postsynaptic potentials within the recorded neuron, spikes elicited by each pulse at a relatively fixed latency were easily distinguished from spontaneous activity, even in units with higher-than-normal spontaneous activity such as the one illustrated.

Differences in recorded potentials did not affect comparison of threshold changes between different populations of cells, provided that populations with comparable resting and spike potentials were sampled (see Woody and Black-Cleworth 1973; Brons and Woody 1980).

Comparison of microampere and nanoampere stimulation technique

As shown in Table 5.1, the results of microampere extracellular and nanoampere intracellular stimulation provide a comparable indication of altered neural excitability in relation to learned behavior. In addition, the classifications of neuronal motor projections obtained by these methods agree, as indicated by the consistency between histograms of spike activity of units classified by each technique (see Woody 1977).

Either intracellular or extracellular stimulation of single cortical neurons can elicit EMG responses in peripheral musculature by activating polysynaptic, corticofugal pathways. The finding of measurable peripheral EMG responses linked to the discharge of *single* cortical units indicates that each unit can significantly influence muscle contraction in a relatively direct manner. Nonetheless, some potential exists for influencing movements by the introduction of

Table 5.1. *Threshold currents required to discharge units of the motor cortex of awake cats in different behavioral states*

	Extracellular currents (μA)[a]	Intracellular currents (nA)[b]
Naive	14.3	1.0
Conditioned	8.9	0.6
Extinction	11.5	0.8

[a]Averages from Woody et al. (1970, table 1).
[b]Averages from Brons and Woody (1980, table 3).

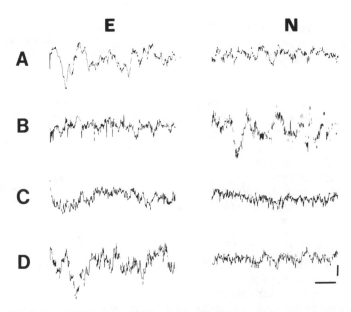

Figure 5.4. Averaged (bipolar) EMG activity evoked by intracellular electrical stimulation (10-msec current pulse of 0.5–6 nA delivered at start of trace) of three different cortical units (*A–C*). E = response in ipsilateral orbicularis oculis muscle, N = response in ipsilateral levator oris muscle. Numbers of responses averaged equal 100, 200, and 180, respectively. Units classified as follows. *A:* Eye muscle projection. *B:* Nose muscle projection. *C:* "Other" projection. *D:* Average of EMG activity (180 responses) synchronous with *spontaneous* discharges of same unit as in *C*. Amplitude calibration = 20 μV; time calibration = 10 msec. (From Woody and Black-Cleworth 1973)

additional messages at various way stations along the final efferent pathways. Projective classifications determined by the response to intracellular stimulation sometimes differed from classifications determined by the patterns of EMG activity associated with spontaneous spike discharge (Figure 5.4C, D). There was, however, a positive correlation (+.5) between classifications made by the two means. The correlation was similar for spikes evoked by either intracellular or extracellular nanoampere stimulation (Brons et al. 1982).

Extracellular nanoampere stimulation

The minimum "threshold" level of 10 msec, positive-going (depolarizing when intracellular), *extra*cellular nanoampere current pulses required to elicit spike discharges on about 30 percent of the stimulus presentations may also be used as a measure of neural excitability. The pulses are delivered through fluid-filled glass microelectrodes while recording extracellularly from a given unit. The locus of the recording/stimulating electrode may be taken to be extracellular if the size of the recorded action potential and the baseline variation are ≤6 mV or if the action potential is <10 mV and no baseline shift is observed. Units recorded extracellularly with fluid-filled glass microelectrodes should also meet the usual criteria for recordings from extracellular units made with metal microelectrodes (i.e., unitary spike height varying no more than baseline variation; unitary action potentials with refractory periods that do not summate).

Determination of threshold in an extracellular unit is compared with that in an intracellularly recorded unit in Figure 5.5. Note the tendency toward multiple spiking in the intracellular recording. In our studies, the latency of spike initiation by extracellulary injected nanoampere current averaged 1.6 msec longer than that by intracellular nanoampere current injection (Woody et al. 1976a). Occasionally, with extracellular nanoampere stimulation at higher currents, there was a decrease in latency approaching the onset of the current pulse. This was attributable to excitation via capacitative current. Levels of applied currents were monitored directly and cross-calibrated outside the preparation by calculating current–resistance drops from current–voltage traces using known resistances. The input resistance of the amplifier was 10^{12} Ω. The same equipment was used for extracellular and intracellular studies performed with fluid-filled glass microelectrodes.

Some extracellularly stimulated units responded to both polarities of injected current at threshold level. Further data may be found in Table 5.2. Generally, the probabilities of discharge increased by 50–1000 percent during passage of the positive-going, extracellular current pulses. The percentage of units responding preferentially to positive or negative extracellular nanoampere currents is shown in Table 5.3.

Movement of the electrode and differences in the position of the electrode from cell to cell could contribute to error in the measurement of threshold excitability. In order to assess the degree to which this factor might have contributed to our experimental results, we plotted the sizes of measured action potentials as a function of the different threshold currents required to discharge

Table 5.2. *Percent of 0.2-nA electrical stimuli eliciting a discharge in a single, extracellularly recorded and stimulated unit of the motor cortex*

	During stimulus pulse	Immediately after stimulus pulse	20 msec after stimulus pulse
+V	50	32	10
+C	20	2	2
−V	20	26	12
−C	12	6	0

Note: V, percentage of stimulus deliveries (0.2 nA) eliciting discharges of any variable latency within the 10 msec. C, percent of stimulus deliveries (0.2 nA) eliciting discharges varying no more than 1 msec in latency during the 10-msec period. +, positive-going (current). −, negative-going (current).
Source: Woody et al. (1976b).

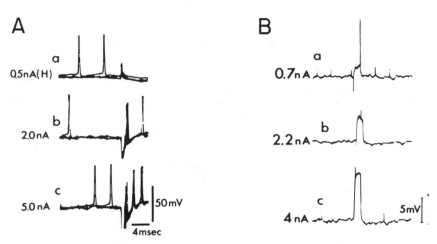

Figure 5.5. *A:* Intracellular nA stimulation of a unit of the coronal pericruciate cortex with hyperpolarizing and depolarizing current pulses of 10-msec duration. Current onset shown by artifact 6 msec prior to end of each sweep. *a:* 0.5-nA hyperpolarizing pulses suppress spike initiation. *b:* 2-nA depolarizing pulses repeatedly initiate a spike discharge. *c:* 5-nA depolarizing pulses repeatedly initiate repetitive spike discharges. All traces three sweeps each. *B:* Extracellular recordings of extracellular nA stimulation of a unit of the coronal pericruciate cortex. Illustration shows determination of current level required to produce unit discharge within 10-msec current pulse delivered in middle of each trace. At 0.7 nA no consistent discharge is produced. At 2.2 nA a discharge is seen. At 4 nA the latency of the initiated discharge is shorter. Usually 20–50 percent of suprathreshold stimulations initiate discharges. The trailing stimulus artifact has been removed from *b* and *c* to facilitate visualization of the evoked spike. (From Woody 1977)

Table 5.3. *Percentage of extracellular units activated by extracellular nA stimulation in conditioned animals*

	Click	Hiss	Both	$\bar{0}$
+	50 (1.6)	42 (2.3)	39 (2.8)	41 (2.2)
−	5	0	5	0
=	45 (3.0)	58 (4.9)	56 (4.7)	59 (5.2)

Notes: +, Responding preferentially to positive-going electrical current pulse; −, responding preferentially to negative-going pulse; =, responding without polarity preference.
Units divided into click-responsive (Click), hiss-responsive (Hiss), click- and hiss-responsive (Both), and unresponsive ($\bar{0}$) units.
Mean threshold response currents (nA) are shown in parentheses for + and = data.
Source: Woody et al. (1976b).

the extracellular units. The results are shown in Figure 5.6. In contrast with the logarithmic plots of current and distance found with microampere stimulation (see Stoney et al. 1968), these results showed a relatively flat relationship between spike size (an indicator of distance from the recorded/stimulated cell) and current, except at the very lowest current values. This result is consistent with observations of Gustaffson and Jankowska (1976) on stimulating spinal motor neurons with currents of less than 5 μA (see their fig. 7). Their measurements were made with two electrodes, one intracellularly and one extracellularly.

Intracellular and extracellular nanoampere current thresholds reflect different features of neural excitability. Intracellular currents provide an index of postsynaptic cellular change. The weak extracellular currents may not. This may be inferred from studies in which differences in excitability found by *extra*cellular stimulation of neurons of the auditory association cortex were not found by *intra*cellular stimulation (Woody et al. 1976b). Also, it is questionable whether extracellularly delivered current of less than 1nA could cross high-resistance plasma membranes in amounts sufficient to produce repeatable spike initiation following repeated current pulses. The magnitudes, polarities, and ranges of extracellular nanoampere currents satisfactory for producing spike discharge also weigh against a chemical effect such as the release of potassium from the electrode into the extracellular space being the factor responsible for spike activation. Activation of presynaptic terminals is also thought to be unlikely since facilitation resulting from presynaptic facilitation would be expected to be associated with hyperpolarization of the terminals, and hyperpolarized terminals might be expected to require more rather than less current to produce spike discharge, the opposite of the correlation found between unit excitability and unit activity (see Woody et al. 1976b). In addi-

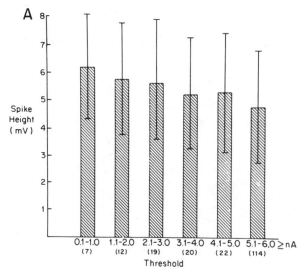

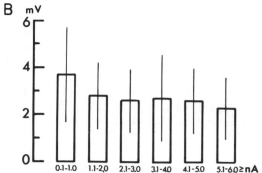

Figure 5.6. Mean ($\pm SD$) sizes of action potentials (ordinate) associated with different threshold currents (abscissa) required to discharge extracellular units at motor cortex (*A; n* as shown in parentheses) and at auditory association cortex (*B; n* = 462). Note slight increase in spike sizes with lower threshold currents, suggesting closer proximity of recording–stimulating electrode to these units than to others requiring higher threshold currents to be discharged. (*A* from Buchhalter 1979; *B* from Woody et al. 1976b)

tion, activation of presynaptic terminals should give rise to excitatory postsynaptic potentials (EPSPs). The latter were not found in further studies performed with triple-barrel electrodes designed to permit simultaneous intracellular and extracellular recording while passing current at either location (Woody and Gruen 1977). Reversible decreases in spike height were observed with passage of depolarizing currents, intracellularly, but not with delivery of extracellular currents, even though they were as much as 50–100 times greater. Repeatable PSP activity was not detected when spike initiation

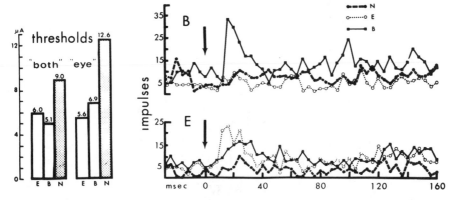

Figure 5.7. Thresholds and total activity of unit areas in blink-trained cats with CR in both eye and nose muscles (B) versus blink-trained cats with pure eye muscle CR (E). E: Units projective to eye muscles. N: Units projective to nose muscles. B: Units projective to both eye and nose muscles. Left: Averaged thresholds (bar graph) of unit areas to extracellular μA stimulation. Right: Unit activity to click CS. Impulses = the total number of impulses found in each 4-msec epoch in all low-spontaneous-discharge-rate units encountered. Times of CS delivery indicated by arrows. (From Woody and Engel 1972)

by extracellular current was prevented by superimposed steady hyperpolarizing current, delivered intracellularly. The results indicated that extracellular nanoampere stimulation sufficient to promote spike discharge was not associated with measurable levels of depolarizing current intracellularly.

CHANGES IN THE EXCITABILITY OF SINGLE CELLS WITH LEARNED BEHAVIOR

Latent facilitation

Changes in the excitability of neurons of the motor cortex of cats can be related to associative conditioning and to nonassociative presentations of an unconditioned stimulus (see Table 5.1 and Figures 5.7 and 5.8) as well as to direct local stimulation of the cortex (Tchilingaryen 1963; Krnjevic et al. 1966a, b; Lippold 1970; Woody and Black-Cleworth 1973; Woody 1974; O'Brien et al. 1977) or the pyramidal tract (Tzebelikos and Woody 1979; Bindman et al. 1979). With conditioning, changes in the thresholds to electrical stimulation of units of the motor cortex are correlated with changes in the patterns of spike activity evoked by the CS (Figure 5.7). In the example shown in Figure 5.7, two different facial motor responses were conditioned. In one, the EMG responses showed a compound blink movement ("both"); in the other, a simple eye blink was found ("eye"). Patterns of activity and thresholds to electrical stimulation changed as a function of training in the units projecting cortico-

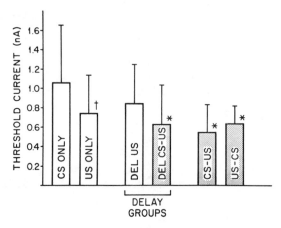

Figure 5.8. Bar graph representations of the mean $\pm SD$ intracellular threshold currents (nA) required to discharge neurons projective (polysynaptically) to muscles performing a compound blink CR. Data are given for six behavioral groups receiving presentations of the stimuli indicated. (Del = >4 weeks delay after final training before testing.) Shaded bars refer to groups of cats that at some time had received paired CS and US trials (including the extinction group). Unshaded bars refer to groups that received only unpaired CS or US trials. Asterisks indicate significant decreases ($p < .05$) in thresholds compared to the CS-only group. The US-only group exhibited a moderate decrease in thresholds ($p < .10$). (From Brons and Woody 1980)

fugally to the muscles performing the CR. Similar correlations between unit projection, activity, excitability, and conditioning have been found for three different conditioned facial movements, each distinguished by the patterns of EMG activity recorded from the involved musculature (see Woody et al. 1970; Woody and Engel 1972; Woody and Black-Cleworth 1973; Brons and Woody 1980; Woody 1982c). The changes in unit excitability provide as sensitive and precise an indication of changes in motor function as changes in unit activity.

The persistence of excitability changes after conditioning and extinction coincides with learning savings of the CR that lasts for periods of a year or longer. Changes in neural excitability produced nonassociatively by presentations of the US alone (in the absence of conditioning) do not last for more than 4 weeks unless the US is associatively paired with a CS (Figure 5.8). Nonassociative stimulation does not produce an overt CR, but is reflected by an increased rate of acquisition of subsequently learned conditioning (Matsumura and Woody 1982). The acquisition of learned motor performance similar to that produced as a UR to the repeatedly presented US is promoted by this means. Repeated presentations of the US alone may therefore be responsible for latent facilitation, just as strong electrical stimulation of the motor cortex may be responsible for an analogous facilitation or reflex dominance.[2]

The possibility that the effects of latent facilitation are mediated, in part, by cholinergic synaptic transmission has been discussed elsewhere (Woody 1982b,

Table 5.4. *Extracellular nA currents required to initiate spike discharges*

	Click	Hiss	Both	$\overline{0}$
C	1.0 (2.3)	3.6 (3.8)	2.6 (3.6)	2.7 (4.0)
R	2.0 (3.1)	2.5 (3.4)	1.9 (2.8)	3.5 (3.7)
N	5.7 (5.1)	5.9 (5.1)	5.7 (5.3)	5.5 (5.2)

Notes: Values are medians, with means in parentheses, expressed in nanoamperes. Units are from conditioned (C), randomization (R), and naive (N) animals. Medians are the average of median scores of each cat; means are the average threshold of all units in the category.

Units and their corresponding means and medians have been separated according to sensory receptive property of the unit: Click, increased discharge to click; Hiss, increased discharge to hiss; Both, increased discharge to both click and hiss; $\overline{0}$, unresponsive to click or hiss.

Source: Woody et al. (1976b).

c), as has the ability to produce persistent increases in excitability of pyramidal cells of the motor cortex by applications of acetylcholine or cyclic GMP paired with current-induced spike discharge (Woody et al. 1974a, 1976a, 1978; Swartz and Woody 1979; Woody and Gruen 1980).

Changes in neural excitability similar to those found in the motor cortex have also been found in facial motoneurons. Matsumura and Woody (1980) reported that thresholds of intracellular currents needed for eliciting spikes in motoneurons of cats were reduced after conditioning and after presentations of US alone. The time course of these effects was comparable to those found in the motor cortex (Matsumura and Woody 1982).

For conditioning to occur in which the acquired (or facilitated) motor response is performed selectively to a particular stimulus such as a CS, discriminative changes may also be needed. Neural correlates of such changes are described next.

Sensory discrimination

When a CS and US are paired associatively as in Pavlovian conditioning, changes in neural excitability supporting discrimination of the CS are found in neurons of the auditory association cortex (Woody et al. 1976b). The excitability to weak extracellular stimulation is altered selectively in neurons receptive to the CS. When the CS and US are presented in random temporal order, smaller increases in excitability to such currents are found in neurons of this region irrespective of their receptivity to click. The results from these studies are summarized in Table 5.4.

Latent inhibition

The excitability of neurons of the motor cortex to weak *extra*cellular nanoampere stimulation also changes as a function of repeated US presentation. Brons et al. (1978, 1982) reported that more extracellular current was required to fire motor-projective units of the motor cortex following repeated stimulations of a US such as glabella tap than was required to fire units in animals not given these USs. The changes lasted longer than 4 weeks and were thought to contribute to latent inhibition. These decreases in excitability to weak extracellular current are correlated with decreases in the rate of spontaneous discharge of the involved neurons (see Figure 5.9).

Latent inhibition is a well-known behavioral phenomenon in which repeated presentation of some CSs results in retardation of subsequently learned motor performance (also see unconditioned loci, Table 5.6, and later discussion). Some USs may produce similar effects. Many USs may be viewed as CSs that produce conspicuous motor responses, and USs may be used as CSs in Pavlovian conditioning (Gormezano and Tait 1976). Four weeks after US presentations, when the latent facilitatory changes described earlier are lost and these inhibitory changes remain, the acquisition of conditioning is retarded (Brons et al. 1982; Matsumura and Woody 1982; Woody 1982b). The finding of latent *inhibition* as well as latent *facilitation* after repeated presentation of a US is not without precedent. In the experiments of Graham-Brown and Sherrington involving stimulation of the cortex, either facilitatory *or* inhibitory (or *both*) changes could occur depending on the nature of the stimulus that was presented.

Sensory preconditioning

The amount of weak extracellular nanoampere current required to excite auditory receptive neurons of the *motor* cortex of the cat (cf. Sakai and Woody 1980) is reduced after a paradigm in which two sensory stimuli are paired as in sensory preconditioning (Buchhalter et al. 1978). A peripheral motor response contingent upon delivery of either stimulus is not required to produce this effect. Behaviorally, the facilitatory effects of pairing two sensory stimuli are well recognized (Brogden 1939; Hoffeld et al. 1958, 1960; Seidel 1959; Kendall and Thompson 1960). First, paired presentations of S_1 and S_2 are given. Then, S_1 is paired with a US to produce conditioning. Thereafter, the ability of the animal to acquire the same conditioned response to S_2 is facilitated. In contrast to the latent facilitation of motor performance described earlier, this represents a transfer of the ability to respond to a particular stimulus. In the physiologic studies of the cortical effects of pairing two auditory stimuli, no significant differences in thresholds to *extra*cellular nanoampere current were found as a function of motor projection. (The number of pairings was small.) The projections were established by extracellular nanoampere stimulation. However, as shown in Table 5.5, the auditory-receptive cells of the motor cortex required less current to excite than did the nonreceptive cells ($t = 1.9$, $df = 46$, $p \leq .06$). In addition, the auditory-receptive cells in the sensory preconditioned cats required significantly less current for excitation than did the receptive cells of naive cats ($t = 2.4$, $df = 55$, $p < .02$). An analysis

Table 5.5

Receptivity	R	NR
SP	3.6 ± 1.4 (24)	4.4 ± 1.5 (24)
N	4.4 ± 1.1 (33)	4.0 ± 1.2 (29)
Projection	P	NP
SP	3.6 ± 1.5 (45)	4.1 ± 1.4 (23)
N	3.9 ± 1.2 (58)	4.8 ± 1.0 (33)

Notes: Values in parentheses represent the number of units in that group. The other numbers indicate the mean extracellular current (nA) (± *SD*) required to produce spike discharges in cells of that category. R, receptive; NR, nonreceptive; P, projective; NP, nonprojective. SP, sensory preconditioned; N, Naive. Further explanations in Buchhalter et al. (1978).

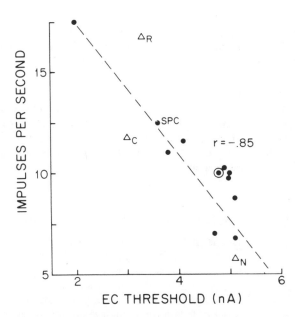

Figure 5.9. A plot of spontaneous unit activity versus extracellular nA current thresholds for spike elicitation in cells of the motor cortex of conditioned cats, CR-extinguished cats, and control cats that received CS only (solid dots). Circled dot indicates overlap of two data points. Additional data supporting an inverse linear relationship ($r = -.85$; line drawn by eye) between baseline activity and thresholds were drawn from other studies employing extracellular nA stimulation. Triangles refer to data obtained from the posterior association cortex in cats that received CS trials only (N), conditioning pairs of CS and US trials (C), or randomly presented CS and US trials (R) (see Woody et al. 1976b). SPC refers to data obtained from the motor cortex of cats presented with 10 consecutive pairs of CS (click) and DS (hiss) trials in a sensory preconditioning type of paradigm (see Buchhalter et al. 1978). (From Brons et al. 1982)

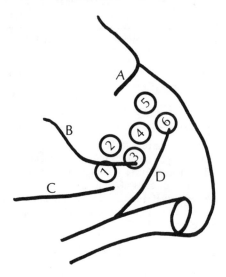

Figure 5.10. Location of electrodes (1–6) at coronal precruciate cortex. Electrodes are approximately 1–1.5 mm apart. A: Cruciate sulcus. B: Coronal sulcus. C: Diagonal sulcus. D: Presylvian sulcus. (From Woody and Yarowsky 1972)

of variance of the median currents per animal also showed a significant difference ($F = 6.4$, $df = 1, 13$, $p < .05$), excluding bias from individual cats. The auditory receptivity of each unit was determined by comparing rates of spike discharge in the periods 40 msec before and after the onset of the auditory stimulus. If the firing frequency in any of 10 4-msec poststimulus periods was greater than the largest response in any of the prestimulus periods, the unit was classified as being receptive to that stimulus.

ELICITATION OF LEARNED BEHAVIOR BY ELECTRICAL CS

Electrical stimulation of the motor cortex of the cat has been shown to be an effective CS for producing eye blink conditioning (Woody and Yarowsky 1972). The 12-msec latency of the conditioned blink responses is so brief (4 msec longer than efferent pathway conduction time) that central processing of transmissions of the CS brought about by learning may be assumed to occur in or near the motor cortex. In addition, such conditioned responses persist after extensive removal of cortical gray matter posterior to the motor cortex (Woody and Yarowsky 1972). The cats were trained to blink by pairing electrical stimulation of the cortex as a CS with glabella tap as a US. The CS was delivered through electrodes implanted chronically in the motor cortex of each cat as shown in Figure 5.10. The tips of the electrodes were approximately 1.5 mm deep in the cortex and approximately 1.5 mm apart. The electrodes were Teflon-coated stainless steel wires of 0.1 mm diameter insulated except at the cut tip.

The threshold for producing an EMG response by electrical microstimula-

Table 5.6. *Changes in threshold for producing an EMG response by electrical stimulation of conditioned and unconditioned cortical loci in six cats after acquisition of 70–80% CR performance levels*

	Number of loci showing		
Type of loci	Decreased threshold	Increased threshold	No change in threshold
Conditioned	9	3	6
Unconditioned	0	15	3

Note: Increase or decrease based on the measurement falling outside the range of thresholds for trials in the same animals in their naive state.
Source: Woody and Yarowsky (1972).

tion was measured before and after training. The results of training are shown in Table 5.6. There was a reduced threshold at loci at which electrical stimulation was presented associatively as the CS and a slightly increased threshold at adjacent loci (at which the CS was not delivered). The latter are termed unconditioned loci in Table 5.6 and in the descriptions that follow. Examples of the myographic responses elicited at the respective loci after conditioning are shown in Figure 5.11.

It is important to recognize that stimulation of almost any area of the coronal pericruciate (motor) cortex of an awake cat can lead to the production of an eye blink if sufficient current is passed. In the above experiments, careful attention was paid to this problem in order to avoid confusing unconditioned with conditioned eye blinks. Apart from the differences in threshold before (control values) and after conditioning, there also appeared to be a significant difference in the latency between conditioned and unconditioned responses. Many of the conditioned responses showed a latency of approximately 12 msec as opposed to the 8-msec latency of the unconditioned responses. Barbiturates abolished the 12-msec latency responses.

Any locus in the coronal pericruciate cortex appeared to have the potential for establishment of a CR by appropriate pairing of electrical CS with glabella tap. The same loci that served as unconditioned loci in one animal were satisfactory as conditioned loci in other animals. It took several weeks for the animals to learn the conditioned response, but once the CR was established, a reduction in threshold could be detected at loci at which the CS was paired associatively with a US during training. The results are in agreement with those of Doty (1965, 1967, 1969) in which electrical CSs delivered at adjacent cortical loci could be discriminated. In his experiments the regions of cortex involved were mainly visual sensory regions, and the studies were done in monkeys. In the present experiments discriminative conditioning could be demonstrated after lesions of the posterior areas of the cortex.

The experiments using electrical stimulation as CS to produce blink CRs in cats demonstrated stronger inhibitory effects than when a weak auditory stimulus was used. Presumably, the weeks required for establishment of the CRs

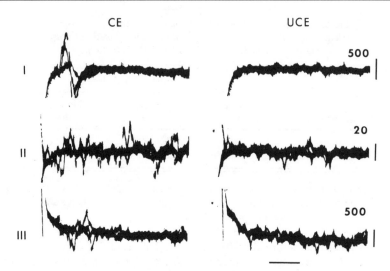

Figure 5.11. Responses to electrical CS at conditioned electrode loci (CE) and unconditioned electrode loci (UCE) in three (I–III) trained animals. Each trace is three
superimposed EMG responses recorded from orbicularis oculi muscles to 0.5-msec
stimulus pulse with current intensity in microamperes as shown by numbers on right.
Stimulus pulse delivery begins (dot) with start of traces; portions of the stimulus artifact were too fast (gap) to be encoded on the storage oscilloscope. The 50-μV and 10-
msec calibrations are as shown. Comparable responses of approximately 12-msec
latency to stimulation at CE are not present following similar stimulation at UCE.
(From Woody and Yarowsky 1972)

led to dissipation of the facilitatory effects of US presentations at unconditioned loci (thresholds for eliciting a blink response were actually increased at
these loci 1.5 mm away from the conditioning regions). Also, after producing
extinction by reversing the order of CS–US presentation, thresholds were
increased at conditioned loci as well. Apparently, presentations of the electrical
CS led to a separate latent (and surround) inhibition consistent with other findings (Graham-Brown and Sherrington 1912; Tzebelikos and Woody 1979; see
esp. Brons et al. 1982). During lengthy extinction a similar phenomenon was
manifested. The experiments also demonstrated that extinction and relearning
of the conditioned behavior could occur after ablation of large portions of the
posterior cortex, thus indicating that cellular changes supporting discriminative
conditioning may occur in other brain regions.

CONCLUSIONS

The following conclusions may be drawn on the basis of the above findings:
    1. Uncertainties of motor representation are real and are attributable to mul-
tiple representations of motor function (due to corticofugal axon collaterals)
and to plasticity.
    2. The plasticity is reflected by changes in levels of neural excitability mea-

sured by, first, extracellular microampere stimulation and intracellular nanoampere stimulation and, second, by extracellular nanoampere stimulation. (Other changes such as presynaptic facilitation and inhibition may be measured by other means.) The techniques provide useful information concerning mechanisms that support conditioning and related phenomena at the cellular level.

3. The changes in neural excitability can be either facilitatory or inhibitory of current-induced spike discharge, as described earlier.

4. The requirements for the production of changes in neural excitability depend on the type of excitability change produced. For the postsynaptic increases in excitability produced in the motor cortex during conditioning, the associative requirements are quite different from what might be expected from the requirements for behavioral conditioning, being somewhat closer to those of pseudoconditioning (Grether 1938). Conspicuous changes in neural excitability may be induced by presentations of USs alone.

Repeated stimulation by an unconditioned stimulus alone suffices to produce postsynaptic increases in excitability in neurons of the motor cortex (Brons and Woody 1980).

The above probably requires cholinergic activation plus the production of repeated doublet or triplet spike discharges in affected neurons (Woody et al. 1976b; Woody 1982b).

Pyramidal tract stimulation is ineffective as a US in the absence of actual antidromic activation (O'Brien et al. 1977).

Cholinergic activation or the production of doublet and triplet discharges alone is ineffective in producing a persistent increase in excitability (Woody et al. 1976a, 1978).

Orthodromic activation produces a transient increase in neural excitability like that seen with application of acetylcholine or cyclic GMP (Woody et al. 1978; Tzebelikos and Woody 1979). (Weak antidromic activation may actually produce a decrease in neural excitability.)

5. Strong antidromic stimulation leading to orthodromic and antidromic activation produces a long-lasting increase in excitability (Bindman et al. 1979) that may be isomorphic with effects of repeated presentations of a US that lasts for less than 4 weeks (Brons and Woody 1980). The latter increases in excitability have been shown to be produced in facial motoneurons as well as neurons of the motor cortex that project, polysynaptically, to the muscles that serve production of the UR. The duration of these changes is similar to that of behaviorally manifest latent facilitation (Matsumura and Woody 1982; Woody 1982b).

6. The changes in neural excitability last longer than 4 weeks if a CS such as click is presented associatively with the US (Brons and Woody 1980).

7. There is a positive correlation between changes in neural excitability, unit activity, and conditioned behavior – specifically with production of the learned motor response and discrimination of the CS.

8. The conspicuous changes that are acquired associatively are those in the auditory association cortex supporting auditory discrimination. Remarkably, these changes may not be as essential for some forms of learning as the less conspicuous effects of association that govern the duration of the changes in excitability of neurons of the motor cortex and facial nucleus.

9. The discriminative aspect of conditioning is supported by a different mechanism than supports development of the learned motor response.

10. The changes in neural excitability at the posterior auditory receptive cortex supporting discrimination of an auditory CS are unnecessary for conditioning to an electrical CS of the motor cortex to be manifest. Direct electrical stimulation of the motor cortex with microampere currents is a successful CS for the production of discriminatively conditioned eye blink CRs (Woody and Yarowsky 1972). Loci at which the CS is delivered associatively with the US are distinguished from other loci 1.5 mm apart with respect to whether or not electrical stimulation will elicit the CR. Although other regions outside the motor cortex may be involved in the production of conditioning, the CR is still produced when regions of the cortex posterior to these areas are removed. The latency of the CRs is nearly that of conduction time from the motor cortex to the periphery (Woody and Yarowsky 1972). Acquisition of eye blink CRs is severely impaired by lesions of the motor cortex (Woody et al. 1974b) and by application of 25 percent KCl to the rostral cortex (Woody and Brozek 1969).

It is suggested that conditioning to the auditory CS is a two-stage process depending on (1) changes in the motor cortex supporting the specificity of the learned motor act and (2) associatively induced changes in the auditory association cortex that permit the response to be elicited selectively by the particular stimulus. The ability to perform this learned motor behavior may depend primarily on the changes that occur in the motor cortex and facial nucleus after presentation of the US. The persistence of the changes supporting the learned motor behavior depends on association of CS and US.

The pyramidal tract cells of the motor cortex may be particularly involved in mediating persistent changes in conditioned motor function. These are the cells of the motor cortex that respond selectively to an auditory stimulus with a response latency sufficiently short to mediate short-latency eye blink conditioning (Sakai and Woody 1980). These cells are responsive to acetylcholine and to cyclic GMP (Woody and Gruen 1980). These cells show long-lasting excitability changes after pyramidal tract stimulation (Bindman et al. 1979), and antidromic activation of these cells is a satisfactory US for conditioning to occur (O'Brien et al. 1977).

Although, overall, our results show promising isomorphisms with conditioning, latent facilitation, latent inhibition, and sensory preconditioning, they also show a complexity of the underlying processes that mediate these behaviors. Just as habituation has multiple facets that are unlikely to be explained by a single mechanism, so are these behaviors unlikely to be explained in toto by the mechanisms described herein.

ACKNOWLEDGMENTS

I acknowledge with pleasure the collaboration of J. Brons, J. Buchhalter, E. Gruen, M. Matsumura, and many others at NIH and UCLA in these experiments and the support of NSF (BNS-78-24146), NIH (ND05958), AFOSR F49620-83-C-007 and NIA (AGO-1754).

NOTES

1. Stimulation of a few areas such as the deep cerebellar nuclei and the neocerebellar cortex was ineffective as a CS unless the stimulus intensity was increased sufficiently to produce gross movements (see Doty 1961; Donhoffer 1966).
2. Weak, brief neural stimulation (with 10-msec nanoampere currents) may instead produce decreases in neural excitability (Woody et al. 1976b); also see Table 5.6 and later discussion of results of repeated weak microampere stimulation delivered intracortically as a CS.

REFERENCES

Allen, G. I., Oshima, T., and Toyama, K. 1977. The mode of synaptic linkage in the cerebro-ponto-cerebellar pathway investigated with intracellular recording from pontine nuclei cells of the cat. *Exp. Brain Res. 29*:123–136.

Asanuma, H. 1973. Cerebral cortical control of movement. *Physiologist 16*:143–166.

Asanuma, H., and Rosen, I. 1973. Spread of mono- and polysynaptic connections within cat's motor cortex. *Exp. Brain Res. 16*:507–520.

Asanuma, H., Stoney, D. D., Jr., and Abzug, C. 1968. Relationship between afferent input and motor outflow in cat motor sensory cortex. *J. Neurophysiol. 31*:670–681.

Bindman, L. J., Boisacq-Schepens, N., and Richardson, H. C. 1971. "Facilitation" and "Reversal of response" of neurones in the cerebral cortex. *Nature New Biol. 230*:216–218.

Bindman, L. J., and Lippold, O. C. J. 1981. *The Neurophysiology of the Cerebral Cortex*. London: Arnold.

Bindman, L. J., Lippold, O. C. J., and Milne, A. R. 1976a. Long-lasting changes of post-synaptic origin in the excitability of pyramidal tract neurones. *J. Physiol. 258*:71–72P.

 1976b. Prolonged decreases in excitability of pyramidal tract neurones. *J. Physiol. 263*:141–142P.

 1979. Prolonged changes in excitability of pyramidal tract neurones in the cat: A post-synaptic mechanism. *J. Physiol.* (London) *286*:457–477.

Bindman, L. J., Lippold, O. C. J., and Redfearn, J. W. T. 1962a. Long-lasting changes in the level of the electrical activity of the cerebral cortex produced by polarizing currents. *Nature 196*:584–585.

 1962b. The non-selective blocking action of γ-aminobutyric acid on the sensory cerebral cortex of the rat. *J. Physiol. 162*:105–120.

 1964. The action of brief polarizing currents on the cerebral cortex of the rat, (1) during current flow and (2) in the production of long-lasting after effects. *J. Physiol. 172*:369–382.

Brogden, W. J. 1939. Sensory preconditioning. *J. Exp. Psychol. 25*:323–332.

Brons, J. F., and Woody, C. D. 1978. Decreases in excitability of cortical neurons to extracellularly delivered current after eyeblink conditioning, extinction, and presentation of US alone. *Soc. Neurosci. Abstr. 4*:255.

 1980. Long-term changes in excitability of cortical neurons after Pavlovian conditioning and extinction. *J. Neurophysiol. 44*:605–615.

Brons, J. F., Woody, C. D., and Allon, N. 1982. Changes in excitability to weak-intensity extracellular electrical stimulation of units of pericruciate cortex in cats. *J. Neurophysiol. 47*:377–388.

Brooks, V. B., and Asanuma, H. 1965. Recurrent cortical effects following stimulation of medullary pyramid. *Arch. Ital. Biol. 103*:247–278.

Brooks, V. B., Kameda, K., and Nagel, R. 1968. Recurrent inhibition in the cat's cerebral cortex. In *Structure and Function of Inhibitory Neuronal Mechanisms,* (C. von Euler ed.), pp. 327–334. Oxford: Pergamon Press.

Buchhalter, J. 1979. Changes in cortical neuronal excitability and activity after presentations of a compound auditory stimulus. Ph.D. thesis, UCLA.

Buchhalter, J., Brons, J., and Woody, C. 1978. Changes in cortical neuronal excitability after presentations of a compound auditory stimulus. *Brain Res. 156*:162–167.

Burns, B. D. 1957. Electrophysiologic basis of normal and psychotic function. In *Psychotropic Drugs,* (S. Garattini and V. Ghetti, eds.), Amsterdam: Elsevier.

Burns, B. D., Heron, W., and Pritchard, R. M. 1962. Physiological excitation of visual cortex in cat's unanaesthetized isolated forebrain. *J. Neurophysiol. 25*:165–181.

Burns, B. D., and Smith, G. K. 1962. Transmission of information in the unanaesthetized cat's isolated forebrain. *J. Physiol. 164*:238–251.

Chang, H. T., Ruch, T. C., and Ward, A. A. Jr. 1947. Topographical representation of muscles in motor cortex of monkeys. *J. Neurophysiol. 10*:39–56.

Donhoffer, H. 1966. The role of the cerebellum in the instrumental conditioned reflex. *Acta Physiol. Acad. Sci. Hung. 20*:247–251.

Doty, R. W. 1961. Conditioned reflexes formed and evoked by brain stimulation. In *Electrical Stimulation of the Brain* (D. E. Sheer, ed.), pp. 397–412. Austin: University of Texas Press.

1965. Conditioned reflexes elicited by electrical stimulation of the brain in macaques. *J. Neurophysiol. 28*:623–640.

1967. On butterflies in the brain. In *Current Problems in Electrophysiology of the Central Nervous System* (V. S. Rusinov, ed.), pp. 96–103. Moscow: Science Press.

1969. Electrical stimulation of the brain in behavioral context. *Ann. Rev. Psychol. 20*:289–320.

Doty, R. W., Rutledge, L. T., Jr., and Larsen, R. M. 1956. Conditioned reflexes established to electrical stimulation of cat cerebral cortex. *J. Neurophysiol. 19*:401–415.

Ferrier, D. 1875. Experiments on the brain of monkeys – no. I. *Proc. Roy. Soc. 23*:409–430.

Fetz, E. E., and Finocchio, D. V. 1975. Correlations between activity of motor cortex cells and arm muscles during operantly conditioned response patterns. *Exp. Brain Res. 23*:217–240.

Galambos, R., and Sheatz, G. C. 1962. An electroencephalograph study of classical conditioning. *Am. J. Physiol. 203*:173–184.

Gamow, G. 1957. *Mr. Tompkins in Wonderland.* Cambridge: Cambridge University Press.

Garol, H. W. 1942. The "motor" cortex of the cat. *J. Neuropathol. Exp. Neurol. 1*:139–145.

Giurgea, C. 1953. Elaboration of conditioned reflexes by direct excitation of cerebral cortex. (In Romanian.) Bucarest: Acad. Repub. Pop. Romine.

Gormezano, I., and Tait, R. W. 1976. The Pavlovian analysis of instrumental conditioning. *Pavlov. J. Biol. Sci. 11*:37–55.

Graham-Brown, T. 1914. Motor activation of the post-central gyrus. *J. Physiol. 48*: xxx–xxxi.

1915a. Studies in the physiology of the nervous system. 22. On the phenomenon of facilitation: Its occurrence in reactions induced by stimulation of the "motor" cortex of the cerebrum. *Q. J. Exp. Physiol. 9*:81–100.

1915b. Studies in the physiology of the nervous system. 23. On the phenomenon of facilitation: Its occurrence in response to subliminal cortical stimuli in monkeys. *Q. J. Exp. Physiol. 9*:101–116.

1915c. Studies in the physiology of the nervous system. 24. "Secondary facilitation" and its location in the cortical mechanism itself in monkeys. *Q. J. Exp. Physiol. 9*:117–130.

1915d. Studies in the physiology of the nervous system. 25. On the phenomenon of facilitation: Its occurrence in the subcortical mechanism by the activation of which motor effects are produced on artificial stimulation of the "motor cortex." *Q. J. Exp. Physiol. 9*:131–146.

Graham-Brown, T. and Sherrington, C. S. 1912. On the instability of a cortical point. *Proc. Roy. Soc. B. 85*:250–277.

Grether, W. F. 1938. Pseudo-conditioning without paired stimulation encountered in attempted backward conditioning. *J. Comp. Psychol. 25*:91–96.

Grunbaum, A. S. F., and Sherrington, C. S. 1901. Observations on physiology of the cerebral cortex of some of the higher apes. *Proc. Roy. Soc. 659*:206–209.

Gustafsson, B., and Jankowska, E. 1976. Direct and indirect activation of nerve cells by electrical pulses applied extracellularly. *J. Physiol. 258*:33–61.

Hoffeld, D. R., Kendall, S., Thompson, R. F., and Brogden, W. 1960. Effect of the amount of preconditioning training upon the magnitude of sensory preconditioning. *J. Exp. Psychol. 59*:198–204.

Hoffeld, D. R., Thompson, R. F., and Brogden, W. 1958. Effect of stimuli–time relations during preconditioning training upon the magnitude of sensory preconditioning. *J. Exp. Psychol. 56*:437–443.

Jasper, H., Ricci, G., and Doane, B. 1960. Microelectrode analysis of cortical cell discharge during avoidance conditioning in the monkey. *Electroencephalogr. Clin. Neurophysiol. Suppl. 13*:137–155.

John, E. R. 1961. High nervous functions: Brain functions and learning. *Ann. Rev. Physiol. 23*:451–484.

Kendall, S. B., and Thompson, R. F. 1960. Effect of stimulus similarity on sensory preconditioning within a single stimulus dimension. *J. Comp. Physiol. Psychol. 53*:439–442.

Krnjevic, H., Randic, M., and Strough, D. W. 1966a. An inhibitory process in the cerebral cortex. *J. Physiol. 184*:6–48.

1966b. Nature of a cortical inhibitory process. *J. Physiol. 184*:49–77.

Lilly, J. C. 1958. Correlations between neurophysiological activity in the cortex and short-term behavior in the monkey. In *Biological and Biochemical Bases of Behavior* (H. F. Harlow and C. N. Woolsey, eds.), pp. 83–100. Madison: University of Wisconsin Press.

Lippold, O. C. J. 1970. Long lasting changes in activity of cortical neurones. In *Short Term Changes in Neural Activity and Behavior* (G. Horn and R. A. Hinde, eds.), pp. 405–421. Cambridge: Cambridge University Press.

Lippold, O. C. J., Redfearn, J. W. T., and Winton, L. J. 1961. The potential level at the surface of the cerebral cortex of the rat and its relation to the cortical activity evoked by sensory stimulation. *J. Physiol. 157*:7–9P.

Livingston, A., and Phillips, C. G. 1957. Maps and thresholds for the sensorimotor cortex of the cat. *Q. J. Expl. Physiol. 42*:190–205.

Loucks, R. B. 1933. Preliminary report of a technique for stimulation or destruction of tissues beneath the integument and the establishing of conditioned reactions with faradization of the cerebral cortex. *J. Comp. Psychol. 16*:439–444.

Matsumura, M., and Woody, C. D. 1980. Excitability increases in facial motoneurones of the cat after serial presentation of glabella tap. *Soc. Neurosci. Symp. 6*:787.

1982. Excitability changes of facial motoneurons of cats related to conditioned and unconditioned facial motor responses. In *Conditioning: Representation of Involved Neural Function* (C. D. Woody, ed.), pp. 451–7. New York: Plenum.

O'Brien, J. H., Packham, S. C., and Brunnhoelzl, W. W. 1973. Features of spike train related to learning. *J. Neurophysiol. 36*:1051–1061.

O'Brien, J. H., Wilder, M. B., and Stevens, C. D. 1977. Conditioning of cortical neurons in cats with antidromic activation as the unconditioned stimuli. *J. Comp. Physiol. Psychol. 91*(4):918–929.

Oshima, T. 1969. Studies of pyramidal tract cells. In *Basic Mechanisms of the Epilepsies* (H. H. Jasper, A. Ward, and A. Pope, eds.), pp. 253–261. Boston: Little, Brown.

Penfield, W., and Boldrey, E. 1937. Somatic motor and sensory representation in the cerebral cortex of man as studied by electrical stimulation. *Brain 60*:389–443.

Penfield, W., and Rasmussen, T. 1957. *The Cerebral Cortex of Man: A Clinical Study of Localization.* New York: Macmillan.

Phillips, C. G. 1959. Actions of antidromic pyramidal volleys on single Betz cells in the cat. *Q. J. Exp. Physiol. 44*:1–25.

Renaud, L. P., Kelly, J. S., and Provini, L. 1974. Synaptic inhibition in pyramidal tract neurons: Membrane potential and conductance changes evoked by pyramidal tract and cortical surface stimulation. *J. Neurophysiol. 37*:1144–1155.

Rusinov, V. S. 1958. Electrophysiological investigation of foci of stationary excitation in the central nervous system. *Zh. Vyssh. Nerv. Deiat. 8*:473–481.

1959. Long lasting excitation, dominant and temporary connection. *Proc. Int. Congr. Physiol. Sci.* (21st), p. 238.

Sakai, H., and Woody, C. D. 1980. Identification of auditory responsive cells in coronal-pericruciate cortex of awake cats. *J. Neurophysiol. 44*:223–231.

Seidel, R. F. 1959. A review of sensory preconditioning. *Psychol. Bull. 56*:58–73.

Sholl, D. A. 1956. *The Organisation of the Cerebral Cortex.* London: Methuen.

Spehlmann, R. S., and Kapp, H. 1964. Direct extracellular polarization of cortical neurons with multibarrelled microelectrodes. *Arch. Ital. Biol. 102*:74–94.

Stefanis, C. and Jasper, H. 1964. Recurrent collateral inhibition in pyramidal tract neurons. *J. Neurophysiol. 27*:855–877.

Stoney, S. D., Jr., Thompson, W. D., and Asanuma, H. 1968. Excitation of pyramidal tract cells by intracortical microstimulation: Effective extent of stimulating current. *J. Neurophysiol. 31*:659–669.

Swartz, B. E., and Woody, C. D. 1979. Correlated effects of acetylcholine (ACh) and cyclic GMP (cGMP) on membrane properties of mammalian neocortical neurons. *J. Neurobiol. 10*:465–588.

Tchilingaryen, L. I. 1963. Changes in excitability of the motor area of the cerebral cortex during extinction of a conditioned reflex elaborated to direct electrical stimulation of that area. In *Central and Peripheral Mechanisms of Motor Functions* (E. Gutman and P. Hnik eds.), pp. 167–175. Prague: Publishing House of the Czechoslovakian Academy of Science.

Tzebelikos E., and Woody, C. D. 1979. Intracellularly studied excitability changes in coronal–pericruciate neurons following low frequency stimulation of the corticobulbar tract. *Brain Res. Bull. 4*:635–641.

Ukhtomsky, A. 1926. Concerning the condition of excitation in dominance. *Refl. Fiziol. Nerv. Fist. 2*:3–15.

Voronin, L. L., and Solntseva, E. I. 1969. After-effects of polarization of single cortical neurons: Intracellular recording. *Pavlov. J. Higher Nerv. Activ. 19*:828–838.

Woody, C. D. 1970. Conditioned eye blink: Gross potential activity at coronal-precruciate cortex of the cat. *J. Neurophysiol. 33*:838–850.

1974. Aspects of the electrophysiology of cortical processes related to the development and performance of learned motor responses. *The Physiologist 17*:49–46.

1977. Changes in activity and excitability of cortical auditory receptive units of the cat as a function of different behavioral states. *Ann. N.Y. Acad. Sci. 290*:180–199.

1982a. *Memory, Learning, and Higher Function: A Cellular View.* New York: Springer-Verlag.

1982b. Neurophysiologic correlates of latent facilitation. In *Conditioning: Representation of Involved Neural Function* (C. D. Woody, ed.), pp. 233–48. New York: Plenum Press.

1982c. Acquisition of conditioned facial reflexes in the cat: Cortical control of different facial movements. *Fed. Proc. 41*:2160-8.

Woody, C. D., and Black-Cleworth, P. 1973. Differences in excitability of cortical neurons as a function of motor projection in conditioned cats. *J. Neurophysiol. 36*:1104–1116.

Woody, C. D., and Brozek, G. 1969. Conditioned eyeblink in the cat: Evoked responses of short latency. *Brain Res. 12*:257–260.

Woody, C. D., Carpenter, D. O., Gruen, E, Knispel, J. D., Crow, T. W., and Black-Cleworth, P. 1976a. Persistent increases in membrane resistance of neurons in cat motor cortex. *Armed Forces Radiobiol. Res. Inst.* SR76 *1*:1–31.

Woody, C. D., Carpenter, D., Knispel, J. D., Crow, T. J., Jr., and Black-Cleworth, P. 1974a. Prolonged increase in resistance of neurons in cat motor cortex following extracellular iontophoretic application of acetylcholine (ACh) and intracellular current injection. *Fed. Proc. 33*:399.

Woody, C. D., and Engel, J., Jr. 1972. Changes in unit activity and thresholds to electrical microstimulation at coronal–pericruciate cortex of cat with classical conditioning of different facial movements. *J. Neurophysiol. 35*:230–241.

Woody, C. D., and Gruen, E. 1977. Comparison of excitation of single cortical neurons in awake cats by extracellularly and intracellularly delivered current. *Soc. Neurosci. Symp. 3*:166.

1980. Effects of cyclic nucleotides on morphologically identified cortical neurons of cats. *Proc. Int. Union Physiol. Sci. 14*:789.

Woody, C. D., Knispel, J. D., Crow, T. J., and Black-Cleworth, P. A. 1976b. Activity and excitability to electrical current of cortical auditory receptive neurons of awake cats as affected by stimulus association. *J. Neurophysiol. 30*:1045–1061.

Woody, C. D., Swartz, B. E., and Gruen, E. 1978. Effects of acetylcholine and cyclic GMP on input resistance of cortical neurons in awake cats. *Brain Res. 158*:373–395.

Woody, C. D., Vassilevsky, N. N., and Engel, J. 1970. Conditioned eye blink: Unit activity at coronal–precruciate cortex of the cat. *J. Neurophysiol. 33*:851–864.

Woody, C. D., and Yarowsky, P. J. 1972. Conditioned eye blink using electrical stimulation of coronal–pericruciate cortex as conditional stimulus. *J. Neurophysiol. 35*:242–252.

Woody, C. D., Yarowsky, P., Owens, J., Black-Cleworth, P., and Crow, T. 1974b. Effect of lesions of cortical motor areas on acquisition of eyeblink in the cat. *J. Neurophysiol. 37*:385–394.

Woolsey, C. N., Settlage, P. H., Meyer, D. R., Sencer, W., Hamuy, T. P., and Travis, A. M. 1952. Patterns of localization in precentral and "supplementary" motor areas and their relation to the concept of a premotor area. *Res. Publ. Ass. Nerv. Ment. Dis. 30*:238–264.

Yoshii, N. 1957. Principes méthodologiques de l'investigation électroencéphalographique du comportement conditionne. *Electroencephalogr. Clin. Neurophysiol. Suppl. 6*:75–88.

6 · Identification of vertebrate neurons modified during learning: analysis of sensory pathways

DAVID H. COHEN

INTRODUCTION

For some years we have been developing a vertebrate model system for the cellular analysis of long-term associative learning (Cohen 1969, 1974, 1980). This system involves visually conditioned heart rate change in the pigeon, and our approach to developing the model proceeded as follows. First, we established a standardized behavioral paradigm that gives rapid acquisition of a quantifiable, classically conditioned response in a pharmacologically immobilized preparation. Second, we delineated the neuronal pathways that must be intact for normal acquisition of this conditioned response. Third, as the neuronal circuitry was specified, we undertook cellular neurophysiological studies of the more peripheral segments of the identified pathways to specify the temporal properties of the information flow through the system during conditioned responding. Finally, we have most recently been able to initiate studies directed toward identifying specific synaptic fields that undergo training-induced modification – a prelude to exploring the mechanisms of such long-term storage during associative learning.

The focus of this chapter is upon the most peripheral sites of training-induced modification in our model system. As a foundation for this discussion, the behavioral model will be briefly reviewed. No systematic treatment of the relevant circuitry is presented, since that has been reviewed recently (Cohen 1980). However, segments of the identified pathways will be described as they relate to cellular neurophysiological results. Similarly, information regarding the central processing time for the conditioned response has been recently summarized (Cohen 1982), and this will be only briefly described as a foundation for interpreting the temporal response properties of neurons undergoing training-induced modification.

GENERAL BACKGROUND

The behavioral model

The behavioral model is well developed and is remarkably effective for cellular neurophysiological analysis of the relevant neuronal circuitry. The standardized paradigm involves the repeated presentation of a 6-sec pulse of whole-field retinal illumination, the conditioned stimulus (CS), immediately followed by a 500-msec foot shock, the unconditioned stimulus (US). The US elicits marked cardioacceleration as one component of the unconditioned response (UR), and after a sufficient number of light–shock (CS–US) pairings the light alone reliably elicits a conditioned cardioacceleratory response (CR) of predictable dynamics. The conditioned heart rate change largely develops within 10 such pairings, and asymptotic performance occurs in 30 pairings (Figure 6.1.). The CR is highly resistant to extinction, and we have shown that orienting and sensitization responses account for but a small proportion of the overall response.

With respect to the specific response characteristics (see Cohen and Goff 1978), early light presentations, independent of foot shock, elicit a small cardioacceleration (the orienting response) that habituates rapidly. Introducing foot shock then transforms this small rate increase into a sensitization response of different response dynamics, but this too extinguishes rapidly. The CR is a monotonic cardioacceleration with a latency of approximately 1 sec and with maximal values in the fifth or sixth seconds of the 6-sec CS period (Figure 6.2); the properties of this response have now been specified in considerable quantitative detail (Cohen and Goff 1978).

Through various investigations (summarized in Cohen and Goff 1978) we have evaluated a number of experimental variables with respect to their roles in affecting the acquisition, asymptotic level, and dynamics of the CR. We have shown that conditioned heart rate change interacts minimally, if at all, with concomitantly developing CRs, and we have developed an immobilized preparation allowing precise stimulus control and training to asymptotic CR levels in a single session of approximately 2 hr (Gold and Cohen 1981a). Moreover, in this immobilized preparation the quantitative properties of the CR do not differ from those of nonimmobilized animals.

In summary, the behavioral model is a robust and effective system for cellular neurophysiological studies of associative learning. It has numerous technical advantages such as rapid acquisition, consistent response dynamics among animals, and pharmacological immobilization. Furthermore, it potentially allows study of a broad range of learning phenomena in a single system, including habituation, sensitization, conditioning, and differentiation. Finally, it might be noted that use of an avian model could facilitate investigation of the cellular mechanisms of information storage during development.

Central processing time for the conditioned response

As described above, the heart rate CR has a latency of approximately 1 sec and persists for 6 sec. The central processing time for this behavioral response need not, however, be of the same order of magnitude. By characterizing the discharge properties of the "cardiac motoneurons" during conditioned responding, it is possible to understand the CR in a "neurophysiological time domain" that excludes delay time at the motor periphery. If, in addition, one determines the temporal properties of the retinal output, then an estimate of central processing time for the CR could be obtained.

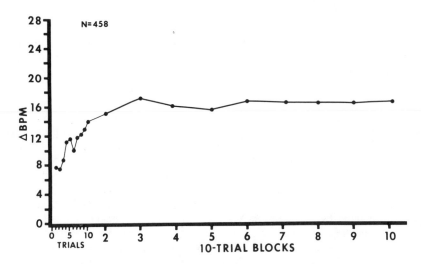

Figure 6.1. Acquisition of conditioned heart rate change. The curve represents mean heart rate changes in beats per minute (BPM) between the 6-sec CS and the immediately preceding 6-sec control periods. Each point represents a group mean for a block of 10 training trials, with the exception of the first block for which individual trial means are shown. (Adapted from Cohen and Goff 1978)

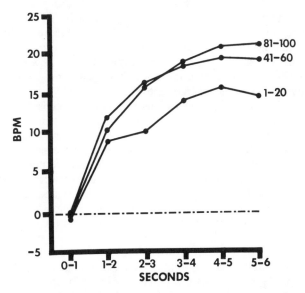

Figure 6.2. CR dynamics at various points in training. The curves show mean heart rate changes from baseline for each succeeding 1-sec interval of the CS period, averaged for trial blocks 1–20, 41–60, and 81–100. Note that stable conditioning occurs in block 1–20 and asymptotic performance by block 41–60. BPM = beats per minute. (From Cohen 1974)

Determining the temporal properties of the information flow through the system is fundamental to gaining some sense of its "analyzability." Moreover, it would establish the temporal boundary conditions for analyzing and interpreting cellular neurophysiological data from more central structures along the implicated pathways.

Discharge properties of the "cardiac motoneurons." We have shown previously that the final common path for the heart rate CR consists exclusively of the cardiac innervation and that both the vagal and sympathetic innervations contribute to the CR (Cohen and Pitts 1968). Furthermore, we have identified the cells of origin of the pre- and postganglionic sympathetic cardiac innervation (Macdonald and Cohen 1970; Cabot and Cohen 1977) and of the preganglionic vagal cardiac innervation (Cohen et al. 1970; Cohen and Schnall 1970). Finally, electrophysiological criteria have been established for identifying these neurons in the unanesthetized, behaving preparation (Cabot and Cohen 1977; Schwaber and Cohen 1978a, b). Based upon this foundation we were able to study the discharge properties of both groups of "cardiac motoneurons" during conditioned responding (Gold and Cohen 1981b; Cabot and Cohen pers. observ.).

With respect to the sympathetic component of the final common path, an important initial observation was that the cardiac postganglionic neurons respond to the visual stimulus prior to any training. The latency of this "orienting response" is approximately 100 msec, and it consists of a short burst of action potentials, followed by a brief depression of discharge and a subsequent return to maintained activity levels or slightly higher before termination of the light. If one continues to present the visual stimulus unpaired with foot shock, then this "orienting response" habituates. However, if light and shock are systematically paired, then both the probability of occurrence and magnitude of the initial light-evoked response increase. This conditioning alters neither the response latency nor the discharge pattern.

A striking feature of the discharge pattern is that it consists primarily of a phasic component (Figure 6.3). The tonic component is minimal, and many cells show no tonic component whatsoever. The phasic response has a duration of only 300–400 msec, suggesting that the central processing time is $\leq$400 msec and that the maintenance of the 6-sec heart rate response occurs at the periphery. A highly nonlinear input–output relation between the sympathetic innervation and the heart is supported by our finding (Cabot and Cohen pers. observ.) that simulating the discharge of the cardiac sympathetic postganglionics with electrical stimulation of the right cardiac nerve for 100 msec evokes a tachycardia of 4–6 sec duration that closely resembles the heart rate CR.

In an analogous study of the vagal cardiac neurons we have found that these cells also respond to the visual stimulus prior to training (Figure 6.4), although in this case the response consists of a decrease in discharge. Also like the sympathetics, the vagal cardiac neurons show both phasic and tonic components in their light-evoked discharge change, although in the vagal cells the tonic component is more prominent than in the sympathetics. The effect of associative training is to amplify the evoked reduction in discharge and to shorten the

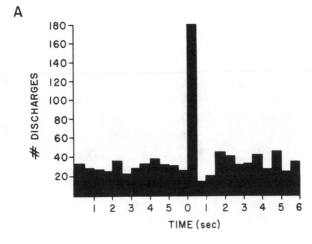

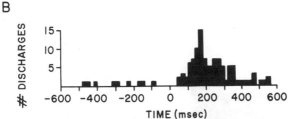

Figure 6.3. Discharge of cardiac sympathetic postganglionic neurons during conditioned responding. *A:* Summary peristimulus time histogram of the responses of nine cardiac postganglionic neurons to 10 CS presentations. The onset of the 6-sec CS is at time 0; bin width is 500 msec. Note that the phasic response is confined to the first 500-msec bin. *B:* Peristimulus time histogram of the discharges of three neurons for the 600 msec immediately before and after CS onset. Note that the phasic component of the response has a latency of approximately 100 msec (first significant deviation from maintained activity) and a duration of 300–400 msec. (Adapted from Cohen 1980)

response latency from 100 msec early in training to 60–80 msec at asymptotic performance (Figure 6.5).

These results for the "cardiac motoneurons" generate a number of conclusions. First, as suggested by earlier denervation experiments (Cohen and Pitts 1968), both the sympathetic and vagal cardiac innervations contribute to the CR in a synergistic manner. Second, the response latencies of the motoneurons are considerably shorter than would have been predicted from the heart rate change. Third, since the sympathetic outflow is the primary contributor to the CR (Cohen and Pitts 1968) and the evoked sympathetic discharge occurs at a latency of 100 msec and persists for only 300–400 msec, the conditioned heart rate change is largely mediated by a short-latency, short-duration burst of

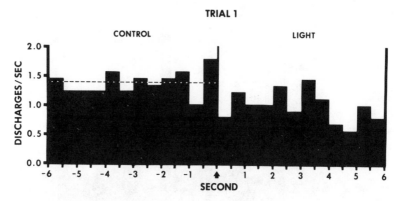

Figure 6.4. Peristimulus time histogram of vagal cardiac neurons during a light presentation before training. The arrow indicates CS onset, and the broken line indicates the mean discharge rate during the 6-sec control period preceding the light. (From Cohen 1982)

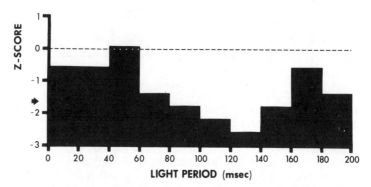

Figure 6.5. Poststimulus histogram showing the latency of the discharge change of vagal cardiac neurons in response to the CS during asymptotic performance. The ordinate indicates the standardized score for each 20-msec bin relative to the baseline distribution, and the arrow indicates the standardized score ($z = -1.65$) below which values differ significantly from baseline at the .05 level. (From Cohen 1982)

motoneuronal discharge. Finally, the effect of associative training is to increase the probability of occurrence and magnitude of the unconditioned response to the light in both sympathetic and vagal cardiac innervations, whereas nonassociative treatment produces attenuation of these initial light-evoked responses.

The retinal output. The retinal ganglion cells of the pigeon behave as a largely homogeneous population in response to whole-field illumination (Duff and Cohen 1975a, b). Their response consists of a short burst at light onset, fol-

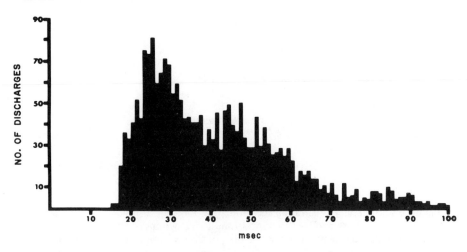

Figure 6.6. Poststimulus histogram of the response of retinal ganglion cells (optic tract fibers) at CS onset. (From Wild and Cohen pers. observ.)

lowed by cessation of discharge during sustained illumination and a more labile burst at stimulus termination. From recordings from single optic tract fibers during conditioning (Wall et al. 1980; Wild and Cohen pers. observ.) we generated a population histogram that reliably characterizes the retinal output during CS presentation (Figure 6.6). This indicates that the discharge at CS onset has a minimum latency of 18 msec and persists for a maximum of 80 msec. This retinal output shows no training-induced modification, and thus this latency range remains invariant over conditioning (Wall et al. 1980; Wild and Cohen pers. observ.).

Estimates of central processing time. From the above data there are various ways in which one could estimate central processing time. One approach is to use the difference between the modes of the histograms for the retinal output and the evoked responses of the "cardiac motoneurons." This gives an estimate of 105 msec for the central processing time for the vagal component of the CR and 135 msec for the sympathetic component (Figure 6.7). The difference of 30 msec between these estimates is consistent with the time required for conduction from the medulla to the preganglionic cell column, plus the relay time from the pre- to postganglionic neurons (Leonard and Cohen 1975).

Another approach to estimating central processing time is to assume that the shortest-latency responses of the retinal ganglion cells are associated with the shortest-latency responses of the "cardiac motoneurons." This gives estimates of 40 msec for the most rapid vagal response and 80 msec for the earliest sympathetic response.

Regardless of what procedure is utilized to estimate the central processing time, the important point is that this time is in the domain of milliseconds

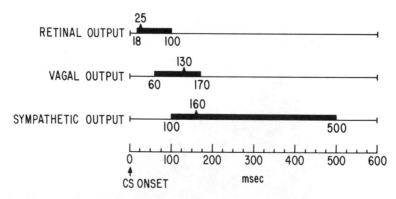

Figure 6.7. Summary illustration for estimating the central processing time for the CR. The bars are extrapolated from poststimulus histograms at CS onset. Below each bar is indicated the temporal range of the phasic response to the CS. The number above each bar indicates the time after CS onset to the mode of the poststimulus histogram.

rather than seconds as might be anticipated from the properties of the behavioral CR. This suggests that the system may be considerably more analyzable than one might initially expect. While certainly not a "simple system," the model might be viewed as a "relatively simple system."

Further comments on the nature of the information flow during conditioned responding

The above neurophysiological findings for the CS periphery and the final common path have further implications for the organization of the neuronal activity mediating the CR. The retinal output in response to the CS is an excitatory wave with a duration of approximately 80 msec. In all likelihood, this wave is the initial "trigger" for the phasic components of the motoneuronal responses. For the vagal cardiac neurons this phasic component has a duration of approximately 80–100 msec, whereas for the cardiac sympathetics the duration is 300–400 msec. Thus, one might view the information flow through the system as being initiated by an 80-msec wave at the retina that is then transmitted rather synchronously along the identified pathways to be expressed as an 80- to 100-msec wave at the vagal cardiac neurons (reflected as decreased discharge) and as a 300- to 400-msec wave at the cardiac sympathetic postganglionics (reflected as increased discharge). This would imply minimal temporal dispersion of the CS-evoked activity for the vagal component of the CR and a temporal dispersion not exceeding a factor of 4–5 for the sympathetic component (Figure 6.7). Further, since the effect of associative training is to augment the initial light-evoked response along the pathway, it might be reasonable to hypothesize a mechanism as straightforward as the facilitation of synaptic transmission at one or more sites along this pathway.

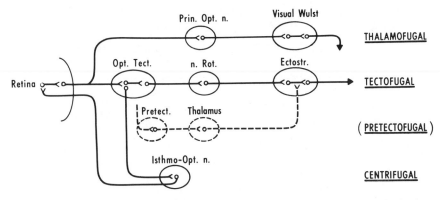

Figure 6.8. Schematic illustration of the major ascending visual pathways. The broken lines indicating the pretectofugal pathway reflect its speculative nature, and the hypothesized pretectofugal projection from the thalamus to the ectostriatum is intended to suggest a termination either upon or in the immediate region of the ectostriatum. Abbreviations: Ectostr. = ectostriatum; Isthmo-Opt. n. = isthmo-optic nucleus; n. Rot. = nucleus rotundus; Opt. Tect. = optic tectum; Pretect. = pretectal region; Prin. Opt. n. = principal optic nucleus of the thalamus (dorsal lateral geniculate). (From Cohen 1974)

SITES OF NEURONAL MODIFICATION

We now address the issue of where along the identified pathways neurons undergo training-induced modification. We have already demonstrated that training-related discharge changes occur at the "cardiac motoneurons," an expected and, indeed, necessary result. However, it is difficult to determine whether such changes involve any local modification at the level of the motoneurons or whether they merely reflect modification of more rostral structures along the pathways. A more favorable approach in this regard is to undertake cellular neurophysiological analysis during conditioning of the visual pathways that transmit the CS information. This allows one to begin at the CS periphery, the retina, and to investigate successively more central cell groups to identify the first structures where discharge patterns change as a function of associative training.

Pathways transmitting the visual CS information

As a foundation for describing the neurophysiological analysis of the visual pathways that transmit the CS information, the identification of these pathways will be briefly reviewed.

We first established that bilateral enucleation prevents development of the CR, excluding the participation of any nonretinal photoreception (Cohen 1974, 1980) (see Figures 6.8 and 6.9) Next we evaluated the possibility of involvement by cell groups that receive a direct retinal projection, the distribution of

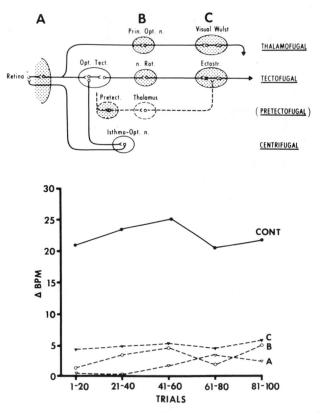

Figure 6.9. Visual system lesions that prevent CR acquisition. Each broken line represents an experimental group, while the solid line shows the performance of control birds (CONT). Curve A illustrates the performance of animals with bilateral enucleation (A in upper panel). Curve B illustrates the performance of animals with a combined lesion of the principal optic nucleus, nucleus rotundus, and the pretectal terminal field of the optic tract (B in upper panel). Curve C illustrates the performance of animals with a combined lesion of the visual Wulst and ectostriatum (C in upper panel). For all curves, each point represents the mean heart rate change in beats per minute (BPM) between the 6-sec CS period and an immediately preceding 6-sec control period. See Figure 6.8 for abbreviations in the upper panel. (From Cohen 1980)

such projections being remarkably similar in avian and mammalian brains (Cohen and Karten 1974). We found no deficits in conditioning consequent to destruction of any single optic tract target (Cohen 1974, 1980). This is not surprising, since each of these cell groups is responsive to whole-field illumination, such that considerable opportunity is available for the parallel transmission of CS information.

In mammals, severe deficits in visual learning follow the interruption of visual pathways to the cortex. Therefore, we turned to an examination of the possible involvement of the homologous pathways of the avian brain. Two such systems have been described, the thalamofugal and tectofugal pathways (Figure 6.8). These are homologous respectively to the mammalian geniculostriate and tectothalamoextrastriate pathways (see Cohen and Karten 1974). Our initial question was whether the interruption of either of these systems would affect CR acquisition. The results were negative. However, we did find that the combined interruption of the two pathways by a lesion including both the visual Wulst and the ectostriatum prevented CR development (Cohen 1974, 1980, pers. comm.) (Figure 6.9). An attempt to cross-validate this finding by combined interruption of the two pathways at the thalamic level produced only a transient deficit. This, in conjunction with various electrophysiological observations in our laboratory, suggested the existence of a third ascending pathway that either converges with the tectofugal pathway upon the ectostriatum or terminates in its immediate vicinity. Given the similarity of avian and mammalian visual pathways, we hypothesized that this pathway may be of pretectal origin (Figure 6.8). We thus conducted an experiment in which the principal optic nucleus, nucleus rotundus, and the pretectal terminal field of the optic tract were destroyed. This lesion produced acquisition deficits comparable to those that follow the combined lesion of the visual Wulst and ectostriatum (Figure 6.9); lesions of the pretectal region alone produced no deficit (Cohen 1974, 1980).

Consequently, it appears that each of these ascending visual pathways is capable of transmitting effective CS information, and it is only with their combined destruction that CR development is precluded. This provides, then, a first approximation to the pathways transmitting the CS information.

Are the visual pathways modifiable or merely input lines?

To address this question we first investigated the retinal output by recording the activity of single optic tract fibers during conditioning (Wall et al. 1980; Wild and Cohen pers. comm.). The results were definitive in indicating no change in either maintained or CS-evoked activity over training (Figure 6.10).

Given an invariant input from the periphery, we undertook analysis of the relevant ascending visual pathways, beginning with the tectofugal system (Figure 6.8). Our strategy was to examine first the telencephalic target of the pathway, the ectostriatum. The rationale was that if the ectostriatal neurons showed no training-induced modification, then one could infer that the subtelencephalic components of the tectofugal pathway are also probably invariant with training. If, on the other hand, the discharge of ectostriatal neurons did, indeed, change over training, then it would be necessary to examine the subtelencephalic components of the pathway to determine the most peripheral site of training-induced modification.

Experiments on the ectostriatum clearly indicated that at least certain classes of ectostriatal neurons show modification of their CS-evoked discharge over training (Wall et al. 1980; pers. comm.) (Figure 6.11). This motivated a study of nucleus rotundus, the thalamic relay of the tectofugal pathway (Figure 6.8). Again, the results indicated substantial training-induced modification

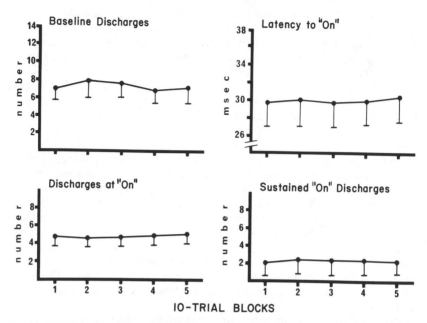

Figure 6.10. Summary of some of the statistics for a sample of approximately 50 single optic tract fibers studied over the course of 10 light presentations (block 1) followed by 40 CS–US presentations (blocks 2–5). The vertical bars indicate standard errors. The upper left panel shows that maintained activity does not change. The upper right panel shows that the latency of the response at CS onset does not change. The lower left panel shows that the number of discharges in the "on response" does not change, and the lower right panel shows that the number of discharges in the tonic response also does not change. (From Wild and Cohen pers. observ.)

of the evoked discharge (Wall et al. 1980, pers. comm.) (Figure 6.12). These findings demonstrate that the tectofugal pathway is involved in information storage and does not behave merely as an input channel for the CS.

To determine the most peripheral site of training-induced modification along the tectofugal pathway we must now study the optic tectum (Figure 6.8). However, rather than undertake these experiments immediately, we chose to investigate the avian homolog of the dorsal lateral geniculate – the principal optic nucleus (Figure 6.8). This decision was based upon two considerations. First, it is known that the superior colliculus in both birds and mammals receives polysensory input, establishing the possibility of CS–US convergence upon collicular neurons contributing to the tectofugal pathway. Therefore, the tectofugal system might be unique among the visual pathways in undergoing training-induced modification. Examination of the avian dorsal geniculate would provide an opportunity to assess the generality of the tectofugal findings. Second, the intratectal circuitry for the tectofugal pathway is not known in detail, and criteria are not available for identifying retinorecipient tectal neurons that

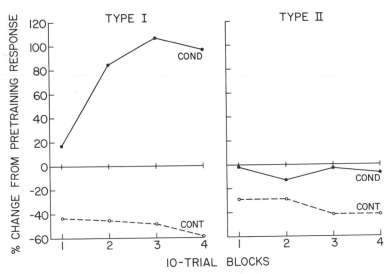

Figure 6.11. Changes in the discharge of ectostriatal neurons over conditioning (solid line) and habituation (broken line). The curves show the mean percentage change from the response to the light prior to training. The results shown are for the phasic responses at CS onset. Type I refers to cells that show increased discharge at CS onset, while type II refers to cells that show decreased discharge at CS onset. It should be noted that for type II cells a negative percentage indicates an attenuation of the inhibitory response to the CS. (From Wall et al., pers. comm.)

contribute to this pathway. By recording from the geniculate neurons we could therefore study a population of retinorecipient cells to determine if training-induced modification occurs as peripherally as the first central synapse of an involved visual pathway.

These experiments have now been concluded, and they show without question that avian geniculate neurons can undergo substantial modification of CS-evoked discharge as a function of training (Figure 6.13) (Gibbs and Cohen 1980; Gibbs et al. pers. comm.).

This generates two important conclusions: (1) At least two of the visual pathways involved in transmitting the CS information show training-induced modification, and (2) such modification can occur as peripherally as the first central synapse of these lemniscal pathways.

CHARACTERISTICS OF THE MODIFICATION OF VISUAL NEURONS

Beyond implicating the thalamofugal and tectofugal pathways in information storage associated with conditioning, the above experiments generated various results on the specific nature of the training-induced neuronal changes along the visual pathways.

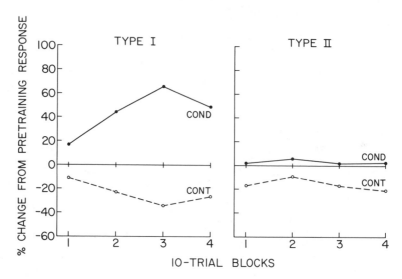

Figure 6.12. Changes in the discharge of rotundal neurons over conditioning (solid line) and habituation (broken line). The curves show the mean percentage change from the response to the light prior to training. The results shown are for the phasic responses at CS onset. Type I refers to cells that show increased discharge at CS onset, while type II refers to cells that show decreased discharge at CS onset. It should be noted that for type II cells a negative percentage indicates an attenuation of the inhibitory response to the CS. (From Wall et al. pers. comm.)

Relationship to visual response properties

Of the three visual structures we have studied, all neurons in the nucleus rotundus and the principal optic nucleus are responsive to whole-field illumination. A small proportion of ectostriatal neurons is unresponsive, and such cells remain unresponsive over conditioning. Thus, all visual neurons that show training-induced modification are responsive to the CS prior to training.

The responsive neurons typically respond at short latency to the CS, one population increasing its discharge at CS onset (type I) and another decreasing its discharge (type II). For example, in the avian geniculate approximately 70 percent of the neurons have type I discharge at CS onset and 30 percent type II discharge (Gibbs et al. 1981). Associative training (conditioning) enhances the initial light-evoked responses of a subset of these neurons. For the type I cells this enhancement consists of a small decrease in response latency and an increase in the number of discharges and duration of the phasic "on response." These response parameters change in the opposite direction during nonassociative treatment, that is, light alone (habituation) or unpaired lights and shocks (sensitization) (Figures 6.14, 6.15, and 6.16). Thus, for the type I neurons associative and nonassociative training have opposite effects on the phasic, CS-evoked discharge.

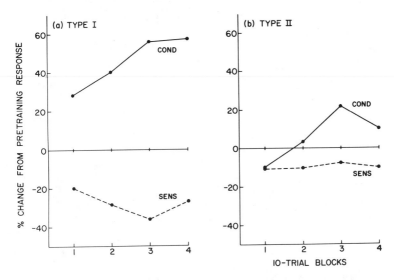

Figure 6.13. Changes in the discharge of dorsal lateral geniculate neurons over conditioning (solid line) and habituation (broken line). The curves show the mean percentage change from the response to the light prior to training. The results shown are for the phasic response at CS onset. Type I refers to cells that show decreased discharge at CS onset. It should be noted that for the type II cells a negative percentage indicates an attenuation of the inhibitory response to the CS. (From Gibbs et al. pers. comm.)

Type II neurons respond at CS onset with a decrease in discharge. Since this decrease frequently approaches a zero response rate, a "floor effect" is created such that the magnitude of the decrease cannot be substantially enhanced by associative training. However, nonassociative treatment can still produce an attenuation of the response, and this consists of successively less inhibition of discharge at CS onset over repeated light presentations (Figure 6.17). Thus, for the type II neurons associative training can perhaps be viewed as "preventing habituation" (Gibbs et al. pers. comm.).

The modifiability of the visual neurons we have studied is independent of the direction of their phasic response to the CS. Moreover, the differential effects of associative and nonassociative training are restricted to the phasic responses to the CS. The discharge later in the CS period, regardless of its characteristics, either does not change or, more frequently, habituates during both associative and nonassociative training (Figure 6.18).

Finally, it is important to appreciate that the maintained activity of the neurons we have studied in the ectostriatum, nucleus rotundus, and principal optic nucleus does not change over the course of training. It may be recalled that the maintained activity of the retinal ganglion cells is also stationary over training. Thus, the enhancement of the discharge of visual neurons by associative training appears to be restricted to their phasic responses at CS onset.

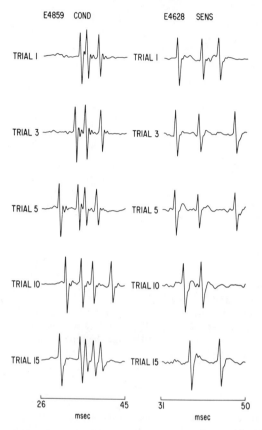

Figure 6.14. Examples of the behavior of single geniculate neurons during conditioning (Cond – left column) and habituation during unpaired lights and shocks (Sens – right column). The training trial from which each record was sampled is indicated to the left of that record. The bar below each column indicates the time period after CS onset for the records that are shown. (From Gibbs et al. pers. comm.)

Temporal properties of the discharge modification

Two temporal features of the discharge changes of visual neurons merit consideration: (1) the phasic response latencies and durations relative to the central processing time for the CR and (2) the time course of the neuronal changes relative to the acquisition rate of the behavioral CR. For this discussion, the results for the type I geniculate neurons will be used as an example.

The minimum response latency of the geniculate neurons is 26 msec, with a mode of 35 msec. The comparable statistics for the retinal ganglion cells are 18 msec and 25 msec. The duration of the "on response" of the geniculate neurons is 97 msec and that of the retinal ganglion cells 80 msec. Conse-

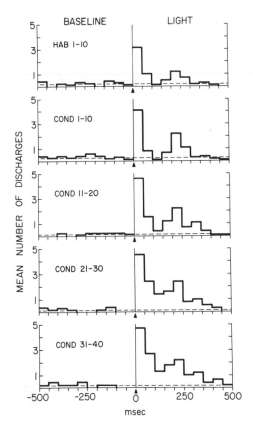

Figure 6.15. Peristimulus time histogram for a geniculate neuron during a block of 10 light presentations (Hab) followed by four 10-trial blocks of CS–US presentations (Cond). For each block the mean number of discharges is indicated for the 500-msec periods preceding and following CS onset (arrow). This is a type I unit, and enhancement of the "on response" over associative training is evident in the mode of the histogram.

quently, the temporal characteristics of the phasic response to the CS at the geniculate are what one would anticipate, given the temporal properties of the retinal ganglion cells. In particular, there is little temporal dispersion of the phasic response generated at the retina. The restriction of the plastic change over conditioning to this narrow time window is also critical, in that it minimizes, if not entirely eliminates, the possibility that the discharge modification of the geniculate neurons reflects changes in other structures that feed back upon the dorsal geniculate.

As to acquisition rate, the training-induced change in the discharge of geniculate neurons is clearly evident within the initial 10 CS–US presentations (Fig-

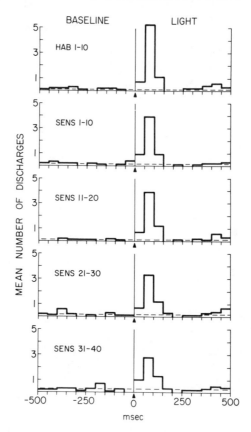

Figure 6.16. Peristimulus time histogram for a geniculate neuron during a block of 10 light presentations (Hab) followed by four 10-trial blocks of unpaired CS and US presentations (Sens). For each block the mean number of discharges is indicated for the 500-msec periods preceding and following light presentation (arrow). This is a type I unit, and attenuation (habituation) of the "on response" over nonassociative training is evident in the mode of the histogram. (From Gibbs et al. pers. comm.)

ures 6.13–6.16), and it is asymptotic by approximately 30 such presentations (Figure 6.13). This accurately parallels the development of the heart rate change (Figure 6.1). Thus, the development of the discharge change of the geniculate neurons can contribute in a causal manner to the development of the behavioral CR.

NECCESSARY CONDITIONS FOR MODIFIABILITY

Most models of neuronal circuitry for classical conditioning postulate at least one site where activity evoked by the CS and US converge. Although other models are possible, this traditional requirement for CS–US convergence is intuitively the most parsimonious. In a preceding section it was stated that the visual neurons showing training-induced modification in our system all initially

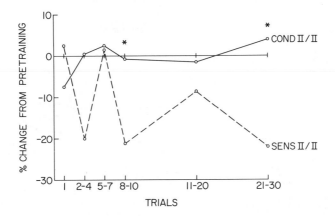

Figure 6.17. Changes in the discharge of a population of type II geniculate neurons over conditioning (Cond) and habituation during unpaired CS and US presentations (Sens). The curves show the mean percentage change from the response to the light prior to training, which for these cells is a decrease in discharge. The results shown are for the phasic response at CS onset. Cells studied during associative training maintain their almost total cessation of discharge at CS onset. In contrast, during nonassociative training this evoked reduction in discharge becomes successively less marked (habituates). (From Gibbs et al. pers. comm.)

responded to the CS; unresponsive neurons in the ectostriatum did not change as a function of training. Therefore, the requirement for CS input is satisfied in the structures we have investigated.

With regard to US input, we have found that a substantial proportion of the visual neurons we have examined also respond in some manner to the US (Gibbs and Cohen 1980; Wall et al. 1980). The exception is the retinal ganglion cells that do not respond at all to the US (Wild and Cohen pers. observ.). Once again using the geniculate neurons for illustration, 88 percent respond to the foot shock (Cohen et al. 1982). Of these US-responsive cells, 58 percent show increased discharge at US onset (type I) and 42 percent decreased discharge (type II) (Figure 6.19). Thus, there is substantial CS–US convergence at the geniculate, establishing the opportunity for some sort of heterosynaptic facilitation.

The necessity for the US input is suggested by our finding that the geniculate neurons that were unresponsive to the foot shock showed no enhancement of their CS-evoked phasic response with associative training. Although CS–US convergence thus appears to be a necessary condition, further analysis of our results indicated that it is not sufficient (Gibbs et al. 1981; pers. comm.). As illustrated in Figure 6.20, the geniculate neurons that had type I responses to the US (increased discharge) showed attenuation (habituation) of their phasic response to the CS over associative training. Indeed, their CS-evoked responses did not differ significantly from those of the nonassociative control neurons. In distinct contrast, the geniculate neurons with type II responses to the US (decreased discharge) showed striking response enhancement during associative training and response attenuation under nonassociative conditions. These

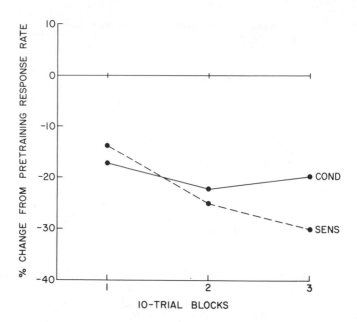

Figure 6.18. Tonic responses of type I geniculate neurons during associative (Cond) and nonassociative (Sens) treatments. Note that these responses are not affected differentially by the treatments, as is the case for the phasic responses. Habituation of the tonic responses occurs in both treatments. (From Gibbs et al. pers. comm.)

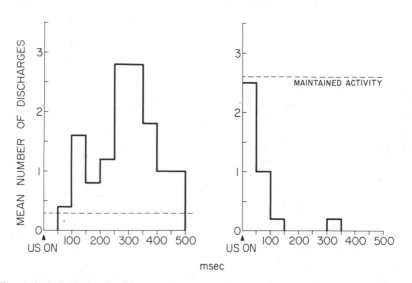

Figure 6.19. Poststimulus histograms illustrating the US responsiveness of geniculate neurons. One population of these neurons shows increased discharge at US onset (type I – left panel). Another population shows decreased discharge (type II – right panel). The level of maintained activity prior to US presentation is shown by the broken lines. (From Gibbs et al. pers. comm.)

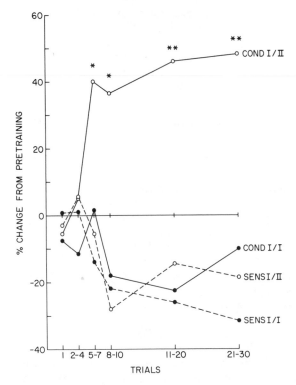

Figure 6.20. Demonstration that training-induced modification of geniculate neurons is a function of the response to the US. The data for the geniculate neurons showing type I responses to the CS (increased discharge) (see Figure 6.13) have now been subdivided on the basis of the response to the US. Thus, "I/II" indicates cells that increase their discharge at CS onset and decrease their discharge at US onset. "I/I" refers to cells that increase their discharge at CS onset and US onset. "Cond" refers to neurons studied during associative training, and "Sens" refers to units studied during nonassociative training. The curves show the mean percentage change from the response to the light prior to training, and the results shown are for the phasic responses at CS onset. Note that the only group showing response enhancement during associative training is that in which the neurons showed decreased discharge at US onset. The associative training group showing increased discharge at US onset does not differ from neurons given nonassociative training (Sens), and all three of these groups show habituation of the CS-evoked response over training. (From Gibbs et al. pers. comm.)

data therefore suggest that there are at least two sources of US input to the lateral geniculate and that the type II input is necessary and sufficient for discharge modification with associative training. While CS responsiveness is also a necessary condition, the nature of the response to the visual stimulus does not appear to be relevant with respect to modifiability.

An interesting implication of these findings is that the directions of the responses to the CS and US need not be the same. Indeed, striking enhancement of the CS-evoked response is found in geniculate neurons that show an excitatory "on response" to the light with decreased discharge in response to the US (Figure 6.20). It is, perhaps, inappropriate to dwell further on this point, since the sign of the response to a given stimulus may be less relevant than the specific transmitter mediating that response.

FURTHER ANALYSIS OF THE US INPUT

As a foundation for more detailed analysis of the plasticity of the geniculate neurons, it is important to gain a further understanding of the nature of the US input. An initial step in this regard is to identify the cell groups that project to the geniculate and are responsive to the US. Using both anterograde and retrograde methods, we have found various nonvisual structures that project to the geniculate (Cohen et al. 1982). Prominent among these is the locus coeruleus, as in mammalian brain (Moore and Bloom 1979). Furthermore, we have recently shown that approximately 80 percent of the locus coeruleus neurons indeed respond to the US, 58 percent with increased discharge and 23 percent with decreased discharge. The median response latency of these neurons is 75 msec. Since the median response latency of the US-responsive geniculate neurons is 100 msec, the locus coeruleus is a reasonable candidate for at least one source of US input.

Based upon these results, we investigated the US responsiveness of geniculate neurons after bilateral destruction of the locus coeruleus. Such a lesion almost entirely eliminates the type II response (decreased discharge) to the foot shock. Consequently, either the locus coeruleus or fibers en passage through that structure appear to mediate the US input that is necessary for training-induced modification of geniculate neurons.

If the locus coeruleus were the only source of US input to the type II geniculate neurons, then one would anticipate that interruption of that input would render these cells unresponsive to the US. This would increase the proportion of unresponsive neurons considerably (from 12 to 49 percent). The proportion of cells with type I responses (51 percent) would not be affected. However, this did not occur. The proportion of unresponsive neurons remained at 12 percent, and the proportion of type I cells increased substantially. This implies that the US-responsive neurons of the geniculate receive both type I and type II inputs, and when the source of the type II input is removed the type II neurons are transformed to type I. We do not yet understand how the type II input dominates one population of US-responsive neurons and the type I input another population.

A further aspect of the results is our finding that destruction of the locus coeruleus increases the latencies and decreases the magnitudes of the type I responses to the US. This suggests that the locus coeruleus contributes to the type I response, as well as mediating the type II response. Supporting this conclusion are ongoing experiments in which we find that electrical stimulation of the locus coeruleus excites some geniculate neurons and inhibits others. If the locus coeruleus in fact contributes a type I input, then excitation of geniculate

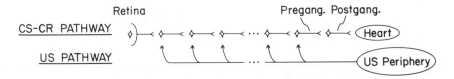

Figure 6.21. Hypothetical model of the organization of the neural circuitry mediating development of the conditioned response. The conditioned stimulus–conditioned response (CS–CR) pathway from the retina to the motoneurons is shown for heuristic purposes as a serial chain of neurons. The pathway transmitting the US information is shown as providing input to all central elements along the CS–CR pathway.

neurons by the US must be mediated by multiple pathways, since type I responses survive interruption of the input from the locus coeruleus.

As in mammals, the avian locus coeruleus consists primarily, if not exclusively, of aminergic neurons (Cabot, pers. comm.). Consequently, the type II input that is necessary for the training-induced modification of geniculate neurons in all likelihood utilizes a biogenic amine as its neurotransmitter. This is intriguing, since in the *Aplysia* sensitization and conditioning may involve a serotonergic input (Kandel, 1976). The avian locus coeruleus has substantial numbers of both noradrenergic and serotonergic neurons (Dubé and Parent 1981), but we have not yet definitively established whether either or both of these cell populations is involved in mediating the type II response of the geniculate neurons.

AN OVERVIEW OF THE ORGANIZATION OF THE INVOLVED CIRCUITRY

To conclude this summary of our ongoing efforts to identify and characterize sites of training-induced modification in our model system, some speculative comments are offered with respect to the overall organization of the circuitry that mediates development of the CR.

For heuristic purposes it is useful to represent the relevant pathways from the retina to the "cardiac motoneurons" as a "serial chain" of neurons (Figure 6.21). As described previously, this pathway is responsive to the CS prior to training. Recall that, for structures studied to date, we have found that all neurons that show training-induced modification are initially responsive to the CS. We have also found that, with the exception of the retinal ganglion cells, the nuclear groups along the identified pathway show substantial US responsiveness. Although we cannot state definitively that each of the cell groups receives direct US input, our present anatomical understanding of the pathway is suggestive of this. We therefore advance the hypothesis that all central structures of the pathway receive direct US input (Figure 6.21).

This model thus proposes CS–US convergence at each central structure along the identified pathway. Such convergence establishes the substrate for some sort of long-term heterosynaptic facilitation at each of these structures, and this is all that would be required to explain the characteristics of training-induced modification that we have observed in our system.

The "distributed" nature of this model, where training-induced modification occurs throughout the participating pathway, allows for parallel changes at each relay. That is, each neuron receiving convergent CS–US input may be capable of training-induced change irrespective of what is occurring at other relays along the pathway. Consequently, the order in which such changes occur need not be serial, and it is possible for neurons more caudally situated along the pathway to show discharge change first. However, the fact that much of the training-induced modification occurs within the first 10 CS–US presentations suggests that the various elements of the pathway would tend to change in parallel. As neuronal modification occurs, a "serial component" could be introduced into the system as well. By way of illustration, the facilitation of the geniculate neurons during conditioning would increase the magnitude of the input to their telencephalic target neurons, and this could facilitate the training-induced change of those neurons. Thus, as conditioning develops a "serial amplification" might contribute to the neuronal changes.

The above model has been generated on the basis of our present understanding of the system. It is intended only as a hypothetical working model that parsimoniously accounts for our results to date, both anatomical and physiological. The model does not address the cellular mechanisms of training-induced modification. Indeed, it is not even intended to imply that such mechanisms are necessarily the same for the different segments of the pathway.

CONCLUDING COMMENTS

To review the status of our model system, an effective behavioral preparation is well developed and has been adapted for cellular neurophysiological studies of conditioning. Further, a rigorous "cellular-connectionistic" approach to delineating the involved neural circuitry has proven its effectiveness. The initial neurophysiological analyses of this circuitry have indicated a surprisingly short central processing time for the CR. This, in turn, suggests the system is considerably more analyzable than might have been anticipated. Supporting this contention are our more recent findings on sites along the pathway that undergo training-induced modification. These data have already suggested certain general rules that obtain, such as the necessity for CS–US convergence. Moreover, they raise the intriguing possibility of involvement of aminergic pathways and have established a promising foundation for pursuing the cellular mechanisms of long-term information storage during associative learning in a vertebrate brain.

ACKNOWLEDGMENTS

The research described here was supported by grants from NSF, NIH, and the Alfred P. Sloan Foundation, most recently by NSF Grant BNS-8016396.

REFERENCES

Cabot, J. B., and Cohen, D. H. 1977. Avian sympathetic cardiac fibers and their cells of origin: Anatomical and electrophysiological characteristics. *Brain Res. 131*:73–87.

Cohen, D. H. 1969. Development of a vertebrate experimental model for cellular neurophysiologic studies of learning. *Cond. Res. 4*:61–80.

1974. The neural pathways and informational flow mediating a conditioned autonomic response. In *Limbic and Autonomic Nervous Systems Research* (L. V. DiCara, ed.), pp. 223–75. New York: Plenum Press.

1980. The functional neuroanatomy of a conditioned response. In *Neural Mechanisms of Goal-Directed Behavior and Learning* (R. F. Thompson, L. H. Hicks, and V. B. Shvyrkov, eds.), pp. 283–302. New York: Academic Press.

1982. Central processing time for a conditioned response in a vertebrate model system. In *Conditioning: Representation of Involved Neural Functions* (C. D. Woody, ed.), pp. 517–34. New York: Plenum Press.

Cohen, D. H., Gibbs, C. M., Siegelman, J., Gamlin, P., and Broyles, J. 1982. Is locus coeruleus involved in plasticity of lateral geniculate neurons during learning? *Soc. Neurosci. Abstr. 8*:666.

Cohen, D. H., and Goff, D. G. 1978. Conditioned heart rate change in the pigeon: Analysis and prediction of acquisition patterns. *Physiol. Psychol. 6*:127–141.

Cohen, D. H., and Karten, H. J. 1974. The structural organization of the avian brain: An overview. In *Birds: Brain and Behavior* (I. H. Goodman and M. W. Schein, eds.), pp. 29–73. New York: Academic Press.

Cohen, D. H., and Pitts, L. H. 1968. Vagal and sympathetic components of conditioned cardioacceleration in the pigeon. *Brain Res. 9*:15–31.

Cohen, D. H., and Schnall, A. M. 1970. Medullary cells of origin of vagal cardioinhibitory fibers in the pigeon. II. Electrical stimulation of the dorsal motor nucleus. *J. Comp. Neurol. 140*:321–342.

Cohen, D. H., Schnall, A. M., Macdonald, R. L., and Pitts, L. H. 1970. Medullary cells of origin of vagal cardioinhibitory fibers in the pigeon. I. Anatomical studies of peripheral vagus nerve and the dorsal motor nucleus. *J. Comp. Neurol. 140*:299–320.

Dubé, L., and Parent, A. 1981. The monoamine-containing neurons in avian brain. I. A study of the brain stem of the chicken *(Gallus domesticus)* by means of fluorescence and acetylcholinesterase histochemistry. *J. Comp. Neurol. 196*:695–708.

Duff, T. A., and Cohen, D. H. 1975a. Retinal afferents to the pigeon optic tectum: Discharge characteristics in response to whole field illumination. *Brain Res. 92*:1–19.

1975b. Optic chiasm fibers of the pigeon: Discharge characteristics in response to whole field illumination. *Brain Res. 92*:145–148.

Gibbs, C. M., and Cohen, D. H. 1980. Plasticity of the thalamofugal pathway during visual conditioning. *Soc. Neurosci. Abstr. 6*:424.

Gibbs, C. M., Cohen, D. H., Broyles, J. and Solina, A. 1981. Conditioned modification of avian dorsal geniculate neurons is a function of their response to the unconditioned stimulus. *Soc. Neurosci. Abstr. 7*:752.

Gold, M. R., and Cohen, D. H. 1981a. Heart rate conditioning in the pigeon immobilized with α-bungarotoxin. *Brain Res. 216*:163–172.

Gold, M. R., and Cohen, D. H. 1981b. Modification of the discharge of vagal cardiac neurons during learned heart rate change. *Science 214*:345–347.

Kandel, E. R. 1976. *Cellular Basis of Behavior.* San Francisco: Freeman.

Leonard, R. B., and Cohen, D. H. 1975. Responses of postganglionic sympathetic neurons in the pigeon to peripheral nerve stimulation. *Exp. Neurol. 49*:466–486.

Macdonald, R. L., and Cohen, D. H. 1970. Cells of origin of sympathetic pre- and postganglionic cardioacceleratory fibers in the pigeon. *J. Comp. Neurol. 140*:343–358.

Moore, R. Y., and Bloom, F. E. 1979. Central catecholamine neuron systems: Anatomy and physiology of the norepinephrine and epinephrine systems. *Annu. Rev. Neurosci. 2*:113–168.

Schwaber, J. S., and Cohen, D. H. 1978a. Electrophysiological and electron microscopic analysis of the vagus nerve of the pigeon, with particular reference to the cardiac innervation. *Brain Res. 147*:65–78.

1978b. Field potential and single unit analysis of the avian dorsal motor nucleus of the vagus and criteria for identifying vagal cardiac cells of origin. *Brain Res. 147*:79–90.

Wall, J., Wild, J. M., Broyles, J., Gibbs, C. M., and Cohen, D. H. 1980. Plasticity of the tectofugal pathway during visual conditioning. *Soc. Neurosci. Abstr. 6*:424.

7 · Reflex alterations in the spinal system: central and peripherally induced spinal fixation in rats

MICHAEL M. PATTERSON, JOSEPH E. STEINMETZ,
and ANTHONY G. ROMANO

INTRODUCTION

Retention or memorylike processes involving spinal reflex systems were initially described by DiGiorgio and her associates (1929, 1943). In these studies, brain lesions placed in the anterior lobe of the cerebellum were found to induce postural asymmetries in the hindlimbs of decerebrate or anesthetized dogs and rabbits. These asymmetries typically involved flexion of one limb and extension of the opposite limb. More surprisingly, DiGiorgio observed that these postural alterations could be retained after a spinal transection if a certain temporal interval, typically 1–2 hr in these studies, was allowed to elapse between onset of asymmetry and spinal section. This phenomenon has become known as "spinal fixation." The critical time period needed to fixate the postural alteration is referred to as spinal fixation time (SFT).

DiGiorgio's work was extended to the albino rat by Chamberlain et al. (1963). These investigators induced hindlimb asymmetries with electrolytic lesions of the cerebellum and other brain areas and concluded that an SFT of approximately 45 min was required to fixate the alteration. In addition, injections of tricyanoaminopropene, an RNA/protein synthesis stimulant, were found to decrease SFT to 25–30 min, whereas injections of 8-azaguanine, an RNA/protein synthesis inhibitor, increased SFT to 70 min. From these results, the researchers concluded that spinal fixation was dependent on biochemical changes and, in particular, on alterations of RNA synthesis or structure.

Giurgea and his colleagues confirmed and extended the work of Chamberlain et al. By removing half of the anterior cerebellar lobe to produce asymmetry and allowing an appropriate time interval to produce fixation, they were able to conduct an extensive series of studies concerning various features of the fixation paradigm. First, the effects of injecting a variety of drugs were studied. In general, central nervous system stimulants and depressants were found to have no effect on fixation. Drugs that enhance performance in many learning situations generally shortened the SFT, whereas drugs that typically impair performance in many learning situations lengthened the SFT (Giurgea and Mouravieff-Lesuisse 1971; Mouravieff-Lesuisse and Giurgea 1968). In another study, electroconvulsive shock (ECS) prolonged SFT, whereas drugs previously found to shorten SFT partially reversed the effects of ECS (Mouravieff-

Lesuisse and Giurgea 1970). Next, fixation time was consistently found to be prolonged by a cortical spreading depression initiated by application of potassium chloride on the temporoparietal cortex ipsilateral to the cerebellar lesion (Mouravieff-Lesuisse and Giurgea 1973). This effect was not obtained when contralateral spreading depression was produced. These findings suggested that spinal fixation depended on a critical degree of cortical regulation. Lastly, the effects of brain extracts on SFT were assessed (Giurgea et al. 1971). Extracts from brains of donor rats with cerebellar-lesion-induced asymmetries were found to decrease SFT in recipient animals that received intraperitoneal injections of the abstract prior to or at the time of a cerebellar lesion. The researchers concluded that the cerebellar lesion produced postural asymmetry in the limbs and induced biosynthesis of a material normally absent in the CNS. Furthermore, this material was thought to shorten the SFT necessary to fixate the cerebellar-lesion-induced asymmetry.

In another series of studies, Palmer and associates examined temporal parameters as well as the involvement of drugs and protein synthesis on the fixation process (Palmer and Davenport 1969; Palmer et al. 1968, 1970a, 1970b, 1970c). Small stab wounds placed in the anterior cerebellum of newborn rats resulted in contralateral hindlimb extension and ipsilateral flexion that fixated in about 38–45 min. In general, metabolic inhibitors and neurohumoral blocking agents were found to lengthen SFT. Analyses of the role of RNA, proteins, and various other chemical compounds revealed many complex interactions on spinal fixation. It was found that drug-induced disturbances in protein and RNA metabolism of spinal neurons either enhanced or inhibited the fixation process, depending on the dose as well as the drug. These results suggest that the cerebellar lesion may set off a complex biochemical reaction that alters reflex balances in the spinal motor nuclei.

We have recently begun an extensive series of studies involving parametric and neurophysiological features of the spinal fixation phenomenon. Initially, a standard preparation was developed for studying fixation in the rat with traditional cerebellar lesioning techniques. In addition, we have developed a second fixation preparation that involves direct stimulation of the rat hindlimb. A summary of the data we have collected using both asymmetry induction methods and a discussion of the fixation phenomenon as a potential model of spinal plasticity are presented here.

SUBJECTS

Adult, male and female, hooded Long-Evans rats were used in the studies presented below. The rats weighed between 350 and 700 g, received food and water ad lib, and were maintained on light/dark cycles of 12/12 hr. All experimental procedures were conducted during the light portion of the light/dark cycle. In each experiment, the animals were anesthetized to a light surgical level with Nembutal (50 mg/kg, i.p.). If during the course of an experiment an animal reacted too strongly to a tail pinch, supplemental doses of Nembutal were administered to maintain the surgical level of anesthesia.

CEREBELLAR LESION STUDIES

Our initial two experiments were conducted to confirm the existence of spinal fixation in rats and to determine more precisely the minimum spinal fixation time (SFT) required to consistently produce retention of a postural asymmetry (Steinmetz et al. 1981). Cerebellar lesions were used to induce the postural asymmetry in a manner similiar to that used by Chamberlain et al. (1963).

Each rat was anesthetized with Nembutal and placed in a stereotaxic head-holding apparatus. The anterior cerebellum was visually located after removal of the interparietal bone. Next, a laminectomy was performed to expose the spinal cord between T-6 and T-8. While the head remained secure in the stereotaxic instrument, the animal was suspended by the tail and supported by a sling that surrounded the abdomen. This apparatus firmly secured the animal yet allowed unobstructed movement of the hindlimbs. Stereotaxically guided unilateral lesions were next placed in the cortex or underlying nuclear areas of the left anterior cerebellum by delivering 18 sec of 10-mA anodal current through an insulated stainless steel electrode. The lesions resulted in postural asymmetries of the hindlimbs typically characterized by flexion of one limb and extension of the opposite limb. In the initial study, each animal was randomly assigned to one of two groups that were allowed SFTs of either 20 or 50 min. In the second study, animals were randomly assigned to one of five groups and were allowed SFTs of 25, 30, 35, 40, or 45 min. Five minutes prior to the end of the respective fixation times, the degree of asymmetry was assessed by suspending weights from the flexed leg until the hindlimbs were level. In this manner, asymmetry was recorded as the weight necessary to bring the limbs into a symmetrical position. At the end of the respective SFTs, a spinal transection was performed at T-7, 5 min were allowed for spinal shock to subside, and weights were again placed on the flexed leg to assess any asymmetry present after spinal section. Postsection asymmetry was recorded as zero if the limb did not return to a flexed position when the weights were removed.

For the two cerebellar lesion studies, percentages of presection asymmetry retained after spinal section were statistically analyzed with Mann–Whitney U-tests. In the initial study, the 50-min SFT group ($M = 53.06\%$) was found to have a significantly greater percentage of original asymmetry retained after spinal section ($p < .001$) when compared to the 20-min SFT group ($M = .44\%$). In the second study, the 45-min SFT group ($M = 58.2\%$) had a significantly greater percentage retention ($p < .05$) that did the 40-min ($M = 15\%$), 35-min ($M = 11.6\%$), 30-min ($M = 0$), and 25-min ($M = 0$) SFT groups. Considering both studies, consistent retention (i.e., every animal within a group demonstrated retention) was found only when SFTs of 45 or 50 min were allowed. Fixation could be observed, however, in some animals allowed SFTs of less than 45 min. (See Figure 7.1 for a summary of mean amounts of pre- and postsection asymmetry present when various SFTs were allowed to elapse after cerebellar lesion.)

These results demonstrate that it is possible, with the present preparation, to consistently obtain spinal fixation with cerebellar lesions when at least 45 min is allowed to elapse between cerebellar lesion and a spinal cord transection. Since fixation could be obtained in a smaller percentage of animals allowed

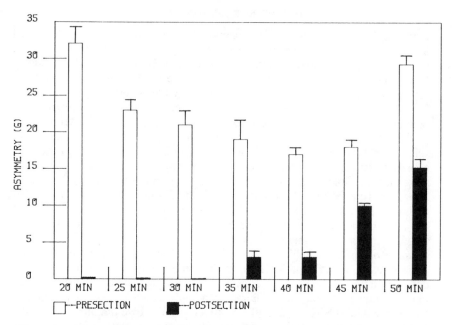

Figure 7.1. Mean amounts of presection and postsection asymmetry present in groups of rats allowed 20, 25, 30, 35, 40, 45, or 50 min of spinal fixation time following a cerebellar lesion. Number of subjects (*ns*) = 15 for 20- and 50-min groups; *ns* = 5 for 25- to 45-min groups; vertical lines = (*SE*).

shorter fixation intervals, it appears that cerebellar lesion-induced fixation may be a gradual progression of events that begins at the time of or after the onset of postural asymmetry, is measureable in some animals at approximately 35 min, and is complete by 45 min.

PERIPHERAL STIMULATION STUDIES

Frankstein (1947) reported that turpentine, when injected into the footpads of cats, caused a limb flexion that lasted for several days. More importantly, he observed that after the animal recovered from the injection and resumed normal locomotion, decerebration resulted in the return of asymmetry. Frankstein interpreted these data to indicate that although the peripheral stress had been removed, long-lasting spinal reflex alterations had occurred and were manifested when cortical compensation was eliminated by decerebration. Considering Frankstein's results, we reasoned that it might be possible to alter reflex patterns of the limbs simply by delivering abnormal, continuous sensory input from the periphery. In this manner, retention of postural alterations could be studied with a method that allowed precise control over onset and offset as well as magnitude of the postural asymmetry used to obtain the fixation. Below is a summary of the series of studies we have conducted utilizing this technique (Steinmetz et al. 1981, 1982b).

The original peripheral stimulation study (Steinmetz et al. 1981) serves to illustrate the general experimental procedures used in all studies that involved peripherally induced fixation. Rats were anesthetized with Nembutal and taped ventral side down to a surgery board, and a laminectomy was performed between T-6 and T-8 to expose the spinal cord. The animals were then suspended by the tail, a position that restrained the animal yet allowed free access to the hindlimbs. Next, each subject was randomly assigned to one of two groups that received either 20 or 45 min of hindlimb stimulation. Stimulation consisted of 1–4 mA, 100-pps, 7-msec repetitive dc pulses delivered through two stainless steel wound clips placed, one on each side, on the upper right hindleg. These stimulation parameters resulted in strong flexion of the stimulated limb. At the end of the respective stimulation periods, the stimulation was terminated and asymmetry measured by suspending weights from the flexed leg until the hindlimbs were level. Next, the spinal cord was sectioned, spinal shock allowed to subside, and persisting postsection asymmetry assessed. Finally, asymmetry was measured 2 hr subsequent to cord section to evaluate the effects of time on any persisting asymmetry. Mann–Whitney U-analysis of the data obtained from this initial study demonstrated that those animals that received 45 min of stimulation retained a significantly greater percentage of asymmetry ($p < .05$) after cord section ($M = 68.33\%$) than did the 20-min stimulation group ($M = 8.28\%$). Only animals in the 45-min group showed asymmetry persistence 2 hr after cord section. These data indicate that postural asymmetry created by direct hindlimb stimulation will outlast a spinal transection if 45 min of constant stimulation are allowed.

The next experiment was conducted to more thoroughly examine temporal parameters involved in peripherally induced fixation. Animals were randomly assigned to four groups that received either 25, 30, 35, or 40 min of hindlimb stimulation prior to spinal section. Mann–Whitney U-analyses of percentage retention scores revealed significantly greater percentages of presection asymmetry present ($p < .05$) in the 40-min group ($M = 47.5\%$) when compared to the 30-min ($M = 22.22\%$) and 25-min ($M = 8.33\%$) groups, for the 35-min group ($M = 46.11\%$) when compared to the 25- and 30-min groups, and for the 30-min group as compared to the 25-min group. Consistent retention was noted only in the 40-min group. When this time series study and the initial stimulation study are combined, a gradual increase in postural alteration retention can be noted as stimulation time increases. Consistent retention was first evident at 40 min. When compared to the cerebellar lesion data, the stimulation data indicate a 5-min downward shift in the minimal SFT necessary to produce fixation. This time shift may be accounted for by differences in intensities of the asymmetry-inducing procedures as well as by differences in cerebellar condition (lesioned vs. intact) during the time interval preceding spinal section. (See Figure 7.2 for a summary of pre- and postsection asymmetries present when various periods of hindlimb stimulation were delivered prior to spinal section.)

The relative importance of maintaining intact spinal–hindlimb sensory and motor connections during the peripheral stimulation procedure was also examined in three separate studies. In the first study, rats underwent complete dorsal and ventral root rhizotomies between T-10 and L-6 prior to 10, 20, 30, 40, or

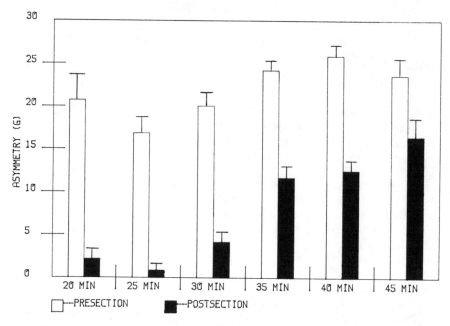

Figure 7.2. Mean amounts of presection and postsection asymmetry present in groups of rats (*ns* = 7) given 20, 25, 30, 35, 40, or 45 min of hindlimb stimulation. Vertical lines = *SE*.

50 min of hindlimb stimulation. The stimulation initially produced strong flexion. However, none of the animals could maintain appreciable levels of flexion for the duration of the various stimulation periods. These results were interpreted to show that observed retention of peripherally induced asymmetry was not merely the product of local stimulation effects on muscle tissue. In a second study, complete dorsal and ventral root rhizotomies were performed on rats after they received 10, 20, 30, 40, or 50 min of hindlimb stimulation. Again, no appreciable levels of postrhizotomy asymmetry could be discerned. This suggested that activity at the hindlimb neuromuscular junction, in the absense of spinal connection, was not sufficient to maintain peripherally induced postural asymmetry. The third study was conducted to examine the possibility of obtaining persistence of asymmetry via direct stimulation of ventral roots in the absence of sensory feedback from hindlimb muscles. First, rats underwent complete dorsal and ventral root rhizotomy. Strong flexion was produced for a 50-min period by direct stimulation of the cut ventral roots. After stimulation was terminated, no appreciable persisting asymmetry could be found. Thus, ventral root stimulation was not effective in producing asymmetry persistence when the hindlimbs were isolated from the spinal cord. The three rhizotomy studies demonstrate that neural connections between hindlimb and spinal cord must remain intact to obtain as well as maintain a fixated postural asymmetry.

Moreover, these results provide substantial evidence for an active role played by spinal reflex centers in the fixation process.

The role of higher centers in producing the fixation effect has been unclear. DiGiorgio (e.g., 1943) was able to produce the effect when a decerebrate preparation was utilized, thus suggesting that the cerebral cortex was not necessary for obtaining fixation. Conversely, Giurgea and colleagues (e.g., 1973) have ascribed an important function to the cerebral cortex and cerebellum in the fixation process. Using spinalized rats, we determined the importance of maintaining intact higher brain structures during peripherally induced fixation. In this study, 20 rats underwent spinal transections before peripheral stimulation was administered. Half of the rats received 45 min of hindlimb stimulation, while the remaining animals were stimulated for 20 min. At the end of stimulation, asymmetry was assessed by suspending weights. The results of this study showed that persistence of asymmetry could be obtained in the spinal rat. Moreover, the 45-min stimulation group ($M = 18$ g) demonstrated a significantly greater amount of persisting asymmetry ($p < .01$) than did the 20-min group ($M = 5$ g). Analysis of temporal parameters involved in post-stimulation asymmetry persistence has revealed that consistent retention is possible when as little as 30 min of stimulation is delivered to the spinalized rat (Steinmetz et al. 1982a). Apparently, cerebral cortex and cerebellar centers are not necessary for producing retention of peripherally induced postural asymmetries. Moreover, spinal fixation appears to develop more rapidly in the absence of these higher brain structures.

When cerebellar lesions are used to induce postural asymmetry, a constant source of descending input to spinal centers from the injured cerebellum is available to produce fixation. A study was undertaken to examine the importance of maintaining constant hindlimb stimulation for the production of peripherally induced fixation. Rats were randomly assigned to five groups and received either 10, 20, 30, 40, or 50 min of stimulation. Following stimulation, a waiting period was allowed before a spinal cord section was performed. The respective waiting periods for the 10-, 20-, 30-, 40-, and 50-min stimulation groups were 40, 30, 20, 10, and 0 min. Thus stimulation times plus waiting period totaled 50 min for all animals. All animals maintained their previously stimulated leg in a flexed position for the duration of the respective waiting periods. An analysis of percentage retention scores revealed no significant differences between the five groups. Only two rats, one a member of the 20-min group and the other a member of the 30-min group, failed to show persistence of asymmetry after spinal section. (See Figure 7.3 for mean amounts of pre- and postsection asymmetries recorded for the five groups of rats.) From these data we concluded that constant delivery of peripheral stimulation was not required for fixation to occur. Stimulation appeared to be necessary only for initiation of the fixation process. In the stimulation–wait procedure, it is likely that stimulation initiated higher-center influences that in turn impinged upon the spinal reflex system during the time interval between stimulation offset and cord section. It should be noted, however, that higher-center influences are not necessary for fixation to occur, since fixation can be produced in the spinal rat. Apparently, adherence to basic temporal parameters is more important for the

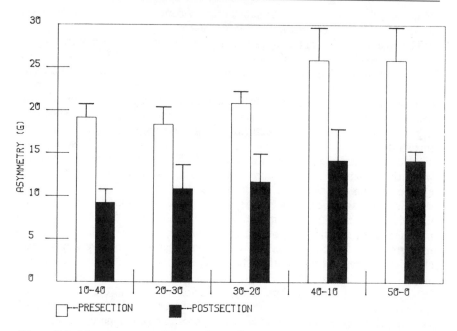

Figure 7.3. Mean amounts of presection and postsection asymmetry present in rats that received 10, 20, 30, 40, or 50 min of hindlimb stimulation (*ns* = 6) followed by 40-, 30-, 20-, 10-, or 0-min waiting periods, respectively, prior to spinal section. Vertical lines = *SE*.

production of retention than is the method used to alter spinal reflex mechanisms.

PRELIMINARY RESULTS

At the present time, several fixation projects are underway in our laboratory. Presented below are preliminary results from three of these studies. First, we have begun exploring the possibility of obtaining fixation in the rat through cerebellar stimulation. In this preparation, postural asymmetries are induced through stimulation of anterior cerebellar cortex. Preliminary results show that retention of postural alterations will occur if 45 min of central stimulation are delivered. Twenty minutes of stimulation does not appear to be adequate for retention.

Another ongoing project involves long-term retention characteristics of the peripheral stimulation preparation. Rats are anesthetized, then given either 45, 60, or 90 min of hindlimb stimulation and returned to their home cages for either 24, 48, or 72 hr. At the end of the waiting periods, the animals are returned to the laboratory, where they undergo a spinal transection. Our preliminary data indicate that even though the animals have regained normal locomotive patterns, flexion of the previously stimulated limb returns after spinal transection. This effect has been noted for all combinations of stimulation times and waiting periods.

Finally, we have gathered preliminary data from a study designed to evaluate the role of cutaneous irritation in promoting postsection flexion in the peripherally stimulated rat. Although previous data suggest an active involvement of the spinal cord in producing alterations associated with postsection retention of asymmetry, an alternative explanation involving nonspinal, peripheral, neural or nonneural components still exists. It is possible that 45 min of peripheral stimulation induces a long-lasting irritation of cutaneous tissue in the general area of stimulation. This focus of irritation could convey constant afferent input to the spinal cord, thus perpetuating flexion of the afflicted limb after spinal section. Peripherally induced fixation would therefore not be a product of altered spinal outflow but rather a product of continuous, injury-based sensory input. To test this possibility, we are blocking the hindlimb cutaneous tissue with a local anesthesia (lidocaine) subsequent to the delivery of 45 min of stimulation. To find the dose of lidocaine effective in blocking the sensory input to the spinal cord that arises from stimulation, we subcutaneously injected varying amounts of lidocaine into the hindlimbs of several pilot animals. Fifty minutes of stimulation was then delivered and amounts of persisting flexion were measured upon termination of stimulation. These procedures revealed that the effects of peripheral stimulation were blocked when 1.25 cc/kg of lidocaine (10 mg/ml) was injected under each of the wound clips. More specifically, flexion could not be observed in these animals either during or subsequent to the delivery of stimulation. Results of this study show that persisting asymmetry can be observed after previously stimulated tissue has been effectively anesthetized with appropriate doses of lidocaine. Apparently, retention of a stimulation-induced asymmetry can be obtained in the absence of ongoing peripheral irritation.

DISCUSSION

The original studies involving cerebellar-lesion-induced spinal fixation confirmed the existence of the phenomenon and provided additional insight into temporal aspects of the retention process. Apparently, cerebellar lesions result in alterations in the tonic descending neural activity that impinges on spinal components of the reflex arc. Our data indicate that over a 45-min period, these asymmetrical descending influences gradually alter reflex activity in the spinal centers. Once the spinal reflex activity has been adequately changed, it becomes self-sufficient in maintaining postural asymmetry in the absence of descending input. Our preliminary findings regarding the use of cerebellar stimulation concur with the lesion data. The cerebellar stimulation method should prove to be valuable for studying the relative role of descending input in the fixation process.

The majority of our research has been concerned with the use of hindlimb stimulation for induction of postural alterations. Peripheral stimulation provides abnormal, tonic input to spinal centers that apparently induces long-lasting alterations in neural conduction through the spinal reflex circuitry. Parametric data concerning temporal aspects of peripherally induced fixation indicate that it is a gradual process of spinal reflex alteration that is complete after 40 min. Thus, it is similar to the cerebellar preparations in that critical time periods must be allowed to elapse for the effect to be seen.

Our data have demonstrated the importance of the spinal cord in promoting and maintaining the postural alteration. Previous research, however, has concentrated on the importance of higher centers in obtaining the fixation effect (e.g., Giurgea et al. 1971). It is likely that higher structures modulate ongoing spinal activity during the peripherally induced fixation process, but this influence is apparently secondary to spinal events. Data obtained from spinalized rats clearly show that fixation can be obtained in the absence of descending influences. Also, studies involving rhizotomized animals have revealed the importance of spinal activity in the fixation process. Stimulation either prior to or after complete ventral and dorsal root rhizotomy did not produce appreciable asymmetry retention. Likewise, appreciable asymmetry persistence was not obtained when direct stimulation of ventral roots was delivered in the absence of dorsal root–spinal cord connections. Intact flexor spinal reflex arcs are apparently necessary for demonstrating the fixation phenomenon. These studies also provide indirect evidence against the possibility that peripherally induced fixation is a product of ongoing peripheral influences arising from the hindlimb. First, stimulation after rhizotomy should have produced retention if fixation was a product of direct stimulation of muscle tissue. Second, flexion should have been evident after poststimulation rhizotomy if flexion was being maintained solely by activity at hindlimb neuromuscular junctions. Finally, preliminary data demonstrating the persistence of postsection flexion after previously stimulated cutaneous tissue is effectively blocked with a local anesthesia rules out the possibility that retention is an artifact of ongoing peripheral sensory input.

The above studies strongly suggest that peripherally induced fixation is not dependent on higher-neural-structure activity and not propagated by ongoing peripheral neural events. Fixation appears to depend heavily on processes that occur at the spinal level. We propose that spinal fixation is a result of long-lasting alterations of reflex patterns in the spinal cord. The various asymmetry induction methods (i.e., cerebellar lesions, cerebellar stimulation, and hindlimb stimulation) are successful at initiating these spinal reflex alterations. Apparently, the various forms of spinal input induce a focus of spontaneous activity in the spinal reflex centers that eventually, over time, become self-sufficient in maintaining postural asymmetry independently of other sources of nervous influence.

The reflex alteration is relatively long-lasting since preliminary data have shown that the postural alteration will reappear upon spinal section when as long as 72 hr is allowed to elapse between the termination of stimulation and spinal section. The findings that postural alterations are long-lasting and occur only when strict temporal parameters are adhered to suggest that fixation may be a spinal analog of memory processes in the intact organism. Traditional "consolidation hypotheses" suggest that experience is first encoded in a labile form for some time and only later becomes "consolidated" in a more permanent form. (See McGaugh and Herz 1971 for review.) In the fixation preparation, the time period between onset of asymmetry and spinal cord section can be regarded as the "labile" phase. If spinal events are allowed to continue for an adequate length of time, the reflex alterations will "consolidate" and become somewhat permanent. Thus, the fixation process appears to share many of the formal properties of memory consolidation in the intact organism, yet is produced in a much more simple system.

The process of spinal fixation appears to be somewhat similar to the phenomenon of the dominant focus described by Ukhtomsky (e.g., 1926) and of the mirror focus described by Morrell (1961). These have been summarized by Kandel and Spencer (1968). In the dominant focus work, Ukhtomsky found that repeated elicitation of any response by a strong stimulus caused that response to become dominant over other responses. He utilized the concept of facilitated neural pathways due to the repeated usage to explain the altered reflex state. Morrell's mirror focus was developed from his clinical observations that epileptic patients often had a primary seizure focus in one hemisphere and a secondary or "mirror" focus in the opposite hemisphere at the homologous site. Morrell's work in primates showed that ethyl chloride spray applied to a small cortical area on one cortex produced seizure activity in the homolateral area of the opposite cortex. Finally, excision of the original lesioned area did not abolish the seizure activity in the mirror cortical site. The activity had become autonomous or fixed.

Kandel and Spencer (1968) summarized several possible mechanisms for such persistent alterations of discharge patterns in neural networks. Included were reverberating electrical signals, transmitter persistence, rebound excitation after inhibition, and direct alteration of pacemaker activity. They differentiated between dynamic (circulation of impulses) and enduring (functional property changes) alterations of neural networks. Although the present studies of spinal fixation have yet to address mechanisms of the effect, it is tempting to postulate that the process is one encompassing both processes: Initially, a dynamic alteration of the neural activity produced by the fixation stimulus occurs, which in turn produces an enduring change in (presumably) interneuron properties that is strong enough to cause a persistence of neural activity after the initiating stimulus is removed. Thus, the fixation process, from our data, seems to produce a spinal autonomous focus of activity that drives the flexor motor units and that remains active for hours following the removal of the initiating stimulus.

The standard preparations described above have provided us with measurable and quantifiable methods for studying the fixation phenomenon. In addition, the preparations are amenable to neurophysiological analysis and should allow us to study spinal events that underlie long-term retention of postural alterations. The spinal preparations should prove to be especially useful for this purpose since spinal reflex activity can be explored with minimal confounding influences from other nervous centers.

ACKNOWLEDGMENTS

The research presented here was supported by NINCDS Grants NS10647 and NS14545, the American Osteopathic Association Bureau of Research Grant 81-80-023, and the Ohio University College of Osteopathic Medicine.

REFERENCES

Chamberlain, T. J., Halick, P., and Gerard, R. W. 1963. Fixation of experience in the rat spinal cord. *J. Neurophysiol. 26*:662–673.

DiGiorgio, A. M. 1929. Persistenza nell'animale spinale, di asimmetrie posturali e motorie di origine cerebellare. Nota I–III. *Arch. Fisiol. 27*:518–580.

1943. Richerche sulla persistenza die fenomeni cerebellari nell'animale spinale. *Arch. Fisiol. 43*:47–63.

Frankstein, S. I. 1947. One unconsidered form of the part played by the nervous system in the development of disease. *Science 106*:242.

Giurgea, C., Daliers, J., and Rigaux, M. L. 1971. Pharmacological studies on an elementary model of learning – The fixation of an experience at spinal level. II. Specific shortening of the spinal cord fixation time (SFT) by a brain extract. *Arch. Int. Pharmacodyn. Ther. 191*:292–300.

Giurgea, C., and Mouravieff-Lesuisse, F. 1971. Pharmacological studies on an elementary model of learning – The fixation of an experience at spinal level. *Arch. Int. Pharmacodyn. Ther. 191*:279–291.

Kandel, E. R., and Spencer, W. A. 1968. Cellular neurophysiological approaches in the study of learning. *Physiol. Rev. 48*:65–134.

McGaugh, J. L., and Herz, M. J. 1971. *Memory Consolidation*. San Francisco: Albion.

Morrell, F. 1961. Lasting changes in synaptic organization produced by continuous neuronal bombardment. In *CIO MS Symposium on Brain Mechanisms and Learning* (J. F. Delafresnaye, A. Fessard, R. W. Gerard, and J. Konorski, eds.). pp. 375–92. Oxford: Blackwell.

Mouravieff-Lesuisse, F., and Guirgea, C. 1968. Pharmacological reactivity of an experimental model of memory: The spinal cord fixation. *Arch. Int. Pharmacodyn. Ther. 176*:471–472.

Mouravieff-Lesuisse, F., and Guirgea, C. 1970. Influence of electroconvulsive shock on the fixation of an experience at spinal level. *Arch. Int. Pharmacodyn. Ther. 183*:410–411.

Mouravieff-Lesuisse, F., and Guirgea, C. 1973. Cortical regulation of the fixation of an experience at spinal level. Paper presented at the meeting of the Association des Physiologists, Bordeaux, France, May 1973.

Palmer, G. C., and Davenport, G. R. 1969. Involvement of phospholipids in the fixation of abnormal spinal reflexes in newborn rats. *Brain Res. 13*:394–396.

Palmer, G. C., Davenport, G. R., and Ward, J. W. 1970a. The effects of neurohumoral drugs on the fixation of spinal reflexes and the incorporation of uridine into the spinal cord. *Psychopharmacology 17*:59–69.

1970b. Involvement of protein synthesis in the fixation of spinal reflexes in the newborn rat. *Brain Res 17*:372–375.

1970c. Involvement of RNA synthesis in the fixation of abnormal reflexes in the newborn rat. *Exp. Neurol. 26*:263–264.

Palmer, G. C., Ward, J. W., and Davenport, G. R. 1968. Biochemical and pharmacological investigation of abnormal spinal reflex activity. *Anat. Rec. 160*:405.

Steinmetz, J. E., Beggs, A. L., Cervenka, J., Romano, A. G., and Patterson, M. M. 1982a. Temporal parameters involved in retention of reflex alterations in spinal rats. *J. Comp. Physiol. Psychol. 96*:325–327.

Steinmetz, J. E., Cervenka, J., Dobson, J., Romano, A. G., and Patterson, M. M. 1982b. Central and peripheral influences on retention of postural asymmetry in rats. *J. Comp. Physiol. Psychol. 96*:4–11.

Steinmetz, J. E., Cervenka, J., Robinson, C., Romano, A. G., and Patterson, M. M. 1981. Fixation of spinal reflexes in rats by central and peripheral sensory input. *J. Comp. Physiol. Psychol. 95*:548–555.

Ukhtomsky, A. A. 1926. Concerning the condition of excitation in dominance. *Nov. Refl. Fiziol. Nerv. Sist. 2*:3-15.

8 · The use of simple invertebrate systems to explore psychological issues related to associative learning

T. J. CAREW, T. W. ABRAMS, R. D. HAWKINS, and
E. R. KANDEL

INTRODUCTION

The primary goal in the study of learning in animals with simple nervous systems is to use their simplicity to analyze the *biological* mechanisms underlying learning and memory. A less obvious but important secondary goal is to use cellular approaches to gain insights into traditional *psychological* issues that center on learning. In this chapter we focus on the second of these goals. Based on our studies of the mechanism of classical conditioning in *Aplysia,* we will examine three psychological issues that have required cellular approaches for their solution: (1) What is the relationship of alpha conditioning to classical conditioning? (2) What is the relationship of sensitization to classical conditioning? and (3) Is impulse activity in a postsynaptic neuron (as first proposed by Hebb) a prerequisite for associative changes to occur at a synapse?

Before taking up each of these questions, we will briefly review our current understanding of the cellular mechanisms of classical conditioning in *Aplysia.*

DIFFERENTIAL CONDITIONING OF THE SIPHON WITHDRAWAL REFLEX IN *Aplysia*

The key to our recent ability to analyze classical conditioning in *Aplysia* has been the use of a discriminative learning task. In their initial studies of classical conditioning of the siphon and gill withdrawal reflex, Carew et al. (1981) used a weak tactile conditioned stimulus (CS) to the siphon, which elicits a reflex contraction of the gill and siphon of very small amplitude. The unconditioned stimulus (US) was a strong electric shock to the tail, which produces a very strong withdrawal reflex. In animals that received specific temporal pairing of the CS and US, the CS acquired the ability to trigger significantly enhanced gill and siphon withdrawal, whereas in animals that received the CS and US either specifically unpaired, randomly presented, or alone, the response to the CS after training was significantly less than in paired animals. The conditioning was rapidly acquired (within 15 trials with an intertrial interval of 5 min) and persisted for several days.

More recently, Carew et al. (1983) have found that this reflex can be differentially conditioned. To produce discriminative conditioning, two sites were chosen as CS pathways in the same animal: (1) the siphon (as in Carew et al. 1981), and (2) the mantle shelf. Stimulation of one site (the CS+) was specifically paired with the US, stimulation of the other (the CS−) was specifically unpaired (Figure 8.1*A*, *B*). Conditioning was assessed in each animal by examining the responses to the CS+ and CS− after training. We found that each site was competent to mediate conditioned siphon withdrawal, and that animals could indeed acquire a differentially conditioned response (Figure 8.1*C*). The differential conditioning can be acquired in only a single trial, and is retained for more than 24 hr. The strength of the learning increases progressively with increased trials. The discriminative capabilities of *Aplysia* are surprisingly good. Not only can two sites on anatomically separate structures (the siphon and the mantle) serve as discriminative stimuli, but even two separate sites on the siphon skin can serve as conditioned stimuli (Carew et al. 1983). These two sites are within the field of innervation of a single homogeneous sensory neuron cluster (the LE cluster).

The demonstration that differential conditioning can be produced in an elementary reflex mediated by a small number of neurons indicates that this relatively advanced form of associative learning is not an exclusive feature of behaviors having complex neural circuitry. Moreover, the finding that independent afferent inputs that activate a common set of motor neurons and interneurons can be differentially conditioned restricts the possible cellular loci involved in the associative learning.

These behavioral observations raise several important questions for further study. For instance, what is the interstimulus interval function of the learning? Can backward conditioning be produced? Answers to these questions will permit comparison with vertebrate learning studies and also set limits on possible cellular mechanisms that might produce the associative effect. Moreover, the important studies of Sahley et al. (1981a, b; see also Chapters 11 and 12) make clear that it is also possible to use simple animals to examine higher forms of learning such as contingency effects, second-order conditioning, blocking, sensory preconditioning, and conditioned inhibition.

The neural circuits for the pathways involved in the learning (the CS, US, CR and UR pathways) have been worked out in some detail and are quite simple. Since the behavioral experiments involved differential conditioning of the siphon withdrawal component of the defensive withdrawal reflex, we focused our cellular experiments on the neural circuit for siphon withdrawal. In the CS and CR pathways, identified siphon mechanosensory neurons make monosynaptic excitatory synaptic connections onto identified central siphon motor neurons (Perlman 1979), peripheral siphon motorneurons (Bailey et al. 1979), and identified interneurons (see Figure 8.2). Sensory neurons that convey the US input from the tail have also been identified (Walters et al. 1981b). To investigate the cellular mechanism of the differential conditioning described above, Hawkins et al. (1983) used a training procedure analogous to that used in conditioning of the withdrawal reflex, involving direct stimulation of the siphon sensory neurons that mediate the reflex. Specifically, we attempted to facilitate synaptic transmission differentially from two individual siphon sensory neurons to a common siphon motor neuron by pairing spike activity in one of the sensory neurons with the US (thus substituting intracellularly produced

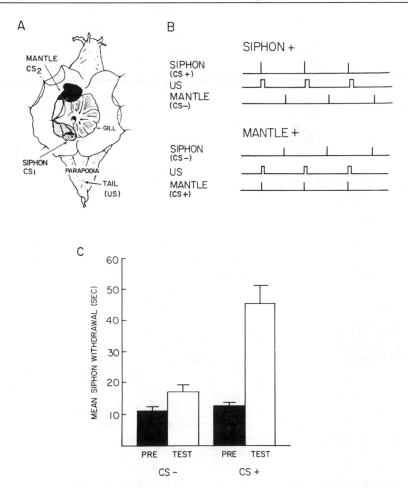

Figure 8.1. *A:* Dorsal view of *Aplysia* illustrating the two sites used to deliver conditioned stimuli: (1) the siphon (stipled) and (2) the mantle shelf (black). The unconditioned stimulus (US) was an electric shock delivered to the tail. For illustrative purposes, the parapodia are shown intact and retracted. However, the behavioral studies
were all carried out in freely moving animals whose parapodia were surgically
removed. *B:* Paradigm for differential conditioning: one group (Siphon+) received the
siphon CS paired (CS+) with the US and the mantle CS specifically unpaired
(CS−) with the US; the other group (Mantle+) received the mantle stimulus as
CS+ and the siphon stimulus as CS−. The intertrial interval was 5 min. *C:* Pooled
test scores from the unpaired (CS−) pathway ($n = 12$) and from the paired (CS+)
pathway ($n = 12$) are compared to their respective prescores. The CS+ test scores
are significantly greater than the CS− test scores ($p = .005$), demonstrating that
differential conditioning has occurred. (Data from Carew et al. 1983)

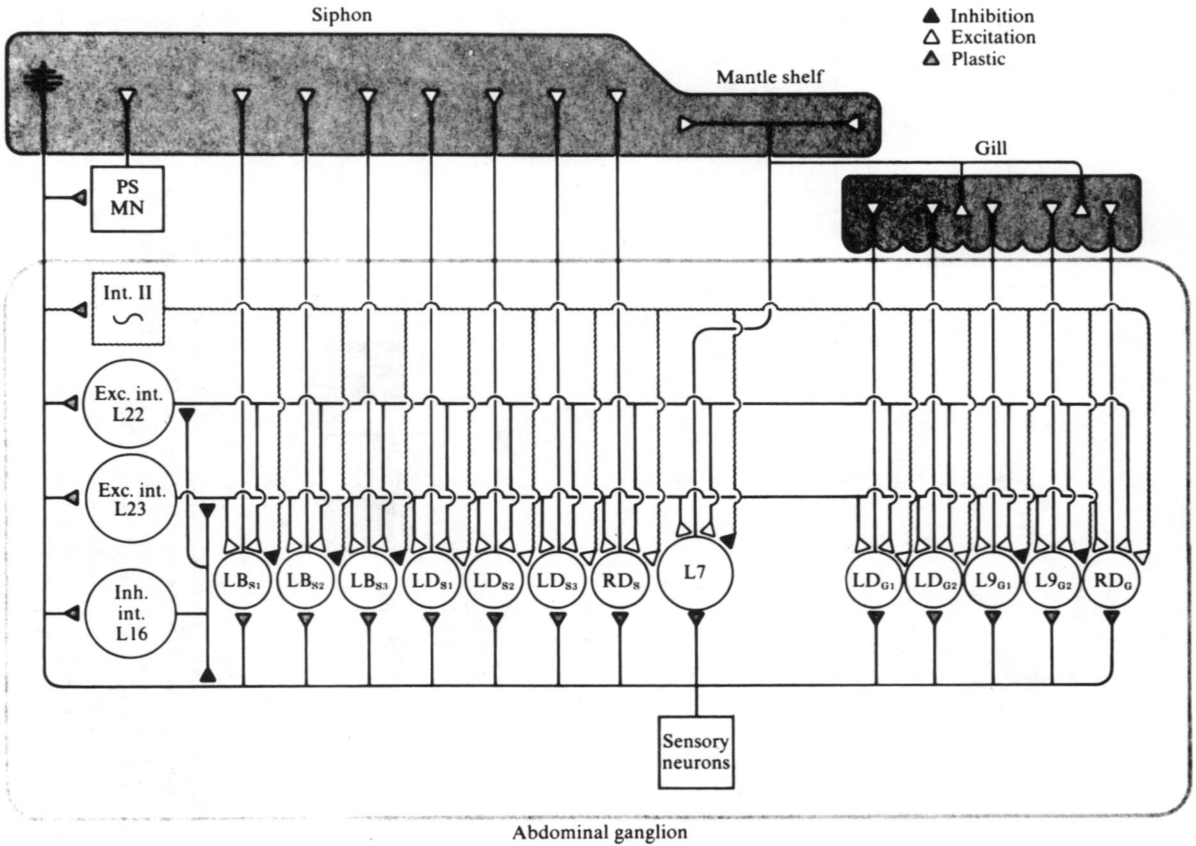

Siphon

Mantle shelf

Gill

▲ Inhibition
△ Excitation
◭ Plastic

PS
MN

Int. II

Exc. int.
L22

Exc. int.
L23

Inh.
int.
L16

LB_{S1} LB_{S2} LB_{S3} LD_{S1} LD_{S2} LD_{S3} RD_S L7

LD_{G1} LD_{G2} $L9_{G1}$ $L9_{G2}$ RD_G

Sensory
neurons

Abdominal ganglion

spike activity for the tactile CS+ input) and producing spike activity in the other sensory cell specifically unpaired with the US (substituting this activity for the CS− input) (Figure 8.3*A*). We found that 15 min after five training trials, the excitatory postsynaptic potential (EPSP) from the paired sensory neuron was dramatically facilitated (approximately 200 percent of pretraining EPSP amplitude), whereas the EPSP from the unpaired sensory neuron actually tended to decrease slightly, to approximately 90 percent of pretraining amplitude, probably due to homosynaptic depression (Figure 8.3*B*$_1$). In other experiments we examined the effects of stimulation of one sensory neuron paired with the tail shock US, compared to another sensory neuron that was not stimulated during training (thus receiving "US-alone" or sensitization training). Both EPSPs showed facilitation after training, but the EPSP from the paired sensory neuron exhibited significantly greater facilitation than that from the US-alone sensory neuron (Figure 8.3*B*$_2$).

These results demonstrate that facilitation of the EPSP from a sensory neuron to a follower neuron is selectively enhanced if the sensory neuron is active just before the facilitating stimulus (the US) is presented. The experimental protocol and parameters that we have used are identical to those that were used in the behavioral studies described above. A direct comparison of the results from the cellular experiments with the results from behavioral experiments with a comparable number of training trials is shown in Figure 8.3*D*. There is quite a good fit between the behavioral and cellular results, not only qualitatively but even quantitatively. The congruence between the behavioral and cellular data, together with the fact that the sensory neurons examined in this study are known to mediate the withdrawal reflex, suggest that activity-dependent enhancement of facilitation is likely to contribute significantly to the differential conditioning of this reflex.

There is probably nothing unique about the siphon sensory neurons that enables them to undergo pairing specific changes. Other neurons (for example, the interneurons in this reflex) may undergo similar changes if two requirements are met: (1) They must be capable of being modulated by the input from a facilitatory system, and (2) they must be active just prior to their receiving the facilitating input. This mechanism is likely to operate throughout the nervous system wherever these two requirements are met. Indeed, Walters and Byrne (1983) have independently found a very similar mechanism of pairing specificity in another group of neurons in *Aplysia*. Thus there is reason to believe that this simple mechanism for learning might be quite general.

Figure 8.2. Schematic neural circuit of the defensive withdrawal reflex, indicating the sensory, interneuronal, and motoneuronal components of the total reflex. The population of sensory neurons innervating the siphon skin consists of about 24 cells. These neurons have direct connections to the motor neurons and indirect connections via several interneurons, 3 of which are illustrated. The 3 interneurons and the 13 central motor cells in the abdominal ganglion are all unique, identified cells. The peripheral siphon motor cells (PS), although identified as a population, are not uniquely identified. The same population of sensory neurons activates both the siphon and gill motor neurons. This accounts for the fact that the two acts of the reflex pattern occur simultaneously. (From Kandel 1976)

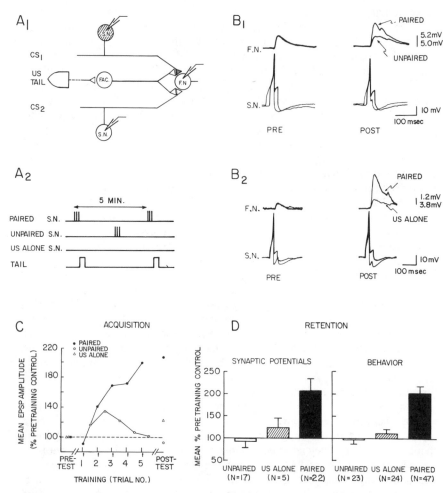

Figure 8.3. Differential facilitation of monosynaptic EPSPs from siphon sensory neurons that mediate the withdrawal reflex. A_1, A_2: Experimental arrangement and training protocol. Activity in the shaded neuron is paired with the US. See the text for details. *B:* Examples of the EPSPs produced in a common follower neuron (F.N.) by action potentials in two sensory neurons (S.N.) before (Pre) and 5 to 15 min after training (Post). The spikes and EPSPs from the two sensory neurons have been superimposed and the gains adjusted to match the amplitudes of the EPSPs before training. B_1: Example of greater facilitation of an EPSP from a paired sensory neuron than of an EPSP from an unpaired sensory neuron in the same experiment. B_2: Example of greater facilitation of an EPSP from a paired sensory neuron than of an EPSP from a sensory neuron that was not stimulated during training (US Alone). *C:* Average acquisition of differential facilitation in 22 experiments (17 with paired and unpaired neurons and 5 with paired and US-alone neurons). The arrows indicate the times at which the US (tail shock) was delivered. *D:* Comparison of retention of differential facilitation of EPSPs and differential conditioning of the withdrawal reflex. (Data from Hawkins et al. 1983)

Table 8.1. *Different types of conditioned responses generated by a classical conditioning paradigm*

Specificity of CR to CS-US pairing	Response to CS	
	Establishment of a new response	Evolution of a preexisting (alpha) response
Specific	Associative classical conditioning	Associative alpha conditioning
Nonspecific	Nonassociative classical conditioning (psuedoconditioning)	Nonassociative alpha conditioning (sensitization)

Source: Kandel and Spencer (1968).

We can now use the insights provided by our analysis of the mechanism of associative learning in this simple withdrawal reflex to consider the distinction between (1) alpha and classical conditioning, (2) classical conditioning and sensitization, and (3) presynaptic and postsynaptic activity in producing classical conditioning.

WHAT IS THE RELATIONSHIP BETWEEN ALPHA CONDITIONING AND CLASSICAL CONDITIONING?

Alpha conditioning refers to the enhancement, through classical conditioning procedures, of a *preexisting* response to a CS. The notion of alpha conditioning was first introduced by Clark Hull (1934), who attempted to distinguish two types of responses that were available to a CS after conditioning: (1) a change in a response to the CS that was initially present, prior to conditioning (this Hull called *alpha* conditioning), and (2) the emergence of a new response to the CS that was acquired through reinforcement (*beta* conditioning). Hull considered beta responses "true conditioned responses" and characterized alpha responses as "a mere sensitization or augmentation of the original unconditioned reaction to the CS." There are two interrelated yet distinct issues of importance for understanding alpha conditioning: (1) Is alpha conditioning a true instance of associative learning? and (2) Are alpha and beta responses fundamentally different? We will discuss each of these issues in turn.

In his writing Hull emphasized the possibility that alpha conditioning – the enhancement of a previously existing response to the CS – is likely to represent nonassociative learning (sensitization). This had the unfortunate consequence that it confounded the difference between sensitization and classical conditioning with the difference between alpha and beta conditioning (see Table 8.1). Hull's basic notion that alpha conditioning must represent sensitization and therefore be nonspecific has influenced psychological theorizing over the last 50 years. However, experiments over the last two decades (reviewed by Kandel and Spencer 1968; Kandel 1976) make it quite clear that alpha and beta conditioning represent different dimensions than do nonassociative and associative conditioning. Both alpha and beta responses can be associatively conditioned in some instances and sensitized in others. This has led many psychologists to suggest that the critical dimension in associative conditioning is not whether the conditioned response is an alpha or a beta response, but rather that the change in response to the CS *is specific to pairing* of that CS with a US (see for example Kimble 1961; Kimmel 1964; Kandel and Spencer 1968; Hilgard and Bower 1975).

Whereas the distinction between nonassociative and associative learning is mechanistically quite fundamental (as we shall show below), the distinction between alpha and beta responses – that is, between enhancement of a previously existing response or development of a new response – may not be fundamental at all (for discussion see Kandel 1976). Although the behavioral response to the CS may show a discontinuity between alpha and beta responses, on a cellular level this distinction is likely to represent two points along a single continuum. The generation of an overt behavioral response requires the generation of action potentials in the motor neurons that cause contraction of the muscles responsible for the behavior. Thus, the action potential, which has a discrete threshold, introduces a discontinuity. However, our data show that a fundamental change produced by associative learning involves a change in synaptic strength, as indicated by a graded change in the EPSP (Figure 8.3*C*). The generation of an action potential and the manifestation of a behavioral response are all-or-none consequences of a continuous process – a graded increment in the amplitude of the EPSP. Thus, in many cases of classical conditioning in which no *behavioral* alpha response is manifest, such as the nicitating membrane response of the rabbit (Gormezano 1972), the CS still can produce a clear *neural* alpha response at the level of the motor neurons (Young et al. 1976; Harvey and Gormezano 1981). Because of these considerations, we suggest that in most CS pathways one could choose a CS that, before conditioning, produces no overt (alpha) behavioral response but, by appropriate manipulation, could be shown to manifest one. Moreover, if one recorded synaptic potentials in interneurons or motor neurons in that pathway, there would be a neural alpha response (see Figure 8.4). The presence of an alpha response thus is likely to depend upon a number of factors such as the sensitivity of the behavioral response measure and the motivational state of the animal.

This issue can be examined in the conditioning of the siphon withdrawal reflex of *Aplysia* described above. In our behavioral studies, we chose a CS of very weak intensity, such that it produced a small reflex response prior to conditioning. Therefore, this would be called alpha conditioning. Yet we would predict that if we adjusted the stimulus intensity so that no initial behavioral response to the CS was observed, a CR would subsequently develop reflecting apparent beta conditioning. In this reflex system, there would be no fundamental difference between alpha and beta conditioning: The basic underlying mechanism of the conditioning (differential enhancement of the excitatory synaptic input in the CS+ pathway) would be the same whether or not the CS+ was below or above behavioral threshold prior to conditioning.

Although we are stressing that prior to conditioning many more responses to the CS might be detected if cellular changes are used as the response measure, we should emphasize that this does not imply that every potential stimulus is prewired to every potential motor response in the nervous system. There is certainly a range of likelihoods that a particular stimulus will come to elicit

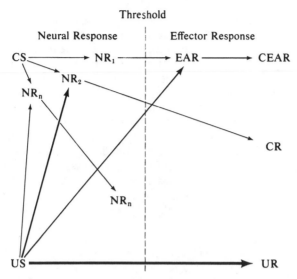

Figure 8.4. A facilitation model comparing a variety of neural and effector (behavioral) responses during conditioning. According to the model, the CS gives rise to a family of neural responses, each of which is designated NR_1, NR_2, etc. The efficacy of each is indicated by the length of the arrow originating from the CS. Some of the neural responses reach effector threshold (NR_1) and produce an effector alpha response (EAR). Others remain below threshold; of these, some might be capable of producing responses if appropriately facilitated (NR_2), others might never reach behavioral threshold or might not involve effector systems (NR_n). The US also produces a family of neural responses; these interact with the neural responses to the CS, facilitating some and inhibiting others. In this scheme only the simplest case of facilitation is considered. The varying ability of the US to facilitate different neural responses is indicated by the thickness of the arrow. When NR_1 is facilitated, it gives rise to a specific conditioned effector alpha response (CEAR). Similarly, when NR_2 is facilitated, it produces an effector response. Since this response to the CS was not previously evident, it constitutes a new response, a CR. (From Kandel 1976)

a particular conditioned response. This is apparent when one considers the issue of biological constraints on learning (Seligman and Hager 1972) where certain stimulus–response relations are readily learned (especially those that are "biologically appropriate" to the animal), while other relationships are very difficult or impossible to learn. The point we wish to make here is not that all stimuli are preconnected to all responses, but rather that many more stimuli have access to particular response systems than might be appreciated from purely a behavioral approach.

WHAT IS THE RELATIONSHIP OF SENSITIZATION TO CLASSICAL CONDITIONING?

We have previously suggested that two associative forms of conditioning (alpha and beta conditioning) represent a single process (much as dishabituation and sensitization do; see Carew et al. 1971). But we can also ask: To what degree do associative and nonassociative learning processes resemble one another? Specifically, to what degree does sensitization resemble classical conditioning? Both sensitization and classical conditioning involve the enhancement of one set of (test) responses by activity in another (reinforcing) pathway. But sensitization differs from classical conditioning in that there is no requirement for temporal association of activity in the two pathways in sensitization, whereas temporal specificity is a defining characteristic of classical conditioning. Because of these similarities, sensitization has often been thought to constitute an elementary building block or behavioral precursor from which classical conditioning may have evolved (Razran 1971; Kandel 1976). However, specification of the exact relationship between these two processes awaited the possibility of comparing them in mechanistic terms.

We have compared sensitization and classical conditioning in the same reflex, both behaviorally and neurophysiologically. On a behavioral level, sensitization is significantly less in magnitude, but parallels classical conditioning both in its time course and in its progressive increment with increased trials (Carew et al. 1981, 1983). On a cellular level, the mechanism of sensitization of the defensive withdrawal reflex is presynaptic facilitation of synaptic transmission from the sensory neurons of the reflex (Castellucci and Kandel 1976). This facilitation is due to a cyclic AMP–mediated decrease in a K^+ conductance that leads to broadening of the action potentials in the sensory neurons, which in turn results in an increase in Ca^{2+} influx and thus more transmitter release (Klein and Kandel 1978, 1980). We have recently found that classical conditioning involves the same type of mechanism (see also Chapter 15), but in amplified form. The action potentials in sensory neurons whose activity had been paired with a US are significantly more broadened than those in sensory neurons whose activity was unpaired (Figure 8.5). These results indicate that the activity-dependent facilitation that underlies conditioning in this reflex is presynaptic in origin and involves a differential increase in Ca^{2+} influx in the paired and unpaired sensory neurons. Thus, at least in this system, the mechanism of classical conditioning appears to be simply an extension of the mechanism of sensitization: presynaptic facilitation caused by an increase in Ca^{2+} influx during each action potential. The pairing specificity characteristic of classical conditioning results from the presynaptic facilitation being amplified or augmented by preceding spike activity in the sensory neurons. These observations render more plausible the idea that classical conditioning might in fact have evolved from the more general facilitatory process that underlies sensitization. A more specific molecular hypothesis is considered in the companion paper (Chapter 9).

IS POSTSYNAPTIC IMPULSE ACTIVITY REQUIRED FOR ASSOCIATIVE LEARNING TO OCCUR?

In 1949 D. O. Hebb proposed a creative hypothetical model of how learning might occur at a synapse. According to this model, the strength of a particular synaptic connection was posited to increase if the use of that synapse contributed to the occurrence of action potentials in the postsynaptic neuron (Figure

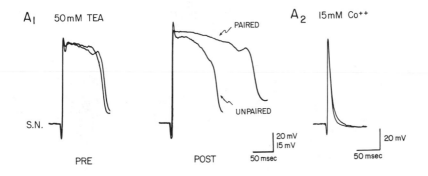

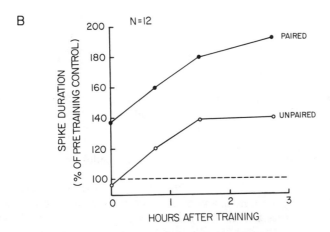

Figure 8.5. Differential broadening of the action potentials in two sensory neurons in the presence of 50-mM tetraethylammonium (TEA). A_1: Examples of the action potentials in a paired and unpaired sensory neuron before (Pre) and 3 hr after training (Post). The action potentials in the two neurons have been superimposed. A_2: The action potentials in paired and unpaired neurons with 15-mM $CoCl_2$ added to the bath after the 3-hr posttests (from another experiment with action potentials similar to those shown in A_1). B: Average differential spike broadening and time course of retention in 12 experiments in which both neurons were held for at least 3 hr after training. (Data from Hawkins et al. 1983)

8.6*A*). This model was extended by Stent (1973), who suggested that activity in a postsynaptic neuron that was paired with a CS input might alter the receptor sensitivity of the postsynaptic cell and thus produce an increase in responsiveness to synaptic input from the presynaptic (CS) neurons after training. The essential feature of both Hebb's original postulate and Stent's extension of it is that action potentials *must* occur in the postsynaptic neuron for the associative changes to occur. These models have received a good deal of attention and popularity despite the fact that there has been little solid experimental

A   B

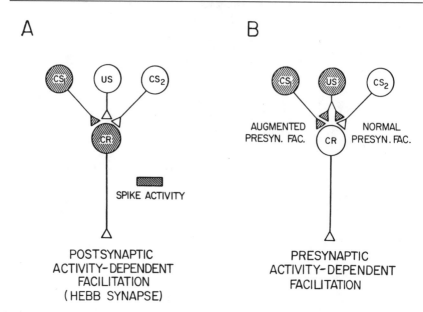

POSTSYNAPTIC
ACTIVITY-DEPENDENT
FACILITATION
(HEBB SYNAPSE)

PRESYNAPTIC
ACTIVITY-DEPENDENT
FACILITATION

Figure 8.6. Diagram of a cellular mechanism of conditioning that requires either (*A*) postsynaptic activity (the Hebb postulate) or (*B*) presynaptic activity.

evidence to support them (for negative results, see Kandel and Tauc 1965; Wurtz et al. 1967).

Having gained some insight into possible cellular mechanisms of classical conditioning in *Aplysia,* we were in a position to test directly the Hebb hypothesis at identified synapses that show a temporally specific associative change that contributes to the conditioning. We found that postsynaptic activity was *neither necessary nor sufficient* to produce the associative synaptic change. That postsynaptic activity was not *necessary* was shown by hyperpolarizing the postsynaptic cell, preventing it from firing action potentials during training; this did not prevent associative synaptic changes from occurring. That postsynaptic activity was not *sufficient* was shown by substituting for the US the direct firing of the postsynaptic neuron, and this did not produce the associative change. Moreover, at those same synapses, subsequent training with tail shock as the US *did* produce the associative synaptic effect (Carew et al. in press). Thus, postsynaptic activity is not critical for associative learning at these synapses.

By contrast, our evidence strongly supports the idea that *presynaptic activity* is required for associative learning (Figure 8.6*B*). Like the Hebb model, this cellular mechanism of learning requires very little special circuitry, because the mechanism of pairing specificity resides in individual neurons. However, unlike the Hebb model, it involves specific neuronal mechanisms (cAMP-dependent protein phosphorylation and decreased ionic conductances; see Chapter 9) that have been described across a variety of phyla and appear to be highly conserved. It is attractive to at least entertain the possibility that, just as these fundamental properties of neuronal function have been conserved across a wide variety of animals, so too might at least some of the basic cellular mechanisms of learning also have been conserved. It will therefore be of interest to see whether this simple mechanism for temporal specificity in *Aplysia* underlies simple forms of associative learning in other animals as well.

CONCLUSIONS

In this chapter we have attempted to illustrate that cellular approaches in simple systems can be used not only to elucidate the mechanisms of a given form of learning, but also to gain insights into issues of pyschological relevance. Several invertebrate animals have now been shown to be capable of a wide variety of both simple and complex forms of associative learning (Mpitsos and Davis 1973; Mpitsos and Collins 1975; Mpitsos et al. 1978; Davis et al. 1980; Gelperin 1975; Sahley et al. 1981a,b; Crow and Alkon 1978; Farley and Alkon 1980; Hoyle 1979; Walters et al. 1979, 1981a; Carew et al. 1981, 1983). As cellular studies progress in these systems, it may be possible to address a number of questions that are of continuing interest on a psychological level. For instance, what is the relationship between contiguity and contingency of stimuli in associative learning? What is the relationship between latent inhibition and habituation? How does extinction relate to forgetting, and how do both of these processes relate to habituation? Are short- and long-term forms of memory fundamentally different? What is the relationship between classical and operant conditioning? These kinds of questions have been of interest to psychologists for many years, and several creative behavioral approaches to these issues have been advanced. It is reasonable to hope that cellular approaches to learning in a variety of simple animals may provide another source of insight into these complex psychological issues.

REFERENCES

Bailey C. H., Castellucci, V. F., Koester, J., and Kandel, E. R. 1979. Cellular studies of peripheral neurons in siphon skin of *Aplysia californica. J. Neurophysiol. 42*:530–557.

Carew, T. J., Castellucci, V. F., and Kandel, E. R. 1971. An analysis of dishabituation and sensitization of the gill-withdrawal reflex in *Aplysia. Int. J. Neurosci. 2*:79–98.

Carew, Y. J., Hawkins, R. D., Abrams, T. W., and Kandel, E. R. In press. A test of Hebb's postulate at identified synapses which mediate classical conditioning in *Aplysia. J. Neurosci.*

Carew, T. J., Hawkins, R. D., and Kandel, E. R. 1983. Differential classical conditioning of a defensive withdrawal reflex in *Aplysia. Science 219*:397–400.

Carew, T. J., Walters, E. T., and Kandel, E. R. 1981. Classical conditioning in a simple withdrawal reflex in *Aplysia californica. J. Neurosci. 1*:1426–1437.

Castellucci, V. F., and Kandel, E. R. 1976. Presynaptic facilitation as a mechanism for behavioral sensitization in *Aplysia. Science 194*:1176–1178.

Crow, T., and Alkon, D. L. 1978. Retention of an associative behavioral change in *Hermissenda*. *Science 201*:1239–1241.

Davis, W. J., Villet, J., Lee, D., Rigler, M., Gillette, R., and Prince, E. 1980. Selective and differential avoidance learning in the feeding and withdrawal behavior of *Pleurobranchaea californica*. *J. Comp. Physiol. 138*:157–165.

Farley, J., and Alkon, D. L. 1980. Neural organization predicts stimulus specificity for a retained associative behavioral change. *Science 210*:1373–1375.

Gelperin, A. 1975. Rapid food-aversion learning by a terrestrial mollusk. *Science 189*:567–570.

Gormezano, D. 1972. Investigations of defense and reward conditioning in the rabbit. In *Classical Conditioning*, vol. 2, *Current Research and Theory* (A. H. Black and W. F. Prokasy, eds.), pp. 151–181. New York: Appleton-Century-Crofts.

Harvey, J., and Gormezano, D. 1981. Effects of haloperidol and pimozide on classical conditioning of the rabbit nictitating membrane response. *J. Pharm. Exp. Ther. 218*:712–719.

Hawkins, R. D., Abrams, T. W., Carew, T. J., and Kandel, E. R. 1983. A cellular mechanism of classical conditioning in *Aplysia*: Activity-dependent amplification of presynaptic facilitation. *Science 219*:400–405.

Hebb, D. O. 1949. *The Organization of Behavior: A Neuropyschological Theory*. New York: Wiley.

Hilgard, E., and Bower, G. 1975. *Theories of Learning*. Englewood Cliffs, N.J.: Prentice-Hall.

Hoyle, G. 1979. Instrumental conditioning of the leg lift in the locust. In *Cellular Mechanisms in the Selection and Modulation of Behavior* (E. R. Kandel, F. B. Krasne, F. Strumwasser, and J. W. Truman, eds.), pp. 577–86. Neuroscience Research Program Bulletin 17. Cambridge: MIT Press Journals.

Hull, C. L. 1934. Learning. II. The factor of the conditioned reflex. In *A Handbook of General Experimental Psychology* (D. Murchison, ed.), pp. 392–455. Worcester, Mass.: Clark University Press.

Kandel, E. R. 1976. *Cellular Basis of Behavior: An Introduction to Behavioral Neurobiology*. San Francisco: Freeman.

Kandel, E. R., and Spencer, W. A. 1968. Cellular neurophysiological approaches in the study of learning. *Physiol. Rev. 48*:65–124.

Kandel, E. R., and Tauc, L. 1965. Mechanism of heterosynaptic facilitation in the giant cell of the abdominal ganglion of *Aplysia depilans*. *J. Physiol. 181*:28–47.

Kimble, G. 1961. *Hilgard and Marquis' Conditioning and Learning*, 2nd ed. New York: Appleton-Century-Crofts.

Kimmel, H. 1964. Further analysis of GSR conditioning: A reply to Stewart, et al. *Psychol. Rev. 71*:150–165.

Klein, M., and Kandel, E. R. 1978. Presynaptic modulation of voltage-dependent Ca^{++} current: Mechanism for behavioral sensitization in *Aplysia californica*. *Proc. Nat. Acad. Sci. 75*:3512–3516.

1980. Mechanism of calcium current modulation underlying presynaptic facilitation and behavioral sensitization in *Aplysia*. *Proc. Nat. Acad. Sci. 77*:6912–6916.

Mpitsos, G. J., and Collins, S. D. 1975. Learning: Rapid aversion conditioning in the gastropod mollusc *Pleurobranchaea*. *Science 188*:954–957.

Mpitsos, G. J., Collins, S. D., and McClellan, A. D. 1978. Learning: Model system for physiological studies. *Science 199*:497–506.

Mpitsos, G. J., and Davis, W. J. 1973. Learning: Classical and avoidance conditioning in the mollusc *Pleurobranchaea*. *Science 180*:317–320.

Perlman, A. 1979. Central and peripheral control of siphon-withdrawal reflex in *Aplysia californica*. *J. Neurophysiol. 42*:510–529.

Razran, G. 1971. *Mind in Evolution: An East–West Synthesis of Learned Behavior and Cognition.* Boston: Houghton Mifflin.

Sahley, C., Gelperin, A., and Rudy, J. W. 1981a. One-trial associative learning modifies food odor preferences of a terrestrial mollusc. *Proc. Nat. Acad. Sci. 78:*640–642.

Sahley, C., Rudy, J. W., and Gelperin, A. 1981b. An analysis of associative learning in a terrestrial mollusc. I. Higher-order conditioning, blocking, and a transient US pre-exposure effect. *J. Comp. Pyysiol. 144:*1–8.

Seligman, M. E. P., and Hager, J. L. 1972. *Biological Boundaries of Learning.* New York: Appleton-Century-Crofts.

Stent, G. S. 1973. A physiological mechanism for Hebb's postulate of learning. *Proc. Nat. Acad. Sci. USA 70:*997–1001.

Walters, E. T., and Byrne, J. H. 1983. Associative conditioning of single sensory neurons suggests a cellular mechanism for learning. *Science 219:*405–408.

Walters, E. T., Carew, T. J., and Kandel, E. R. 1979. Classical conditioning in *Aplysia californica. Proc. Nat. Acad. Sci. 76:*6675–6679.

1981a. Associative learning in *Aplysia:* Evidence for conditioned fear in an invertebrate. *Science 211:*504–506.

1981b. Identification of sensory neurons involved in two forms of classical conditioning in *Aplysia. Soc. Neurosci. Abstr. 7:*353.

Wurtz, R. H., Castellucci, V. F., and Nusrala, J. 1967. Synaptic plasticity: The effect of the action potential in the postsynaptic neuron. *Exp. Neurol. 18:*350–368.

Young, R. A., Cegavske, C. F., and Thompson, R. F. 1976. Tone-induced changes in excitability of abducens motor neurons and of the reflex path of nictitating membrane response in rabbit. *J. Comp. Physiol. Psychol. 90:*424–434.

9 · Cellular and molecular correlates of sensitization in *Aplysia* and their implications for associative learning

J. S. CAMARDO, A. S. SIEGELBAUM, and
E. R. KANDEL

Elementary forms of learning are divided into two general categories – non-associative learning and associative learning. This distinction is based on whether or not the learning requires a specific temporal association between stimuli or between a stimulus and a response. A particularly important question concerns the relationship of sensitization (which is nonassociative) and classical conditioning (which is associative), two forms of learning that are thought to be quite similar. Both involve the effect of one stimulus on the response to another stimulus. Classical conditioning, however, requires associative pairing of the two stimuli whereas sensitization does not. Does classical conditioning partake of the molecular machinery involved in sensitization, or does associative learning recruit a completely new family of cellular mechanisms?

These questions cannot be answered in purely behavioral terms; their answer requires the development of preparations in which each form of learning can be studied on the cellular level and the underlying alterations in synaptic connections can be elucidated and compared. As illustrated in the preceding as well as in other chapters in this volume, it has recently become possible to carry out this type of dual analysis in several invertebrate animals. Chapter 8 described recent studies on classical conditioning of siphon and gill withdrawal reflex in *Aplysia*. In this chapter we describe studies in *Aplysia* that explore sensitization of these reflexes on the cellular and molecular levels. We first review our studies on the mechanisms of sensitization, suggesting that the short-term form of this nonassociative form of learning is due to the enhancement of transmitter release that results from depression of a novel K^+ channel. We next describe biochemical and pharmacological studies indicating that the modulation of this current involves cyclic-nucleotide-dependent protein phosphorylation. The studies of sensitization provide the first direct evidence that protein phosphorylation mediated by cyclic AMP (cAMP) can serve as a mechanism for modulating synaptic strength. Finally, we outline some thoughts on the possible relationship between the molecular mechanisms of sensitization and those of classical conditioning.

SENSITIZATION IS DUE TO PRESYNAPTIC FACILITATION OF TRANSMITTER RELEASE

To explore the mechanism of sensitization, we have used the gill withdrawal reflex in *Aplysia*. As outlined in Chapter 8, the neural circuit of this reflex consists of only 24 sensory cells, 6 motor cells, and several interneurons. Habituation and sensitization, both nonassociative forms of learning, result from alterations in the amount of chemical transmitter released by the terminals of the sensory neuron (Castellucci et al. 1970). This fact is technically fortunate because it has allowed us to trap these forms of nonassociative learning in a series of monosynaptic connections – an essential prerequisite for a detailed analysis.

At the cellular level, sensitization involves an enhancement of synaptic transmission at the same locus as that which is involved in habituation: synapses between the sensory neuron and their central target cells, the interneurons, and the motor neuron (Figure 9.1). The physiologic mechanism for sensitization is presynaptic facilitation (Castellucci and Kandel 1976). Specific interneurons make excitatory connections onto the terminals of the sensory cells (Bailey et al. 1981; Hawkins et al. 1981). Stimulation of these interneurons or addition of exogenous serotonin facilitates release of transmitter from the sensory cell terminals. These postsynaptic effects have been shown to be mediated by a cAMP-dependent protein phosphorylation.

PRESYNAPTIC FACILITATION RESULTS FROM INCREASED CALCIUM INFLUX DUE TO DELAYED REPOLARIZATION OF THE ACTION POTENTIAL

How does serotonin increase the release of transmitter from the sensory neuron terminals? Klein and Kandel (1978) observed that facilitating input to the sensory cells as well as serotonin, both of which increased transmitter release, also produced a long-lasting decreased-conductance, excitatory synaptic potential in the sensory neurons and delayed repolarization of their action potential. Klein and Kandel (1980) suggested that both these effects resulted from transmitter modulation of a membrane ionic current. They analyzed the slow postsynaptic potential (PSP) using the voltage clamp technique and found that, in the cell body, 5-HT produces a decrease in conductance and an inward movement of net current. The reversal potential of this decreased-conductance excitatory postsynaptic potential (EPSP) is sensitive to changes in extracellular K^+. Moreover, the inward movement of the net current is blocked by high concentrations of K^+ channel blockers or by substituting Cs^+ for intracellular K^+ (using the monovalent inophore Nystatin). Based on these findings, Klein and Kandel concluded that both the slow EPSP and the increase in duration of the action potential result from a common process, a decrease in K^+ current. The delay in repolarization increases the duration of Ca^{2+} influx and thereby increases transmitter release (for review see Klein et al. 1980).

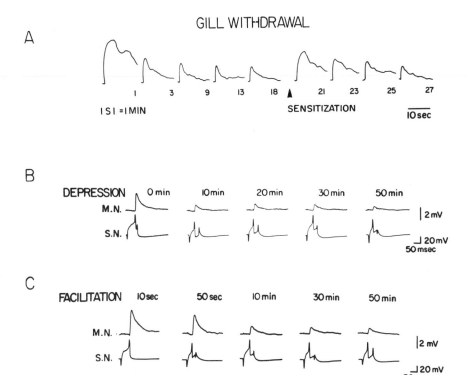

Figure 9.1. Habituation and sensitization occur at the same single synapse. *A:* Contraction of the gill decreases with repeated stimulation of the siphon by a water jet and increases following a shock to the head. A stimulus of the same intensity is presented before and after the head shock. *B:* Habituation results from homosynaptic depression. In the isolated abdominal ganglion, the postsynaptic potential (PSP) elicited onto the gill motor neuron by a single spike in a siphon sensory cell decrements with repeated stimulation (once every 10 sec). *C:* Sensitization results from heterosynaptic facilitation. The PSP is increased in size by repetitive stimulation of the left connective, one of the nerves that carries information to the abdominal ganglion from the head.

SEROTONIN MODULATES A NEW SPECIES OF K⁺ CHANNEL

Four types of K^+ currents have been found in vertebrate and invertebrate neurons: (1) an early K^+ current, (2) a delayed K^+ current, (3) a Ca^{2+}-dependent K^+ current, and (4) the "M" current (see Brown and Adams 1980; Thompson 1978). Which of these is modulated by serotonin? Using voltage clamp analysis in combination with pharmacological blocking agents, Klein et al. (1982) found that serotonin did not affect any of these known currents. Rather, serotonin modulates a novel current, which we have called "S" current by analogy with "M" current.

In the following section we will briefly review the evidence that distinguishes the "S" current from other K^+ currents.

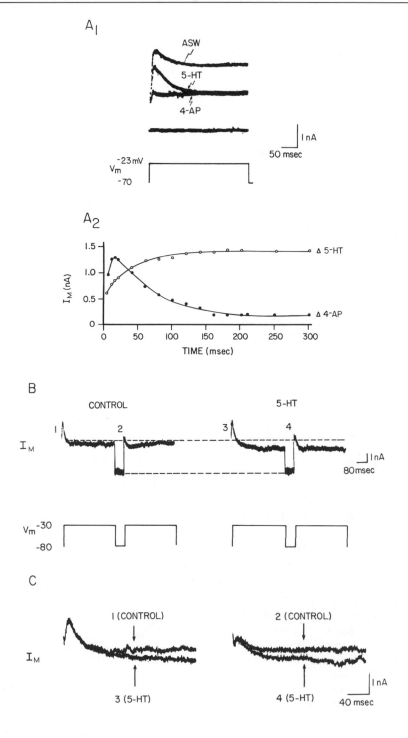

THE 5-HT-SENSITIVE CURRENT DIFFERS FROM THE EARLY K^+ CURRENT

The K^+ currents in the sensory cells can be distinguished by differences in their kinetics, voltage dependence, and susceptibility to pharmacological blocking agents. The early K^+ current (I_A) is a rapidly activating and rapidly inactivating outward current that reaches a peak near the beginning of a voltage step, then decays to a low steady-state level (Connor and Stevens 1971). By contrast, the 5-HT–sensitive current has little voltage-dependent inactivation. As a result, serotonin decreases outward current throughout the entire duration of a long depolarizing command. In fact, reduction of outward current by 5-HT is greater after I_A has largely inactivated (Figure 9.2). The outward current sensitive to 4-AP (obtained by subtracting the current elicited in a solution containing 5-HT and 4-AP from the current in 5-HT alone) has an early onset and decreases rapidly. By contrast, the current modulated by 5-HT (obtained by subtracting the current in 5-HT from that in normal seawater) persists throughout the 300-msec pulse (Figure 9.2A_2).

We also inactivated I_A by presenting repeated depolarizing commands or by holding at a depolarized level (-30 mV) and found that inactivation of I_A does not decrease the response of the outward current to 5-HT. Figure 9.2B is representative of three paired pulse experiments in which the average peak current (I_A) was inactivated by 35 percent. In no case was the response to 5-HT dimin-

Figure 9.2. Kinetic and pharmacological distinction between 5-HT–sensitive outward current and early K^+ currents. A_1: A 300-msec step from -70 activates two overlapping outward currents, a fast current that inactivates within 200 msec and a slower current. Addition of 10^{-4} M 5-HT reduces the slow current and leaves the early K^+ current unaffected. Subsequent addition of 5 mM 4-AP (an early K^+ channel blocking agent) eliminates the early K^+ current. The bottom trace in the current record is the holding current. Solution contained 60 μM TTX. A_2: The time course of the 5-HT–sensitive current and 4-aminopyridine–sensitive current. Open circles represent the difference between the currents elicited in normal seawater before and after 5-HT. Closed circles represent the difference between currents in the presence of 5-HT, before and after application of 4-AP. These points were calculated from currents in part A_1. B: Inactivation of the early K^+ current (I_A) does not affect responsiveness to 5-HT. Pairs of depolarizing commands from -80 to -30 mV were given, the first to activate I_A and allow it to inactivate, and the second to demonstrate its continued reduction by prior inactivation. I_A is activated in steps 1 and 3 and inactivated by approximately 30 percent in steps 2 and 4 of the pairs. The pairs of voltage clamp pulses to -30 mV were given every 5 sec, in ASW with 50 μM TTX. Each step of the pair was 400 msec long, and the interval between steps in the pair was 65 msec. Following addition of 5-HT, the total outward current is decreased, as is evident in the currents in steps 3 and 4. The decrease is not smaller in step 4 as compared to step 3, despite inactivation of I_A. C: Superimposition of the traces illustrated in part B at a faster sweep speed. Steps 1 and 3 (I_A-activated) and steps 2 and 4 (I_A-inactivated) have been superimposed. Despite significant inactivation of I_A, the decrease in outward current in response to 5-HT is not affected. This is expected, since the 5-HT–sensitive current inactivates very little during the short period between the voltage steps. (From Klein et al. 1982)

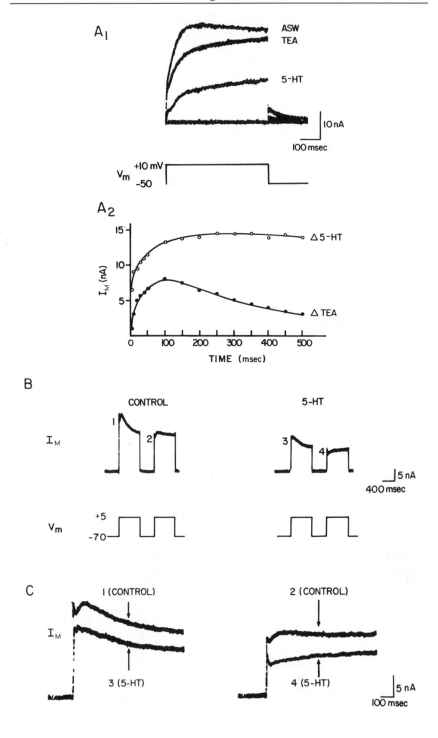

ished during the second step, where I_A was inactivated, indicating that 5-HT is as effective when I_A is maximal as when it is reduced. This finding, together with the difference in kinetics between the 5-HT–sensitive current and I_A, indicates that 5-HT acts on a current different from I_A.

THE 5-HT-SENSITIVE CURRENT DIFFERS FROM THE DELAYED K$^+$ CURRENT

We used a similar strategy to that described above to distinguish the 5-HT–sensitive current from delayed rectification, which can be elicited by a voltage step to $+10$ mV from a holding potential of -50 mV. This current (I_K) peaks in 100–200 msec and decays to a steady-state level. Addition of tetraethylammonium (TEA), which is thought to block I_K somewhat selectively, decreases a current with this time course. By contrast, 5-HT–sensitive current does not decay with time. The action of 5-HT is also independent of the inactivation of I_K. When delayed rectification is reduced by inactivation, the absolute 5-HT response is unchanged (Figure 9.3).

SEROTONIN DOES NOT DECREASE THE Ca^{2+}-ACTIVATED K$^+$ CURRENT

Ca^{2+} entry into molluscan and other neurons activates a third K$^+$ current, the Ca^{2+}-activated K$^+$ current, I_C (Meech 1974), that can be decreased by substitution of Ba^{2+} for Ca^{2+} (Hermann and Gorman 1979). Figure 9.4 shows current–voltage curves of cells bathed in either normal or Ba^{2+}-substituted seawater before and after application of 5-HT. Serotonin reduced the residual

Figure 9.3. Kinetic and pharmacological distinction between 5-HT–sensitive and delayed K$^+$ currents. A_1: Pharmacological blockage of currents before and after sequential addition of 10 mM Et$_4$N$^+$ (TEA) and 5-HT. Et$_4$N$^+$ reduces outward current decreasingly with time; 5-HT reduces an outward current that increases with time. Addition of these agents in reverse order gives the same result. There is also a reduction of outward tail currents on repolarization after 5-HT, but not after Et$_4$N$^+$ application. A small residual early K$^+$ current is discernible near the beginning of the traces. Seawater contained 60 μM TTX. A_2: The time course of the differences in outward current as a result of the application of Et$_4$N$^+$ (closed circles) and 5-HT (open circles). The points are derived from the currents illustrated in part A_1. B: Inactivation of delayed rectification does not decrease the 5-HT response. The cell was held at -70 mV, and pairs of voltage commands were given to $+5$ mV, to elicit delayed rectification. The commands were 800 msec long, separated by 500 msec. In the second step of each pair, the delayed current was significantly inactivated (steps 2 and 4). Despite the inactivation of the delayed current, the 5-HT response is not decreased. Steps 1 and 2 show the control, steps 3 and 4 the decrease in outward current following 5-HT. The experiment was carried out in ASW with 50 μM TTX. C: Superimposition of currents illustrated in part B with (2 and 4) and without (1 and 3) inactivation of I_K, before (1 and 2) and after (3 and 4) 5-HT application. Despite the inactivation of delayed rectification, the response to 5-HT is not reduced. (From Klein et al. 1982)

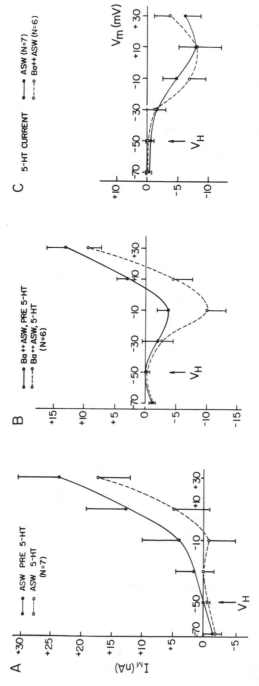

Figure 9.4. Effects of 5-HT in normal seawater and in Ca^{2+}-free seawater with 11 mM Ba^{2+}. *A*: Current–voltage curves obtained with 25-msec steps from a holding potential of -50 mV in normal seawater before (closed circles) and after (open circles) addition of 10^{-4} M 5-HT (mean ± *SD*). *B*: Current–voltage curves similar to *A*, in Ba^{2+}-substituted seawater. *C*: 5-HT response in normal and Ba^{2+}-substituted seawater. Differences between curves before and after 5-HT application were computed for cells in normal or Ba^{2+}-substituted seawater. The curves do not differ significantly in shape or magnitude. (From Klein et al. 1982)

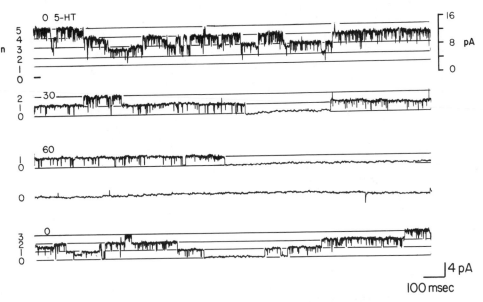

Figure 9.6. Action of serotonin on single-channel current. Patch clamp current with a patch membrane potential of 0 mV (defined as the potential inside the cell minus the potential in the patch pipette). Patch pipette potential was set equal to the cell's resting potential of −34 mV, measured with an intracellular microelectrode. *A:* Current recorded in the absence of serotonin. The membrane patch displays a net outward current that fluctuates among five levels (indicated by solid lines and left-hand ordinate) due to simultaneous openings of one to five channels. (In this record the channels never closed to the zero current level; the zero level was determined from another part of the experiment, not shown.) *B:* Current obtained 2 min after addition of 30 μM serotonin to the bath. *C:* One minute after addition of 60 μM serotonin to the bath. The two traces in *C* are a continuous recording showing the closure of the remaining active channel. In this experiment, channel closings were correlated with increases in cell input resistance. *D:* Current records obtained about 5 min after superfusing the cells with serotonin-free ASW. The reduction of channel unit current probably reflects a slight hyperpolarization of the cell's resting potential due to the withdrawal of the intracellular electrode from the cell prior to changing the bath solution. Outward current shown in upward direction. (From Siegelbaum et al. 1982)

ious solutions. Opening of the S channel persists when the patch is exposed to high-K$^+$ (360 mM) and low-Ca^{2+} (10^{-9} M) solution, indicating that its opening is independent of calcium on the inner surface of the membrane. It is therefore unlikely to be gated by intracellular Ca^{2+} and is different from the Ca^{2+}-dependent K$^+$ current (I_C).

PRESYNAPTIC FACILITATION INVOLVES A cAMP CASCADE

An important aspect of the role of serotonin in sensitization and presynaptic facilitation was provided by the demonstration by Cedar et al. (1972) that 5-

HT increases cAMP in the *Aplysia* ganglion. In a recent extension of this work, Bernier et al. (1982) have shown that 5-HT and connective stimulation, but not other neurotransmitters, increase the amount of cAMP in individual sensory cells three-fold, providing direct evidence that cAMP is a key intermediary through which serotonin exerts its effects on transmitter release. This increase in cAMP is specific to 5-HT (it is not stimulated by dopamine, histamine, or octopamine) and persists for several minutes following a brief stimulus. Moreover, Eppler et al. (1982) have recently isolated cAMP-binding proteins in sensory neurons with properties similar to the regulatory units of cAMP-dependent protein kinase, indicating that the cellular machinery for regulating kinase activity exists in these cells. Finally, Castellucci et al. (1980, 1982) have found that intracellular injection of purified catalytic subunit of cAMP-dependent protein kinase prolongs the duration of the action potential and enhances transmitter release, whereas the Walsh kinase inhibitor blocks the spike broadening and the enhancement of transmitter release produced by facilitating stimuli.

DEPRESSION OF THE SEROTONIN-SENSITIVE CURRENT BY cAMP AND THE BIOCHEMICAL LOCUS FOR THE MEMORY FOR SHORT-TERM SENSITIZATION

In light of this biochemical evidence for the role of cAMP in facilitation, we examined the effect of cAMP on the serotonin-sensitive current. We found that intracellular injection of cAMP exerts the same effect on membrane currents as does bath application of 5-HT (Figures 9.7 and 9.8). cAMP decreases the steady-state current at depolarized holding potentials and the transient current elicited by a short depolarizing step from a holding potential of -50 mV. Moreover, the serotonin-sensitive channel can be modulated by cAMP injection: Intracellular cAMP closes the same channel that we have observed to be sensitive to serotonin (Figure 9.9).

The outward current consistently returns to normal soon after termination of the cAMP injection. As a result, the decrease in outward current can be elicited and reversed several times with repeated injection. The ability of cAMP to modulate the current is independent of whether the current is elicited by steady-state depolarization (Figure 9.8) or by transient depolarizing steps from a holding voltage of -50 mV. These results suggest that the exogenous cAMP is rapidly hydrolyzed by the phosphodiesterase within the sensory neurons and that the return of current to control level is due to a decrease in intracellular cAMP following termination of the injection. Since a brief injection of cAMP alone does not produce a long-lasting effect, prolonged facilitation produced by nerve stimulation probably requires continued production of cAMP or inhibition of its breakdown. This is consistent with the finding of Castellucci et al. (1982) that injection of the Walsh inhibitor of protein kinase rapidly reverses the effect of 5-HT. Thus, continued kinase activity appears to be required to maintain facilitation and could be achieved by a sustained elevation

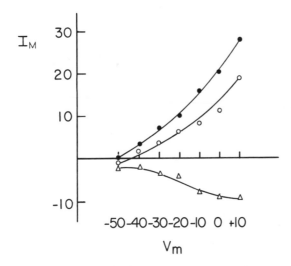

Figure 9.7. Steady-state current–voltage *(I–V)* relation. Cyclic AMP changes the *I–V* relation of the sensory cells by decreasing the outward current in a moderately voltage-dependent manner. Curves were obtained using 9-sec steady-state ramp depolarizations from a holding potential of −50 mV. Solid circles, control; open circles, current after cAMP injection. Difference is plotted by triangles, which show the net cAMP-sensitive current. The effect of cAMP on the steady-state *I–V* relation is indistinguishable from that of serotonin.

Figure 9.8. Cyclic AMP decreases outward currents. This cell was voltage-clamped at 30 mV; every 10 sec the voltage was stepped to −65 mV. The current elicited by this step is shown before and after injection of cAMP. At arrows, cAMP was injected by removing positive current from an intracellular electrode filled with 0.5 M cAMP, allowing cAMP to leak into the cell for 2 sec. The steady-state current moves inward (down), and the current elicited by the voltage step is smaller. Both return to control level 1 min after the injection. A second injection causes a similar decrease, and this also returns to control level. The duration of the cAMP effect can be extended by increasing the amount of time the positive current is turned off.

of cAMP. Since brief exposure to 5-HT produces a prolonged elevation of cAMP that lasts more than 15 min, it is likely that the time course of the memory for short-term sensitization reflects the time course of cAMP elevation in the sensory cells.

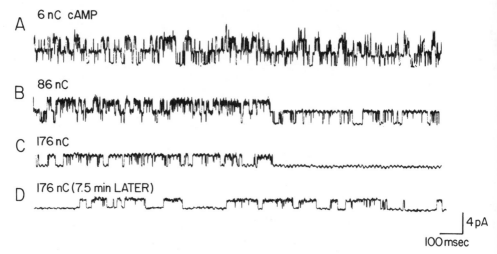

Figure 9.9. Effect of intracellular injection of cAMP on single-channel current. Following establishment of seal between patch electrode and membrane, the cell was impaled with a microelectrode filled with 1 M cAMP (sigma) and cAMP was injected into cells by hyperpolarizing current pulses (0.5–2.5 nA). Total charge injected into cell (nanocoulombs) indicated on top of each trace. *A:* Current record obtained soon after impalement of cell with cAMP electrode. *B, C:* Current records after subsequent injection periods. During the record shown in *C* the remaining active channel closed. The cAMP electrode was withdrawn from the cell and no channel openings were observed for 5 min. Channel activity then partially recovered (trace *D*). Cell membrane potential hyperpolarized somewhat during the course of the experiment. Patch membrane potential was approximately +14 mV in *A*, +11 mV in *B*, +7 mV in *C*, and +2 mV in *D*. (From Siegelbaum et al. 1982)

THE S CHANNEL AS A SPECIFIC TARGET MOLECULE INVOLVED IN SYNAPTIC EFFICACY

The findings that single K^+ channels are modulated by serotonin and cAMP confirm our previous results with voltage clamp studies and correlate well with biochemical studies. Moreover, the K^+ channel modulated by 5-HT is a major K^+ conductance in the sensory cell body and exerts a profound effect on the electrophysiological behavior of these cells. Closing of this channel delays repolarization of the action potential. Since the channel is open at the resting potential, is relatively noninactivating, and remains open with steady-state depolarization, its closing will increase the input impedance of the cell.

One of the most interesting features of this channel is that it is critically involved in the modulation of synaptic transmission. Closing of this channel leads to an increase in the efficacy of the synaptic connection between the sensory neurons and the motor neurons. This increase in efficacy underlies a simple form of learning. Thus, the serotonin-sensitive K^+ channel represents a membrane protein whose activity is altered by learning by means of a cAMP-dependent phosphorylation cascade. In this molecular scheme, the S channel is the "target" protein for the acquisition of this simple form of learning (Figure 9.10). We now need to determine how cAMP-mediated phosphorylation modifies the activity of the K^+ channel. Does it do so by phosphorylating the channel directly, thereby altering its molecular structure? Or does the kinase act on a regulatory protein associated with the K^+ channel? We are now exploring these alternative possibilities using cell-free "inside-out" patches of sensory neuron membrane.

In the next section of this chapter, we will discuss the possibility that the mechanism underlying sensitization is a component of more complex forms of learning.

RELATIONSHIP OF SENSITIZATION TO CLASSICAL
CONDITIONING

Because classical conditioning and sensitization are formally similar, it has often been thought that they may be related in mechanism (Grant 1943; Sokolov 1963a, b; Kandel and Tauc 1965; Wells 1968). The mechanisms that account for sensitization in *Aplysia* could also contribute to classical conditioning. Classical conditioning can be divided into two components (for a discussion of these models see Kandel and Tauc 1965; Kupfermann and Pinsker 1969): (1) a temporal discriminative component that is responsible for the associative specificity and (2) a memory component that is responsible for the time course of the learning. The key point we would like to advance is that the two components could represent distinct neuronal processes: Conventional presynaptic facilitation could provide the neural mechanism for the memory component, the component responsible for maintaining the enhanced strength of the response. The associative component could be provided by amplifying the presynaptic facilitation in an activity-dependent manner in those neurons of the CS pathway where impulse activity precedes facilitating input from interneurons of the US pathway.

As summarized in the preceding chapter, experiments by Hawkins et al. (1983) and Walters and Byrne (1982) provide support for the idea that the associative component also resides in the sensory cells. They have shown that the effect of facilitating input on the sensory cell is greatly increased when it is preceded by a spike or train of spikes in the sensory cell.

The mechanism for this activity-dependent amplification of presynaptic facilitation is at present unknown. Because maintained elevation of cAMP is an important element of conventional presynaptic facilitation (see Figure 9.8), facilitation paired with activity in the sensory neuron might increase the cAMP more than does facilitation unpaired with activity. For example, spikes in the sensory cell could enhance the activity of adenylate cyclase. Since the cyclase is a membrane protein, it might be influenced by changes in membrane voltage. Alternatively, impulse activity may increase intracellular Ca^{2+}, which may increase the activity of the adenylate cyclase or of the cAMP-dependent protein kinase. Klein et al. (1982) observed that full expression of the response to 5-HT requires calcium, and calcium has also been shown to be an important intermediary in many other cAMP-dependent cellular processes (Rasmussen

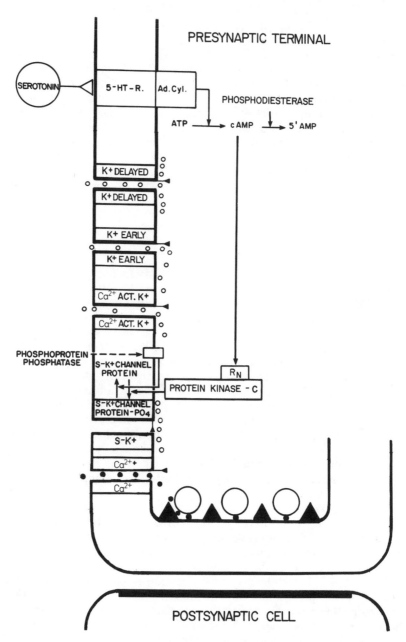

Figure 9.10. Schematic diagram of the sensory neuron terminal depicting the action of 5-HT as limited to a particular K^+ channel, here called S channel to distinguish it from the other K^+ channels in the membrane. This model shows the channel protein itself to be the substrate for protein phosphorylation. There is no direct evidence for this, and it is equally possible that the phosphorylated protein is separate from the channel and affects the channel in an as-yet-undetermined way.

and Goodman 1977). Our observation that 5-HT and cAMP act through a specific membrane protein (S channel) offers another site for an activity-dependent interaction; perhaps the ability of the K^+ channel to be closed by cAMP depends to some extent upon its previous history. A voltage change across the membrane could effect a major change in the affinity of the channel for some "closing" molecule (possibly a phosphate group). The K^+ channel is an attractive site for such an activity-dependent interaction: It is the last step in the cascade of reactions, and therefore the most specific place to effect a change.

AN OVERALL VIEW

We have in this discussion placed great emphasis on presynaptic facilitation. A variety of other plastic mechanisms – either post- or presynaptic – would also serve for both the modulatory and the associative component of associative learning. There are, at this early point, two primary reasons for favoring this presynaptic mechanism over others. The first is that Hebb synapses (1949), Mahr synapses, Stent synapses (1973), and other intriguing constructs that have been repeatedly and popularly invoked to obtain associative specificity are all based upon theoretical postulates about synaptic transmission for which there is no evidence. (For discussion of Hebb synapses, see Chapter 8.) Twenty years ago Kandel and Tauc (1965) and subsequently Wurtz and Castellucci (1967) tested Hebb's postulate and found no support for it at a well-studied synapse that showed long-lasting nonassociative plasticity. Two decades of further study have not yet revealed a convincing supporting example. As Carew et al. have discussed in the preceding chapter, the Hebb-type modification is neither necessary nor sufficient for classical conditioning in *Aplysia*. It therefore seems more profitable to posit known mechanisms capable of memory on the one hand and of temporal specificity on the other than to continue to invoke models for which there is no support. Finally, Quinn and his colleagues (Duerr and Quinn 1982) have found that the learning mutants of *Drosophila* that fail to show associative conditioning are also incapable of sensitization. These mutants show a defect in cyclic nucleotide metabolism. A single-gene mutant of *Drosophila* called dunce, that cannot learn sensitization, lacks cAMP-phosphodiesterase, the enzyme that degrades cAMP, and therefore has abnormally higher levels of this nucleotide. These mutants also cannot learn associative conditioning tasks. Thus genetic, cell biological, and biochemical approaches provide encouragement to thinking that common cellular and molecular mechanisms are responsible for modulating the strength of the behavior in sensitization and classical conditioning.

REFERENCES

Abrams, T., Hawkins, R. D., Carew, T., and Kandel E. R. 1982. Activity-dependent facilitation accounts for pairing specificity in classical conditioning of a defensive withdrawal reflex in *Aplysia. Soc. Neurosci. Abstr.* 8:385.
Bailey, C. H., Hawkins, R. D., Chen, M. C., and Kandel, E. R. 1981. Interneurons

involved in mediation and modulation of the gill-withdrawal reflex in *Aplysia.* IV. Morphological basis of presynaptic facilitation. *J. Neurophysiol. 45*:340–360.

Bernier, L., Castellucci, V. F., Kandel, E. R., and Schwartz, J. H. 1982. Facilitatory transmitter causes a selective and prolonged increase in cAMP in the sensory neurons mediating the gill and siphon withdrawal reflexes in *Aplysia californica. J. Neurosci. 2*:1682–1691.

Brown, D. A., and Adams, P. R. 1980. Muscarinic suppression of a novel voltage-sensitive K^+ current in a vertebrate neurone. *Nature 283*:673–676.

Castellucci, V., and Kandel, E. R. 1976. Presynaptic facilitation as a mechanism for behavioral sensitization in *Aplysia. Science 194*:1176–1178.

Castellucci, V. F., Kandel, E. R., Schwartz, J. H., Wilson, F. D., Nairn, A. L., and Greengard, P. 1980. Intracellular injection of the catalytic subunit of cyclic-AMP dependent protein kinase simulates facilitation of transmitter release underlying behavioral sensitization in *Aplysia. Proc. Nat. Acad. Sci. 77*:7492–7496.

Castellucci, V. F., Nairn, A., Greengard, P., Schwartz, J. H., and Kandel, E. R. 1982. Inhibitor of cAMP-dependent protein kinase blocks presynaptic facilitation in *Aplysia. J. Neurosci. 2*:1673–1681.

Castellucci, V., Pinsker, H., Kupfermann, I., and Kandel, E. R. 1970. Neuronal mechanisms of habituation and dishabituation of the gill-withdrawal reflex in *Aplysia. Science 167*:1745–1748.

Cedar, H., Kandel, E. R., and Schwartz, J. H. 1972. Cyclic adenosine monophosphate in the nervous system of *Aplysia californica.* I. Increased synthesis in response to synaptic stimulation. *J. Gen. Physiol. 60*:558–569.

Connor, J., and Stevens, C. F. 1971. Voltage clamp studies of a transient outward membrane current in gastropod neural somata. *J. Physiol.* (London) *213*:21–30.

Duerr, J. S., and Quinn, W. G. 1982. Three *Drosophila* mutations that block associative learning also affect habituation and sensitization. *Proc. Nat. Acad. Sci. 99*:3646–3650.

Eppler, C. M., Palazzolo, M. J., and Schwartz, J. H. 1982. Characterization and localization of cAMP-binding proteins in the nervous system of *Aplysia. J. Neurosci. 2*:1692–1704.

Gorman, A. L. F., and Thomas, M. V. 1980. Potassium conductance and internal calcium accumulation in a molluscan neurone. *J. Physiol.* (London) *308*:287–313.

Grant, D. A. 1943. Sensitization and association in eyelid conditioning. *J. Exp. Psychol. 32*:201–212.

Hamill, O., Marty, A., Neher, E., Sakmann, B., and Sigworth, F. J. 1981. Improved patch-clamp techniques for high-resolution current recording from cells and cell-free membrane patches. *Pflugers Arch. 391*:85–100.

Hawkins, R. D., Abrams, T. W., Carew, T. J., and Kandel, E. R. 1983. A cellular mechanism of classical conditioning in *Aplysia:* Activity-dependent amplification of presynaptic facilitation. *Science 219*:400–405.

Hawkins, R. D., Castellucci, V. F., and Kandel, E. R. 1981. Interneurons involved in mediation and modulation of the gill-withdrawal reflex in *Aplysia.* II. Identified neurons produce heterosynaptic facilitation contributing to behavioral sensitization. *J. Neurophysiol. 45*:315–326.

Hebb, D. D. 1949. *The Organization of Behavior: A Neurophysiological Theory.* New York: Wiley.

Hermann, A., and Gorman, A. L. F. 1979. Blockade of voltage-dependent and calcium-dependent K^+-current components by internal Ba^{++} in molluscan pacemaker neurons. *Experientia 35*:229–231.

Kandel, E. R., and Tauc, L. 1965. Heterosynaptic facilitation in neurones of the abdominal ganglion of *Aplysia depilans. J. Physiol.* (London) *181*:1–27.

Klein, M., Camardo, J., and Kandel, E. R. 1982. Serotonin modulates a specific potassium current in the sensory neurons that show presynaptic facilitation in *Aplysia. Proc. Nat. Acad. Sci. 79*:5713–5717.

Klein, M., and Kandel, E. R. 1978. Presynaptic modulation of voltage-dependent Ca^{++} current: Mechanism for behavioral sensitization in *Aplysia californica. Proc. Nat. Acad. Sci. 75*:3512–3516.

1980. Mechanism of calcium current moduation underlying presynaptic facilitation and behavioral sensitization in *Aplysia. Proc. Nat. Acad. Sci. USA 77*:6912–6916.

Klein, M., Shapiro, E., and Kandel, E. R. 1980. Synaptic plasticity and modulation of the calcium current. *J. Exp. Biol. 89*:117–157.

Kupfermann, I., and Pinsker, H. 1969. Plasticity in *Aplysia* neurons and some simple neuronal models of learning. In *Reinforcement and Behavior* (J. Tapp, ed.), pp. 356–386. New York: Academic Press.

Meech, R. W. 1974. The sensitivity of *Helix aspersa* neurons to injected calcium ions. *J. Physiol. 237*:259–277.

Rasmussen, H., and Goodman, D. P. 1977. Relationships between calcium and cyclic nucleotides in cell activation. *Physiol. Rev. 57*:421–509.

Siegelbaum, S. A., Camardo, J. S., and Kandel, E. R. 1982. Serotonin and cyclic AMP close single potassium channels in *Aplysia* sensory neurons. *Nature 299*:413–7.

Sokolov, E. N. 1963a. Higher nervous functions: The orienting reflex. *Annu. Rev. Physiol. 25*:545–580.

Sokolov, E. N. 1963b. *Perception and the Conditioned Reflex* (S. W. Wazdenfeld, trans.). New York: Pergamon Press.

Stent, G. S. 1973. A physiological mechanism for Hebb's postulate of learning. *Proc. Nat. Acad. Sci. 70*:997–1001.

Thompson, S. H. 1977. Three pharmacologically distinct potassium channels in molluscan neurones. *J. Physiol.* (London) *265*:465–488.

Walters, E. T., and Byrne, J. 1982. Associative conditioning of single sensory neurons in *Aplysia*. I. Activity-dependent heterosynaptic facilitation. *Soc. Neurosci. Abstr. 8*:386.

Wells, M. J. 1968. *Lower Animals.* New York: McGraw-Hill.

Wurtz, R. H., Castellucci, V. F., and Nusrala, J. M. 1967. Synaptic plasticity: The effect of the action potential in the postsynaptic neuron. *Exp. Neurol. 18*:350–368.

10 · A behavioral and cellular neurophysiological analysis of associative learning in *Hermissenda*

TERRY CROW

INTRODUCTION

One focus of research on the mechanisms of learning and memory is the analysis of relatively simple examples of learned behavior in invertebrates. These animals offer a number of advantages for investigating alterations in identified individual neurons in reasonably well-described neural circuits that mediate modifications in behavior (for a review of this strategy, see Kandel 1976). In a number of mollusks behaviors have been altered by conditioning procedures that meet most of the operational criteria for associative learning (Mpitsos and Davis 1973; Mpitsos and Collins 1975; Mpitsos et al. 1978; Gelperin 1975; Crow and Alkon 1978; Crow and Offenbach 1979; Walters et al. 1979; Sahley et al. 1981a, b; Carew et al. 1981). One preparation that has been particularly useful for studying learned behavior is the Pacific nudibranch *Hermissenda crassicornis*. In this example of associative learning the temporal association of illumination with a presumed aversive stimulus results in a long-lasting change in normal visually guided locomotion. Considerable progress has been made toward an understanding of the learning. For example, the experimental conditions sufficient to produce this learning are now well documented (Crow and Alkon 1978; Crow and Offenbach 1979, 1983; Crow 1983). Neural pathways for both the conditioned stimulus (CS) and unconditioned stimulus (US) have been described, and some cellular as well as subcellular mechanisms have been identified (Crow and Alkon 1980; Neary et al. 1981; Alkon et al. 1982; Crow 1982).

The general strategy in this research has been to identify the relevant sensory pathways that mediate the conditioning, examine the cellular and synaptic organization of the sensory systems with morphological and neurophysiological techniques, identify changes in these systems correlated with behavior, and identify possible cellular mechanisms that explain the learned behavior. The results of these efforts are briefly described in the remainder of the chapter.

GENERAL BEHAVIORAL STRATEGY

The development of this behavioral paradigm used to train *Hermissenda* was guided by the earlier cellular analysis of the visual and gravity-detecting systems (statocysts) by Alkon and his collaborators (for a review see Alkon 1976).

In both the open field and a restricted experimental chamber *Hermissenda* displays robust phototactic behavior (Alkon 1974; Crow and Alkon 1978). As a result of these early studies two sensory systems in *Hermissenda* were well described raising the possibility of developing conditioning procedures that utilize these afferent pathways in an attempt to determine if the behavior of *Hermissenda* could be modified by associative learning. I found that various components of the phototactic behavior can be modified by a conditioning procedure that resembles traditional Pavlovian or classical conditioning. The conditioning paradigm employed in these studies consisted of pairing light with stimulation of the gravity detectors (statocysts) by high-velocity rotation (Crow and Alkon 1978; Crow and Offenbach 1979).

In a strict operational sense light can be defined as the CS and rotation could be viewed as the US, although an unconditioned response (UR) has not been identified. According to traditional views of classical conditioning, the absence of a relationship between the postconditioning response to light and the response properties of rotation (UR) complicates the issue of defining what is actually learned by the application of this conditioning procedure. However, an alternative possibility to the traditional view of a stimulus–response association involves the association between the CS and the stimulus properties of the reinforcing event (US). The results of studies employing response-free reinforcers, such as in the sensory preconditioning procedures, have shown that associations can be formed between stimuli that evoke minimally distinctive responses (Rescorla and Holland 1982). Additional evidence to support the view of sensory–sensory associations comes from studies where the response evoked by the US is eliminated by experimental manipulations. An example of this approach is the conditioned eye blink paradigm (Woody 1974) where the UR normally evoked by the US was eliminated prior to conditioning by crushing the facial nerves innervating the blink musculature. Normal learning still occurs during the period of functional denervation of the muscles, which can be tested following recovery of function several weeks after the crush lesions were performed (Crow and Woody 1973). This issue of the encoding of the stimulus or response properties of the reinforcing event is also present for flavor aversion studies where the conditioned response does not resemble the functional properties of the reinforcing stimulus (US) (Garcia et al. 1972). However, since most of the characteristics of conventional conditioning have also been demonstrated for flavor aversion learning (Revusky 1977), the absence of a distinctive response to the US would not provide evidence that these examples of learning are qualitatively different from traditional Pavlovian conditioning.

In the behavioral studies on *Hermissenda* the selection of rotation as a US was based upon the sensory system that was to be stimulated and the convergence between sensory pathways rather than on any reliable behavior evoked by rotation (UR). Although this experimental strategy left the issue of the functional relationship between the US and the behavior to be modified unresolved, it is interesting to note that there is evidence from the vertebrate conditioning literature that rotational stimulation functions as an aversive US. For example, Braun et al. (1973) reported that for rats distinctive taste cues associated with rotational stimulation led to a specific taste aversion. More recently it was shown that rotation could be used to devalue a previously effective US

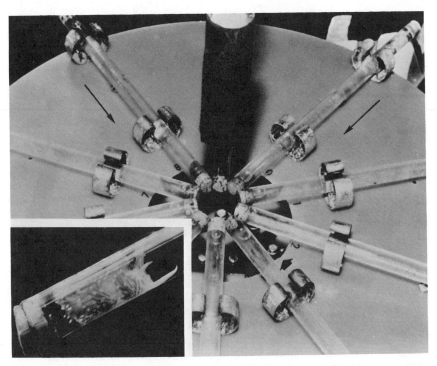

Figure 10.1. Training and testing apparatus. The response latencies to enter a light spot projected onto the center of the turntable by an overhead illuminator were recorded automatically when the *Hermissenda* moved toward the light source (direction of arrows) and interrupted the light between illuminator and photocells (arrowhead). *Hermissenda* were subjected to different behavioral treatments consisting of light and rotation while confined to the end of glass tubes filled with seawater. (From Crow and Alkon 1978)

since following conditioning consisting of a tone (CS) paired with food (US), repeated pairings of the US with rotation resulted in a reduction of the value of the food as an unconditioned stimulus (Holland and Rescorla 1975). The results from the vertebrate conditioning studies notwithstanding, the issue concerning the aversive nature of rotation for *Hermissenda* has yet to be resolved.

A major goal in developing a behavioral paradigm for *Hermissenda* was to produce a reliable and quantifiable change in the behavior of individual animals that could be automated for groups of animals so as to reduce the contribution of handling effects and experimenter bias. The conditioning procedure consisted of pairing light with rotation of a modified turntable. This automated behavioral system has been described in detail by Tyndale and Crow (1979) and thus will be discussed only briefly in this chapter. For conditioning and behavioral measurement the animals were placed in glass tubes filled with seawater that were inserted into clips on the surface of the turntable (see Figure 10.1). A start gate (small foam plug) was inserted into a small

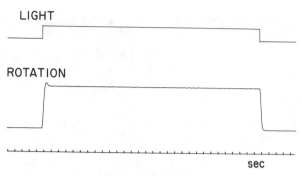

Figure 10.2. Temporal relationship between the light (CS) and rotation (US) for the conditioning procedure. The light trace was activated by a photocell, and the rotation trace represents the output voltage of a tachometer connected to the turntable used in behavioral conditioning.

opening near the outside ends of the glass tubes during training to restrict movement in the tubes and thus insure that all animals were subjected to the same gravitational force during rotation. A test–train–test procedure was used where the time taken by individual animals to locomote from one end of the tube into a center illuminated area at the opposite end of the tube was measured automatically. In subsequent studies the time taken to initiate locomotor behavior in response to light was also measured (Crow and Offenbach 1979, 1983). Following the pretest baseline measurements the animals received 50 trials of light paired with rotation each day for 3 consecutive days. An example of the temporal relationship between the light (CS) and rotation (US) used in the conditioning procedure is shown in Figure 10.2. Immediately following training the animals were again tested to measure the time taken to locomote into the center illuminated area of the turntable. The animals were tested again consecutively 2, 3, 4, and 5 days following training. The behavior of the animals on the retention test days was compared with the initial pretest measurements to assess changes in phototactic behavior produced by the conditioning procedure. The training and testing sequences and control groups used to investigate the contribution of nonassociative factors are summarized in Figure 10.3. The general finding of this initial study was that light paired with rotation produced a long-lasting reduction in the normal positive phototactic behavior of *Hermissenda* that was dependent on the temporal association of light with rotation (see Figure 10.4). Specifically, there were no significant differences between the groups on the baseline measurements of phototactic behavior prior to conditioning. Overall differences between groups were found at the end of 3 training days (Kruskal–Wallis one-way analysis of variance, $p < .001$). Multiple comparisons revealed that the treatment group that received paired light and rotation was significantly different from all of the control groups ($p < .001$), and none of the control groups were significantly different from each other. The changes in behavior produced by the conditioning procedure are

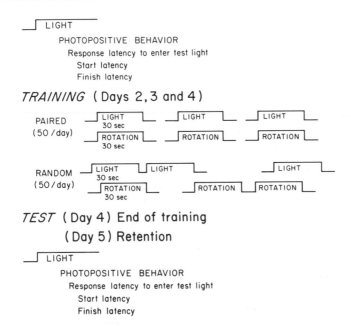

Figure 10.3. Behavioral testing and training sequences. Animals received pretest exposure to the test light to establish baseline responding. Start latencies plus finish latencies equal the response latency to enter the test light. Experimental groups received 150 trials of light paired with rotation, and control groups received an equal number of random presentations of light and rotation. Animals were tested immediately after the last training trial (day 4) and on the retention test day 24 hr after training (day 5). For some of the behavioral experiments the animals were tested 48, 72, and 96 hr after training. (From Crow and Offenbach 1983)

long-term, lasting for several days. This was revealed by the significant differences found between the paired groups and control groups on retention tests presented 2 days ($p < .005$), 3 days ($p < .005$), and 4 days ($p < .05$) following training. These initial studies focused on changes in behavior on retention days since retention would be the initial focus of the cellular neurophysiological studies (see below). However, the change in phototactic behavior does increase with repeated exposure to light–rotation pairings. The results from pilot studies examining changes in behavior at the end of each 50-block training session revealed that the change in behavior increased with each successive session, resembling an acquisition function.

A replication of the behavioral study was done using laboratory-reared *Hermissenda*. In this behavioral study we used laboratory-reared *Hermissenda* because these animals exhibit reduced individual variability of phototactic behavior (Crow and Harrigan 1979). Again we found that the conditioning procedure produced long-term changes in phototactic behavior, and in this case

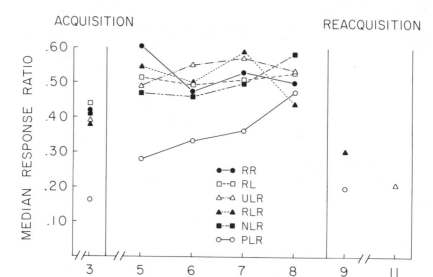

Figure 10.4. Median response ratios for acquisition, retention, and reacquisition of a long-term behavioral change in response to a light stimulus in *Hermissenda* (random rotation, random light, unpaired light and rotation, nothing, and paired light and rotation). The response ratio [in the form of $A/(A + B)$] compared the latency during the test (B) with the baseline response latency (A). Group data consist of two independent replications for all control groups and three independent replications for the experimental group. (From Crow and Alkon 1978)

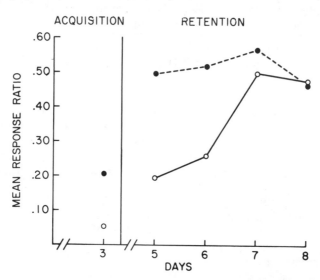

Figure 10.5. Average response ratios for acquisition and retention of the behavioral change in response to light in laboratory-reared *Hermissenda*. Filled circles, random light and rotation; open circles, light associated with rotation. (From Crow and Harrigan 1979)

the differences were detected with small groups ($N = 5$) of animals (see Figure 10.5). The effects of the associative factors were easier to assess because baseline measurements (pretests) for treatment and control groups were virtually identical. These results also indicated that laboratory-reared animals would be useful in cellular studies of the associative learning since changes unique to the temporal association of stimuli would be easier to detect.

DESCRIPTION OF THE CS AND US PATHWAYS

Investigation of the statocyst pathway in mediating the US

The role of the photoreceptors in mediating the light stimuli in this paradigm is now well documented. However, the question arises as to the contribution of the statocysts in mediating the temporally specific change in phototactic behavior. Since rotation does not evoke a clearly identifiable response, the role played by the statocyst in this paradigm is of critical importance. Other sensory systems might be expected to contribute to the behavioral change such as tactile input from mechanoreceptors around the foot and/or body wall.

The evidence for the involvement of the statocyst in mediating the change in behavior during acquisition came from a study of laboratory-reared animals. As reported by Crow and Harrigan (1979), light paired with rotation resulted in a significant change in normal phototactic behavior. Whereas the normalized suppression ratios for the paired groups were still significantly different from the random controls, the absolute magnitude of the latency measurements on retention tests for the conditioned group was less than previously reported for field-collected *Hermissenda*. This indicated that the conditioning procedure may be less effective with laboratory-reared animals. A posttraining dissection of the F1 and F2 generations of laboratory-reared animals used in the study revealed that the statocysts typically contained only one statoconium compared to normal statocysts containing 150–200 statoconia (Figure 10.6). A subsequent study with laboratory-reared animals (F1 generations) compared the effectiveness of the conditioning procedure for reared animals that had statocysts that contained one statoconium with reared animals that had statocysts with several or normal numbers of statoconia (Harrigan et al. pers. comm.). These results showed that following conditioning, animals with statocysts containing one statoconium moved into the light faster than laboratory-reared animals with statocysts containing several or normal numbers of statoconia. These results may reflect a decrease in the effectiveness of the US on the statocysts since intracellular recordings from hair cells during rotation of isolated nervous systems with statocysts containing single statoconia showed smaller hair cell generator potentials evoked by rotation, and thus a consequence of this would be weaker synaptic interactions with the visual system (Harrigan et al. pers. comm.). If tactile input was of critical importance, then one would expect that there could be no difference between preparations with single statoconia and normal numbers of statoconia. Therefore, these results lend support to the hypothesis that the US pathway involves the gravity-detecting organs (statocysts) in *Hermissenda*.

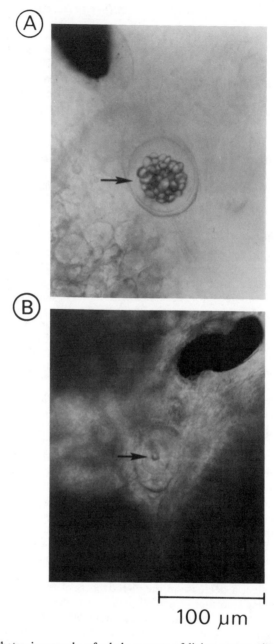

100 µm

Figure 10.6. Photomicrographs of whole mounts of living preparations showing the statocysts of normal field-collected *Hermissenda* (*A*) and laboratory-reared *Hermissenda* (*B*). The statocyst wall enclosed a fluid and cluster of statoconia (arrow) in *A* and a single statoconium (arrow) in *B*. All animals from the F1 and F2 generations and most from the F3 generation had statocysts containing one statoconium. (From Crow and Harrigan 1979)

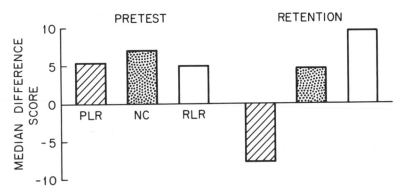

Figure 10.7. Dark–light differences scores for trained ($n = 40$) and control ($n = 19$) animals. Median difference scores on pretest and retention tests for paired group, random group, and normal controls. Difference scores represent latency to move to the end of the tubes in dark minus latency to enter illuminated area for same animals. Positive difference scores indicate faster latencies in light, and negative difference scores show slower latencies in light. Four different orders of testing on pretest and retention were examined for the experimental group: LD, LD; DL, DL; DL, LD; LD, DL. (From Crow and Offenbach 1983)

CS specificity

Are the changes in visually guided locomotion found after conditioning specific to stimulation of the CS pathway with light? Experiments designed to examine this question revealed that the change in phototactic behavior produced by the conditioning procedure is not due to nonspecific changes in locomotor behavior for trained animals. These studies (Crow and Offenbach 1979, 1983) showed that locomotor behavior changed when tested in the presence of light and did not change when tested in the dark following behavioral training (Figure 10.7). However, contextual cues associated with training in the glass tubes might be expected to make a contribution to the change in behavior. Such cues, however, probably do not make a major contribution to the change in behavior since changes in the response to light after training have been detected in the open field as well as in the glass tubes. Thus this example of learning exhibits trans-situational generality, a characteristic of vertebrate learning.

Additional behavioral studies by Crow and Offenbach (1979, 1983) found that the training procedure significantly increases the time to initiate locomotor behavior in response to light. In the dark-adapted animal, most of the variability in the time taken to move into the center illuminated area could be accounted for by the time taken to *initiate* locomotor behavior. We also found that the conditioning procedure changed another component of phototactic behavior. Trained animals remained in the brightest part of the light gradient significantly less time as compared to random control groups ($p < .01$) and pretest measures ($p < .01$) (Figure 10.8). Taken together, these behavioral results on the modification of various components of visually guided behavior

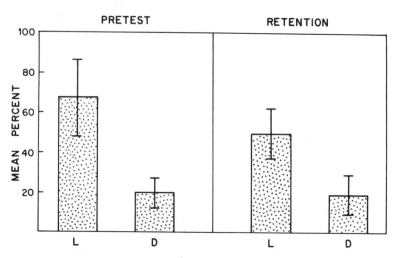

Figure 10.8. Time at end of the tubes in light and in the red light. Mean percentage time ($\pm SD$) in illuminated area at the end of tubes and at end of tubes in the red light for the same animals ($n = 9$) on the pretest and retention test following behavioral training. *Hermissenda* spend significantly more time at the end of the tubes when illuminated for both pretest and retention. However, the animals remain in the light less after behavioral conditioning as compared to pretest measures. (From Crow and Offenbach 1983)

in *Hermissenda* meet most of the operational criteria for associative learning as studied in mammals. The behavioral change is dependent on the temporal association of the conditioning stimuli, is long-lasting, is reversible, and shows stimulus specificity. The random control group actually provides a conservative estimate of nonassociative effects since about 20 percent of the random presentations resulted in some overlap between conditioning stimuli (Crow and Alkon 1978).

Anatomy of the CS pathway

The relative simplicity of the eyes and the well-defined synaptic interactions between the individual photoreceptors within each eye make *Hermissenda* a favorable preparation for cellular neurophysiological analyses. The eyes are situated on the dorsal surface of the circumesophageal nervous system between the cerebropleural and pedal ganglia (see Figures 10.9 and 10.10). Each of the eyes contains five photoreceptors (Dennis 1967), two designated as type A and three as type B (Alkon and Fuortes 1972). This classification of photoreceptor types is based upon electrophysiological and morphological criteria. Type A photoreceptors are approximately 25 μm in diameter and are located in the anteroventral portion of the eye. The type B cells are approximately 40 μm in diameter and are located in the posterodorsal portion of each eye. Both the type A and B photoreceptors can be identified further by their lateral and/or

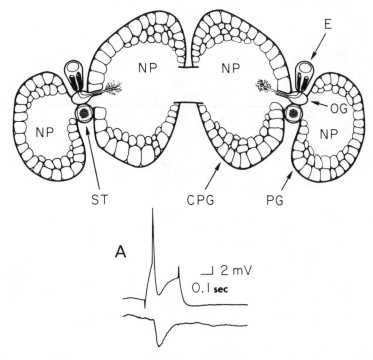

Figure 10.9. Diagram of the dorsal surface of the circumesophageal nervous system of *Hermissenda crassicornis*. Two type B photoreceptor somata, black areas in each eye, their axons and terminal processes drawn schematically. Photoreceptors receive synaptic input from other photoreceptors at their terminal endings in the neuropil. E, eye; OG, optic ganglion; NP, neuropil; ST, statocyst; CPG, cerebropleural ganglia; PG, pedal ganglia. Intracellular recordings from two type B photoreceptors within the same eye are shown in part *A* to illustrate the direct inhibitory interactions between type B photoreceptors. A current pulse delivered to one type B photoreceptor evokes a spike (top trace), which is followed by an inhibitory postsynaptic potential in the second type B photoreceptor (bottom trace). This result is invariant from preparation to preparation. The top of the spike in *A* has been retouched. (From Crow et al. 1979)

medial position in the eye with respect to the cerebropleural ganglion. Type A and B photoreceptors exhibit somewhat different electrophysiological characteristics in response to light. Type B photoreceptors, unlike the type A, are spontaneously active in darkness and are responsive to dim illumination after moderate periods of dark adaptation. In addition, the central B photoreceptor appears to have a lower dark-adapted spontaneous spike discharge frequency than either the lateral or medial type B cells. The axons of the five photoreceptors emerge from the posterior portion of the eye and form the optic nerve (Figure 10.11). The optic nerve courses through the optic ganglion and merges with axons from optic ganglion cells to form the optic tract, which then enters the cerebropleural ganglion and terminates in a cluster of swellings along the

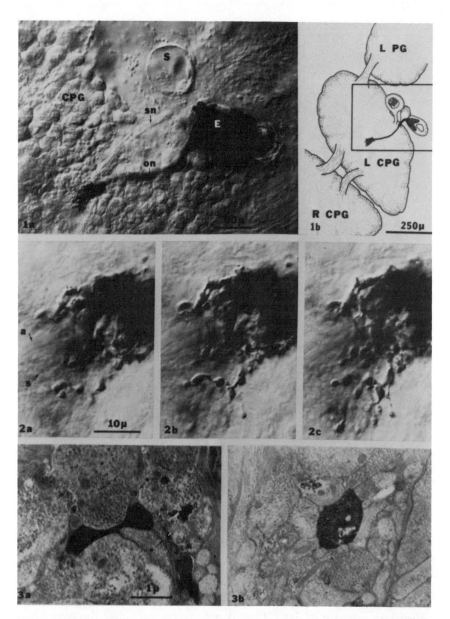

Figure 10.10. 1*a:* 240× optical section of the portion of the circumesophageal ganglia of *Hermissenda crassicornis* indicated (at 90×) by the box in *b*. The primary features are the eye (E) and optic nerve (on) in which one axon has been stained with HRP, and the statocyst (S) and the static nerve (sn). L PG, left pedal ganglion; L CPG, left cerebropleural ganglion; R CPG, right cerebropleural ganglion. 2*a–c:* Three sequential optical sections, separated by 1-μm intervals, of the branched region in part 1, now magnified 1500×. A greater sense of continuity and depth may be obtained by viewing them pairwise in stereo. Note en passant and terminal swellings (see text). (From Senft et al. 1982) 3: Electron micrographic cross sections (13,500×) of swell-

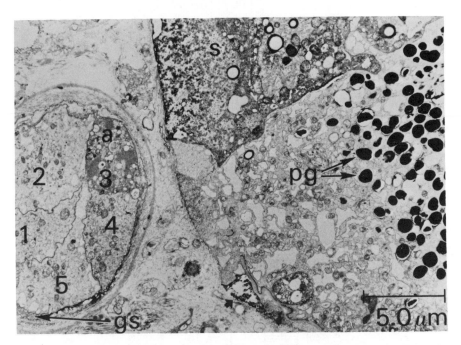

Figure 10.11. Electron micrograph showing HRP-labeled photoreceptor soma (s) and cross section of optic nerve. The pigment cells contain numerous pigment granules (pg). The optic nerve consists of five axons (1–5) surrounded by a glial sheath (gs). A darkly labeled axon (a) corresponds to the labeled soma. A partially filled area indicated by arrow (bottom center) may be the result of some leakage from a previous unstable electrode penetration. (From Crow et al. 1979)

secondary processes and endings (see Figure 10.10). Synaptic interactions between photoreceptors within the same eye and second-order neurons occur at the swellings along the secondary processes in the neuropil (Crow et al. 1979). Transection of the optic nerve eliminates synaptic and spike potentials from the photoresponse since there is a spatial separation of synapses and the area of spike generation from the region of phototransduction.

CELLULAR NEUROPHYSIOLOGICAL CORRELATES

Modification of primary sensory neurons in the CS pathway

The second major goal of these studies was to examine the neural locus of the behavioral modification. As an initial start in the cellular investigation of this

Caption to Figure 10.10 (*cont.*)
ings as seen in part 2. Note that the labeled swellings in both parts 2 and 3 are on the order of 1–2 μm in diameter, but that smaller processes (*) are also seen in the electron micrographs. (From Senft et al. 1982)

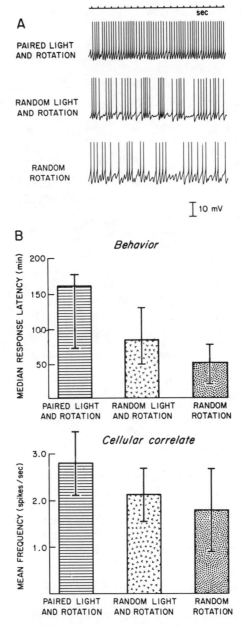

Figure 10.12. Effect of training procedures on the spontaneous activity of dark-adapted type B photoreceptors and photopositive behavior. *A*: Examples of intracellular recordings from dark-adapted B photoreceptors. *B*: Histograms comparing behavioral changes in response to light immediately after the last training trial with the cellular correlate (spike frequency of B photoreceptors) from experimental ($n = 25$) and random control animals (total $n = 25$: random rotation, $n = 7$; random light and rotation, $n = 18$). Behavioral data are expressed as medians ± interquartile ranges. Intracellular recordings were taken after behavioral response. Spike frequency

example of associative learning in *Hermissenda* we investigated changes in the CS pathway by recording intracellularly, from the primary sensory neurons, the photoreceptors in the eyes of the trained animals. The synaptic organization within the eyes and between the visual and statocyst pathways are well documented and thus will not be discussed in detail here (for a review see Alkon 1973a, b, 1976). In order to examine possible neural correlates in the CS pathway, we recorded intracellularly from the type B photoreceptors after 3 days of behavioral training (Crow and Alkon 1980). This was accomplished by training and testing the individual animals, then isolating circumesophageal nervous systems, and recording from B photoreceptors using standard intracellular techniques. Of particular interest was the finding that the spontaneous activity of the dark-adapted type B photoreceptors was significantly increased from trained animals as compared with the various random control groups [$F(2, 47) = 7.32; p < .01$] (Figure 10.12). Since differences in the variance of the samples from treatment and control groups can effect subsequent statistical analyses, we examined this with a Cochran test for homogeneity of variance and found that the sample variances were not significantly different. The a posteriori tests showed that the spontaneous activity of B photoreceptors from the paired group was significantly higher than the two random control groups, which did not differ from each other (Newman–Keuls test, $p < .01$). The results of a replication of the experiment using a smaller sample size ($n = 6$) and a blind procedure yielded similar outcomes ($t = 2.47; p < .025$) (Crow and Alkon 1980). Although the increase in spike frequency was measured after 15 min of dark adaptation, it was possible that the change in spike frequency was triggered by light. A direct test of this assumption has not been carried out because of the difficulties in obtaining stable penetrations of photoreceptors under red illumination. Since we did not have a direct functional relationship between the activity of the B photoreceptor and a behavior, we examined the relationship between the behavior and spike frequency in more detail. If there was a quantitative relationship between the magnitude of the change in phototactic behavior and the activity of B photoreceptors for individual animals, then one might argue that the change in the photoreceptor may be functionally related to the behavior of interest. Such a statistical relationship was indeed found, since the correlation between the type B photoreceptor spike frequency and the time taken to enter the light was significant for animals from the paired light–rotation group (Spearman $p = .63; p < .01$). What could account for this increase in B photoreceptor activity? One possibility is that the increase could be the result of either synaptic disinhibition of the photoreceptors or an increase in excitatory postsynaptic potential (EPSP) frequency in the photoreceptors. Both of these explanations were examined and could be ruled out since inhibitory postsynaptic potential (IPSP) frequency in B photoreceptors was greater, not less, following training, and the frequency of EPSPs in the photoreceptors was not different between the experimental and control groups (Crow and Alkon 1980). The results were more consistent with the hypothesis that following conditioning light evoked a tonic, long-lasting depolarization of

Caption to Figure 10.12 (*cont.*)
data are shown as mean spikes per second ± standard deviations. Cutoff scores of 180 min were used for all groups during behavioral testing. (From Crow and Alkon 1980)

the photoreceptors. Since the occurrence of spike activity and synaptic potentials in the intracellular recordings complicates the cellular analysis, we analyzed the cellular correlates further in axotomized preparations where spike activity and synaptic input was eliminated by cutting the optic nerve. This procedure results in isolated photoreceptor somas since there is a spatial separation between the site of phototransduction and the area of spike generation and synaptic input (Alkon and Fuortes 1972). An analysis of the axotomized preparations immediately after training revealed that the photoreceptors from the trained animals were significantly more depolarized than the random control group ($U = 5, p < .024$). This suggests that training may produce a decrease in a steady-state dark conductance of the photoreceptor membrane, dependent on light and/or voltage. Consistent with this hypothesis was the finding that the input resistance of axotomized photoreceptors was significantly higher for trained animals as compared to the random control group ($n = 4, p = .026$) (see Figure 10.13). In addition to the conductance changes, differences in the waveforms of the light-evoked generator potentials for trained and random control groups were also observed. The amplitudes of the hyperpolarizing dip and tail of the generator potential were significantly increased in the cut nerve preparations from the trained animals as compared with the random controls ($p = .024$ and $p = .053$ respectively), which also suggests that there are primary changes in the photoreceptors' response to light with conditioning. We proposed that the results could be explained by a decrease of a light- and/or voltage-dependent K^+ conductance in the B photoreceptors (Crow and Alkon 1980). This suggestion has been supported by the recent findings of a reduction in an early voltage-dependent K^+ current in B photoreceptors of trained animals (Alkon et al. 1982). The decreased K^+ conductance could be a consequence of long-term alterations of intracellular Ca^{2+} (see Chapter 15, this book), which is important in a number of photoreceptor properties, such as adaptation.

Cellular correlates in putative motor neurons

Having found cellular correlates of learning in the primary sensory cells, the possibility that other neurons in the nervous system of *Hermissenda* may exhibit changes with learning was investigated. As described above, the conditions that are sufficient to produce learning have been well documented. However, little is known regarding how the learning generates changes in phototactic behavior of *Hermissenda*. If changes in the type B photoreceptors associated with learning were the primary locus for learning in this system, alterations in activity in interneurons and motor neurons would be a consequence of these primary changes. Alternatively, the changes in the photoreceptors may reflect primary alterations in other neurons that provide synaptic input to the photoreceptors or parallel input to the central pattern generators and motor neurons. As an initial step in addressing these questions I have recently begun to examine correlates of associative learning in the motor system.

Since one of the changes in phototactic behavior involves an increase in the time taken to initiate locomotion in response to light (CS), it is important to identify motor neurons that may contribute to visually guided locomotor behavior. *Hermissenda*, like a number of gastropods, locomotes by the action

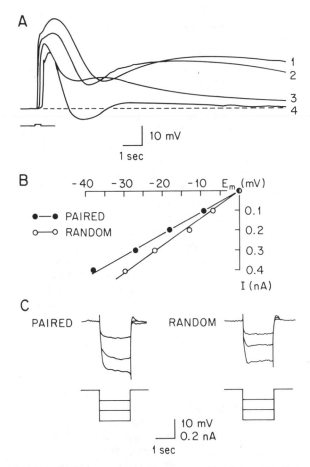

Figure 10.13. Cellular changes in cut nerve preparations from experimental and ran-dom control animals. *A*: Receptor potentials of dark-adapted type B photoreceptor from an experimental animal (light paired with rotation). Responses evoked by brief light flashes of increasing intensity (4), -log 3.0; (3), -log 2.0; (2), -log 1.0; (1), -log 0.5. Dashed line indicates resting membrane potential and lower trace indicates dura-tion of light flash. The absence of spikes and synaptic potentials indicates that the photoreceptor soma was successfully isolated from the area of spike initiation and syn-aptic input. *B*: Representative linear current–voltage relationship of dark-adapted iso-lated (cut nerve) type B photoreceptors from experimental (paired) and random con-trol groups. *C*: Examples of changes in membrane potential of dark-adapted isolated B photoreceptors from experimental and random control animals. Electrotonic poten-tials evoked by hyperpolarizing square current pulses (bottom traces) through a bal-anced bridge. Resistance measurements taken with the single electrode–bridge circuit are consistent with data from experiments in which the photoreceptors were impaled simultaneously with two microelectrodes for current injection and voltage recording. (From Crow and Alkon 1980)

of pedal cilia located on the sole of the foot. During ciliary locomotion the cilia beat against a mucus film between the foot and substratum. The ciliary action is under control of the central nervous system. However, the neurons in the central nervous system controlling or influencing the ciliated cells have not been found in *Hermissenda*. Some light-sensitive neurons have been identified in the pedal ganglion, although those cells were not involved in locomotion (Jerussi and Alkon 1981).

Recently two putative motor neurons in the pedal ganglia that may influence locomotor behavior have been identified (Crow 1981). Using intracellular staining techniques such as lucifer yellow and horseradish peroxidase, it was found that the cells send axons directly out pedal nerve P3 without sending collaterals into the cerebropleural ganglion. These morphological results are consistent with the interpretation that the pedal cells are motor neurons, since they have a major axon in a nerve that innervates the foot and/or body wall (Jerussi and Alkon 1981), although the evidence does not unequivocally prove that they are motor neurons. Electrophysiological studies revealed that these two cells exhibit alterations in cellular activity in response to illumination of the eyes in the isolated nervous system. Light evokes an increase in spike frequency and the frequency of postsynaptic potentials (PSPs). The light-evoked activity is not a direct effect of the light on the pedal cells nor the result of extraocular input to the neurons, since cutting the optic nerve, thus isolating the output of the eye to the rest of the nervous system, eliminates the light response. In addition, the light response is eliminated in solutions that block synaptic transmission.

Neural correlates of the conditioning were examined in one of the identified putative motor neurons, cell P5. In normal animals or random controls, cell P5 exhibits cyclic bursting activity in response to steady illumination. Since trained animals exhibit a significant increase in the time taken to initiate locomotion in response to illumination, the relationship between the behavior and activity of P5 at the onset of illumination was examined initially to determine if there was a correspondence between the behavior and activity in the putative motor neuron. In these studies the start latencies of trained and random control groups were measured and then the circumesophageal nervous systems were isolated as described (Crow and Alkon 1980) and placed in a recording chamber. Intracellular recordings from the motor neuron showed a significantly higher frequency of inhibitory postsynaptic potentials (IPSPs) in the presence of light in cells from trained animals as compared to the random controls ($p < .01$).

This inhibition of motor neuron activity may explain the observed delays in start latency of conditioned animals. The behavior that is modified by the conditioning procedure involves a continuous ongoing activity (locomotion) or a delay in this activity (initiation of locomotion) rather than a modification of reflex activity that is tested by the delivery of a punctate stimulus. Therefore, in this example of learning the change in behavior is manifested during conditions of steady illumination or continuous presentation of the CS. Thus in addition to examining the response of the motor neurons to the onset of light, I also examined their responses under conditions of light adaptation similar to the conditions used to test changes in phototactic locomotor behavior. After 5

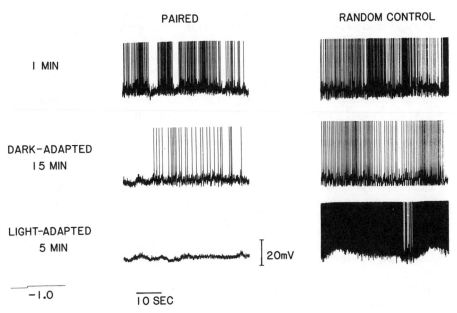

Figure 10.14. Neural correlates of conditioning in a putative motor neuron (P5). Intracellular recordings from pedal neuron P5 in preparations that received paired light and rotation or random presentations of light and rotation. Comparison of the spontaneous activity of conditioned and random controls are after 1 min of dark adaptation (top), after 15 min of dark adaptation (middle), and in the presence of the CS (light) after 5 min of light adaptation (bottom). Log intensity of -1.0 was produced with a neutral density filter that attenuated a quartz-iodide light source ($\sim 2 \times 10^5$ erg cm sec^{-1}). Intracellular recordings taken 24 hr following behavioral conditioning. (From Crow 1981)

min of steady illumination the activity of the pedal motor neurons was examined during successive 1-min periods. The putative motor neuron exhibited a dramatic depression in activity during illumination for the trained animals (Figure 10.14). In contrast to the results from trained animals, the preparations from the random control groups showed an increase in cyclic bursting activity, which is characteristic of the normal response to light in cell P5 (Figure 10.14).

The increase in the frequency of IPSPs in P5 after conditioning could be due to synaptic input from a number of different sources. In order to examine the contribution of visual inputs to the electrical activity of P5 the optic nerve was cut after recording spike activity and IPSP frequency from cell P5 in trained animals. Axotomy thus eliminates the contribution of the visual system to the synaptic activity in pedal motor neuron P5. After the transection of the optic nerve, cell P5 was again repenetrated and dark-adapted, and the preparation was then illuminated. These experiments showed that the IPSPs normally evoked by light in trained animals were now eliminated. This indicates that the source of the inhibition (IPSPs) is from the eye or optic ganglion and the changes found in putative motor neurons with training are presynaptic to the motor neurons. These results support the hypothesis that the primary changes with learning in this system are in the visual (CS) pathway, although they do not unequivocally rule out contributions from other sources in mediating this change in behavior following conditioning. It will be necessary to identify all the motor neurons involved in this type of mucociliary locomotion and input from the central pattern generators before this issue can be resolved.

DISCUSSION AND CONCLUSION

The results of our cellular neurophysiological and biochemical studies (Crow and Alkon 1980; Neary et al. 1981) indicate that there are alterations in primary sensory neurons produced by the conditioning procedure. These findings may be viewed as somewhat surprising with regard to more traditional viewpoints, mainly from the vertebrate conditioning literature that focused on plasticity in association areas or in areas of the nervous system not involved primarily with sensory or motor function (see Weinberger 1982 for discussion). An exception to this research focus in the mammalian nervous system is the work of Woody and his colleagues on cellular correlates of conditioning that are thought to be intrinsic to motor neurons in the cat sensorimotor cortex (Woody 1974). However, our results are not that unusual when one considers that primary sensory neurons in many invertebrates are required to perform a number of integrative functions that are typically performed by secondary cells in vertebrate sensory systems, where considerably more neurons are involved in the processing of sensory information (see Alkon 1976). As an example in *Hermissenda*, lateral inhibition is the result of the mutually inhibitory connections between the three type B photoreceptors in each eye (see Figure 10.9). In contrast, lateral inhibition or surround inhibition involves the synaptic interactions between intrinsic circuit interneurons and secondary sensory cells in the retina and olfactory bulbs of the mammalian nervous system. Another example of primary sensory integration involves phototransduction and spike generation. Spike generation occurs in the photoreceptors of *Hermissenda*, unlike the case of mammalian photoreceptors. Thus, since a number of integrative functions involved with the processing of visual information occurs in the primary photoreceptors of *Hermissenda*, it is not surprising to find plastic capabilities at this level of the sensory systems in relatively simple nervous systems of gastropod mollusks. Additional evidence consistent with this proposal is the recent investigation of cellular analogs of classical conditioning in two different populations of primary mechanoreceptors in *Aplysia* (Abrams et al. 1982; Walters and Byrne 1982). In one study (Hawkins et al. 1983) it was proposed that primary changes in the sensory neurons could account for differential conditioning of a defensive withdrawal reflex (Carew et al. 1983). Conditioned modifications of sensory systems are not unique to invertebrates. Conditioned alterations in neuronal activity have been found in the auditory system of cats (Buchwald et al. 1966; Ryugo and Weinberger 1978) and of rabbits (Gabriel et al. 1976) and in the visual system of pigeons (Cohen 1982).

In a general sense, learning results in a specific change in how the organism operates in the environment. Our work with *Hermissenda* indicates that conditioning results in the development in memory of a relatively permanent representation based on experience (light–rotation pairings) that changes the way that *Hermissenda* responds to an illuminated environment. By placing the memory in the sensory system, a number of ongoing behaviors or activities can be modulated in the presence of the conditioned stimulus (light). This seems to fit the experimental evidence, since conditioning affects a number of different components of phototactic behavior in *Hermissenda*. In the dark-adapted animal light inhibits the initiation of locomotor behavior, and under other conditions light can decrease the rate or speed of locomotion once locomotor behavior has been initiated. Another finding consistent with this hypothesis is that conditioned animals avoid the brightest part of the illuminated gradient and spend a greater amount of time in the adjacent, less illuminated part of the gradient (Crow and Offenbach 1979, 1983).

The conditions sufficient to produce this example of associative learning are well documented. However, the questions of what is actually learned and how the learning generates changes in phototaxic behavior have yet to be answered. These issues require an understanding of the cellular basis of locomotion in *Hermissenda* that we do not presently have. However, it is interesting to make a prediction about how the known sensory correlates might affect the behavior. Changes intrinsic to the type B photoreceptor as a result of conditioning may result in larger steady-state responses to illumination, perhaps due to the increase in the voltage-dependent sustained light-induced Ca^{2+} current as suggested by Alkon et al. (1982). The change in the response of the photoreceptors to light may alter the activity of central pattern generators responsible for bursting activity in motor neurons involved in locomotion. However, it is paradoxical that a greater light response following conditioning should result in such a profound decrease in normal positive phototaxis in an animal that normally exhibits positive responses to brighter illumination. This suggests that other cells in the circuit controlling locomotion may contribute to the response after conditioning in addition to the type B photoreceptors. Understanding how the behavior is generated, how conditioning results in a light-dependent modulation of the burst pattern of a locomotor system, and how learned changes in a sensory system can modulate ongoing patterns of behavior will be the subjects of future research.

ACKNOWLEDGMENTS

I wish to thank Dr. J. H. Byrne for his comments on an earlier draft of this chapter. This work was supported by NIH fellowship (F32 NS05021-12) and NIH research grant HD15793.

REFERENCES

Abrams, T. W., Hawkins, R. D., Carew, T. J., and Kandel, L. R., 1982, Activity-dependent facilitation accounts for pairing specificity in classical conditioning of a defensive withdrawal reflex in *Aplysia. Soc. Neurosci. Abstr. 8*

Alkon, D. L., 1973a, Neural organization of a molluscan visual system. *J. Gen. Physiol. 61*:444–461.

1973b. Intersensory interactions in *Hermissenda. J. Gen. Physiol. 62*:185–202.

1974. Associative training of *Hermissenda. J. Gen. Physiol. 64*:70–84.

1976. The economy of photoreceptor function in a primitive nervous system. In *Neural Principles in Vision.* (J. F. Zettler and R. Weiler, eds.), pp. 410–426. New York: Springer-Verlag.

Alkon, D. L., and Fuortes, M. G. F. 1972. Responses of photoreceptors in *Hermissenda. J. Gen. Physiol. 60*:631–649.

Alkon, D. L., Lederhendler, I., and Shockimas, J. J. 1982. Primary changes of membrane currents during retention of associative learning. *Science 215*:693–695.

Braun, J. J., and McIntosh, H. Jr. 1973. Learned taste aversions induced by rotational stimulation. *Physiol. Psychol. 1*:301–304.

Buchwald, J. S., Halas, E. S., and Schramm, S. 1966. Changes in cortical and subcortical unit activity during behavioral conditioning. *Physiol. Behav. 1*:11–22.

Carew, T. J., Walters, E. T., and Kandel, E. R. 1981. Classical conditioning in a simple withdrawal reflex in *Aplysia californica. J. Neurosci. 1*:1426–1437.

Carew, T. J., Hawkins, R. D., and Kandel, E. R. 1983. Differential classical conditioning of a defensive withdrawal reflex in *Aplysia californica. Science 219*:397–400.

Cohen, D. H. 1982. Central processing time for a conditioned response in a vertebrate model system. In *Conditioning: Representation of Involved Neural Functions* (C. D. Woody, ed.), pp. 517–34. New York: Plenum Press.

Crow, T. 1981. Neurophysiological correlates of conditioning in identified putative motor neurons in *Hermissenda. Soc. Neurosci. Abstr. 7*:352.

1982. Sensory neuronal correlates of associative learning in *Hermissenda. Soc. Neurosci. Abstr. 8*:824.

1983. Conditional modification in *Hermissenda crassicornis:* Analysis of time-dependent associative and nonassociative components. *J. Neurosci. 3*:261–268.

Crow, T., and Alkon D. L. 1978. Retention of an associative behavioral change in *Hermissenda. Science 201*:1239–1241.

1980. Associative behavioral modification in *Hermissenda:* Cellular correlates. *Science 209*:412–414.

Crow, T., and Harrigan, J. F. 1979. Reduced behavioral variability in laboratory-reared *Hermissenda crassicornis* (Eschscholtz, 1831) (Opisthobranchia: Nudibranchia). *Brain Res. 173*:179–184.

Crow, T. J., Heldman, E., Hacopian, V., Enos, R., and Alkon, D. L. 1979. Ultrastructure of photoreceptors in the eye of *Hermissenda* labeled with intracellular injection of horseradish peroxidase. *J. Neurocytol. 8*:181–195.

Crow, T., and Offenbach, N. 1979. Response specificity following behavioral training in the nudibranch mollusk *Hermissenda crassicornis. Biol. Bull. 157*:364.

1983. Modification of the initiation of locomotion in *Hermissenda. Brain Res. 271*:301–310.

Crow, T. J., and Woody, C. D. 1973. Acquisition of a conditioned eyeblink response during reversible denervation of obicularis oculi muscles in the rat. *Brain Res. 64*:414–417.

Dennis, M. J. 1967. Electrophysiology of the visual system in a nudibranch mollusc. *J. Neurophysiol. 30*:1439–1465.

Gabriel, M., Miller, J. D., and Saltwick, S. E. 1976. Multiple unit activity of the rabbit medial geniculate nucleus in conditioning, extinction, and reversal. *Physiol. Psychol. 4*:124–134.

Garcia, J., McGowan, B. K., and Green, K. F. 1972. Biological constraints on conditioning. In *Classical Conditioning,* vol. 2, *Current Research and Theory* (A. H. Black and W. F. Prokasy, eds.), pp. 3–27. New York: Appleton-Century-Crofts.

Gelperin, A. 1975. Rapid food-aversion learning by a terrestrial mollusk. *Science* 189:567–570.

Hawkins, R. D., Abrams, T. W., Carew, T. J., and Kandel, E. R. 1983. A cellular mechanism of classical conditioning in *Aplysia.* Activity-dependent amplification of presynaptic facilitation. *Science 219*:400–404.

Holland, P. C., and Rescorla, R. A. 1975. The effect of two ways of devaluing the unconditioned stimulus after first- and second-order appetitive conditioning. *J. Exp. Psychol. 1*:355–363.

Jerussi, T. P., and Alkon, D. L. 1981. Ocular and extraocular responses of identifiable neurons in pedal ganglia of *Hermissenda crassicornis. J. Neurophysiol. 46*:659–671.

Kandel, E. R. 1976. *Cellular Basis of Behavior.* San Francisco: Freeman.

Mpitsos, G. J., and Collins, S. D. 1975. Learning: Rapid aversive conditioning in the gastropod mollusk *Pleurobranchaea. Science 188*:954–957.

Mpitsos, G. J., Collins, S. D., and McClellan, A. D. 1978. Learning: A model system for physiological studies. *Science 199*:497–506.

Mpitsos, G. J., and Davis, W. J. 1973. Learning: Classical and avoidance conditioning in the mollusk *Pleurobranchaea. Science 180*:317–320.

Neary, J. T., Crow, T., and Alkon, D. L. 1981. Change in a specific phosphoprotein band following associative learning in *Hermissenda. Nature 293*:658–660.

Rescorla, R. A., and Holland, P. C. 1982. Behavioral studies of associative learning in animals. *Ann. Rev. Psychol. 33*:265–308.

Revusky, S. H. 1977. Learning as a general process with an emphasis on data from feeding experiments. In *Food Aversion Learning.* (N. W. Milgram, L. Krames, and T. M. Alloway, eds.), pp. 1–72. New York: Plenum Press.

Ryugo, D. K., and Weinberger, N. M. 1978. Differential plasticity of morphologically distinct neuron populations in the medial geniculate body of the cat during classical conditioning. *Behav. Biol. 22*:275.

Sahley, C., Gelperin, A. and Rudy, J. W. 1981a. One-trial associative learning in a terrestrial mollusc. *Proc. Nat. Acad. Sci. 78*:640–642.

Sahley, C., J. W. Rudy, and Gelperin, A. 1981b. An analysis of associative learning in the terrestrial mollusc *Limax maximus.* I. Higher-order conditioning, blocking, and a transient US–preexposure effect. *J. Comp. Physiol. A 144*:1–8.

Senft, S. L., Allen, R. D., Crow, T., and Alkon, D. L. 1982. Optical sectioning of HRP-stained molluscan neurons. *J. Neurosci. Meth. 5*:153–9.

Tyndale, C. L., and Crow, T. 1979. An IC control unit for generating random and nonrandom events. *IEEE Trans. Biomed. Eng. 26*:649–655.

Walters, E. T., and Byrne, J. H. 1983. Associative conditioning of single sensory neurons suggests a cellular mechanism for learning. *Science 219*:405–408.

Walters, E. T., Carew, T. J., and Kandel, E. R. 1979. Classical conditioning in *Aplysia californica. Proc. Nat. Acad. Sci. 76*:6675–6679.

Weinberger, N. M. 1982. Sensory plasticity and learning: The magnocullular medial geniculate nucleus of the auditory system. In *Conditioning: Representation of Involved Neural Functions* (C. D. Woody, ed.), pp. 697–718. New York: Plenum Press.

Woody, C. D. 1974, Aspects of the electrophysiology of cortical processes related to the development and performances of learned motor responses. *Physiologist 17*:49–69.

11 · A strategy for cellular analysis of associative learning in a terrestrial mollusk

A. GELPERIN, S. J. WIELAND, and S. R. BARRY

The ability of the human brain to store information and modify the meaning of that information through experience is surely one of the most intrinsically interesting phenomena in the natural world. Increasingly, the problem of how our brains accomplish this task is viewed as a tractable problem, whose solution should be attainable given the full range and power of modern methods for the biophysical, biochemical, anatomical, and behavioral analysis of neural networks. A variety of experimental strategies are being used on neural networks drawn from the major vertebrate and invertebrate phyla, as amply documented by the contents of this volume. We have selected a terrestrial gastropod mollusk for physiological study, stimulated by early behavioral data demonstrating several properties of vertebrate taste aversion learning in this animal (Gelperin 1975). More recent behavioral data showing a number of striking parallels between *Limax* learning and vertebrate conditioning data (see Chapter 12) have fueled our conviction that we can study the mechanism of information storage in the human brain using the protohumanoid mollusk *Limax maximus*.

In pursuit of the causative biochemical and biophysical changes underlying associative memory storage, we have developed a preparation of the isolated central nervous system (CNS) of *Limax* that can be trained using procedures analogous to those employed in our whole-animal studies (Sahley et al. 1981b, c). The isolated CNS of *Limax* will perform one-trial, discriminative taste aversion learning in vitro (Chang and Gelperin 1980; Culligan and Gelperin 1983.) We are engaged in several types of cellular and neurochemical analyses aimed at localizing and characterizing the critical synaptic events involved in associative learning and memory storage. We will first describe the properties of the isolated *Limax* CNS preparation that encouraged us to study its learning ability and then describe the learning ability of the isolated brain and the analysis of dopaminergic and cholinergic synapses as related to its learning ability.

THE ISOLATED CNS PREPARATION

An essential step in our neurophysiological analysis of *Limax* taste aversion learning was the development of a preparation of the lips, cerebral ganglia, and buccal ganglia that preserved several of the reflex pathways first studied in

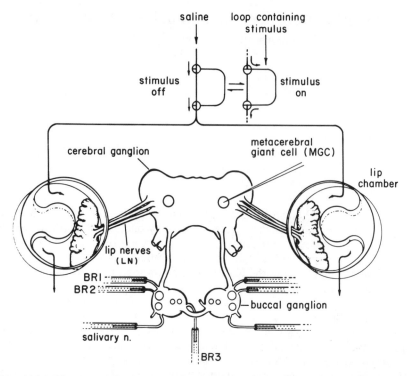

Figure 11.1. The *Limax* lip–brain preparation. A diagram of the isolated lips, cerebral ganglia, and buccal ganglia preparation that allows controlled, selective chemostimulation of the lips while recording intracellularly from cerebral and buccal neurons and extracellularly from buccal nerve roots. BR1, BR2, BR3 = buccal nerve roots 1, 2, 3.

whole-animal experiments. This isolated CNS preparation (Gelperin and Forsythe 1976; Prior and Gelperin 1977) allows controlled repetitive stimulation of the lip chemoreceptors with standardized food extracts while recording intracellularly from identified interneurons and motoneurons (Gelperin 1981) and extracellularly from buccal ganglion motor nerves serving feeding muscles (Gelperin et al. 1978). Figure 11.1 is a diagram of the preparation. The feeding motor program (FMP) is hard-wired in the buccal ganglia and can be reliably activated by repeated 30-sec chemostimuli delivered to the lips at 30-min interstimulus intervals. FMP is the neural substrate for the rhythmic rasping–swallowing movements of ingestion. If FMP is produced by the buccal motor network in response to lip chemostimulation, we conclude that the central decision-making interneurons have accepted the food extract as adequate to trigger ingestion. If FMP is not produced by the buccal motor network in response to lip chemostimulation, we conclude that the central interneurons have "decided" that the lip input is not adequate to trigger ingestion. Unlike some other gastropod preparations (McClellan 1982), the *Limax* buccal motor network produces only one output pattern, that for ingestion.

The lip chemostimulus input–FMP output pathway in the isolated lip–brain preparation of *Limax* shows three forms of reflex modification that mimic the behavior of the intact animal. If the isolated lip–brain preparation is tested with a series of palatable food extracts of increasing concentration, the probability of eliciting a bout of FMP and the length of FMP bouts emitted increase as the concentration of chemostimulus increases (Reingold and Gelperin 1980). Proprioceptive feedbacks from buccal muscles and gut wall have also been shown to operate in a parallel fashion in the intact animal and isolated lip–brain preparation (see Reingold and Gelperin 1980 for details). These results encouraged us to attempt to train the isolated lip–brain preparation using a procedure analogous to the protocol used to train the intact animal.

TRAINING THE ISOLATED BRAIN

The isolated lip–brain preparation of *Limax* can learn in one trial to associate a food extract that initially triggers FMP with a very bitter plant substance such as quinidine. The learning is expressed as a modification of the lip input–FMP output pathway such that the food extract that was initially effective in eliciting FMP does not trigger FMP after pairing. The learning by the isolated brain is discriminative because other food extracts not paired with the bitter taste remain effective at eliciting FMP (Chang and Gelperin 1980).

Our initial demonstration of this phenomenon has recently been extended in studies that allowed us to stimulate the right lip and left lip separately. The isolated cerebral and buccal ganglia were arrayed for recording FMP responses in the same way as before, but the lip perfusion system was modified so that we could train the brain using one lip and test for learning using the other lip (Gelperin et al. 1981; Culligan and Gelperin 1983). FMP responses were scored by a naive observer who was given the filmed records of the buccal nerve root responses to lip chemostimuli and a measuring algorithm for recognizing bites or rasps in terms of spike frequencies and phase relations of spike bursts between buccal nerve roots. Graphs of FMP responses were derived from these blindly scored data by plotting the reciprocal of the interbite interval versus time since beginning of the experiment. An example of cross-brain associative learning by an isolated *Limax* CNS is shown in Figure 11.2*A, B*. Training was carried out using the left lip only. The learned suppression of FMP response was expressed using test stimuli applied to *both* the left *and* right lip. Other preparations that experienced the same stimuli but with a 3-hr delay between potato and quinidine application did not learn; that is, they continued to give FMP responses to both potato and carrot applied to both right and left lips (Culligan and Gelperin 1983). These results make it very unlikely that the modification produced by the learning involves the peripheral chemoreceptor array.

Our most recent training experiments with the isolated brain preparation have attempted to build a stronger bridge between the whole-animal and iso-

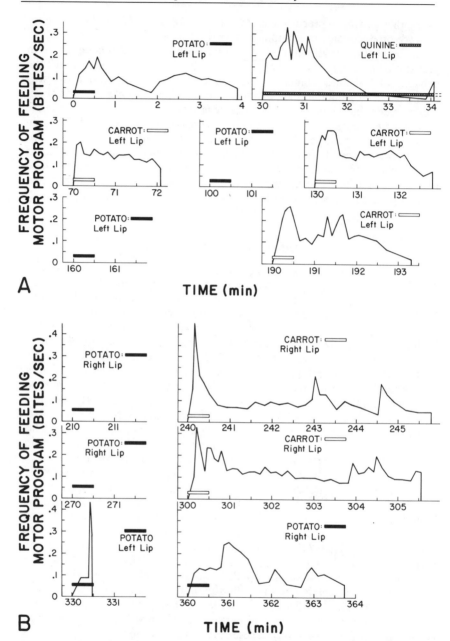

Figure 11.2. Feeding motor program responses of an isolated lip–brain preparation during training and testing. *A*: Training was carried out to associate potato and quinidine using the left lip only. A selective suppression of the FMP response to potato but not carrot was produced by a single training trial. *B*: The selective suppression of FMP response to potato is evident when test stimuli are applied at the naive (right) lip.

lated-brain training and testing procedures. We have trained a group of intact animals to associate an initially attractive taste with a very bitter taste, and then dissected the lips and cerebral and buccal ganglia from these same animals for testing as in our previous experiments. Of 14 lip–brain preparations derived from slugs trained to avoid a standard food extract, 7 showed a clear and selective FMP response decrement to that food extract when it was applied to their isolated lips in vitro (Gelperin and Culligan in press). We are encouraged to believe that the whole-animal training and the isolated-brain training procedures involve the same central processing and memory storage networks.

DOPAMINE AND FMP

To approach a cellular basis for feeding control and associative learning, we are delineating the neurons in the cerebral ganglia that receive chemosensory input and those that project to the buccal motor ganglia to initiate, modulate, or suppress expression of the feeding motor program.

By using retrograde labeling techniques with cobalt chloride and fluorescent dyes, we have located approximately 15 cerebral neurons that project to each buccal ganglion and that are thus candidates for feeding control interneurons. Of these, 6 medium-sized neurons (approximately 50 μm diameter) lie in clusters that have been shown by histochemical fluorescence to contain catecholamines (Osborne and Cottrell 1971; Wieland and Gelperin 1983). Using double-labeling techniques, we are pursuing a direct confirmation that some of these interneurons contain catecholamines. In addition, one neuron that projects to the buccal ganglion, the metacerebral giant cell (MGC), is known to contain serotonin (Cottrell and Osborne 1970; Gelperin 1981).

Cerebral and buccal ganglia were extracted with acid, and these extracts were chromatographed on reversed-phase columns using high-performance liquid chromatography (Mefford 1981) (Figure 11.3). Elution times were compared with authentic compounds and recoveries were estimated by inclusion of internal standards. In a series of four determinations, the cerebral ganglia were found to contain 380 pmol serotonin and 120 pmol dopamine per animal; the buccal ganglia contained 30 pmol serotonin and 19 pmol dopamine. No norepinephrine nor oxidized metabolites of monoamines were detected. Using ^{3}H-tyrosine as a prescursor, we detected accumulation of ^{3}H-dopamine, but little, if any, ^{3}H-norepinephrine in both ganglia.

Exogenous serotonin flowing over the lip–brain preparation produced general excitation of neurons with axons in the buccal motor roots (Figure 11.4A). Concentrations of serotonin greater than 10^{-5} M produced excitation with synchronized bursting, but seldom feeding motor program. These results are consistent with the role of the MGC as a positive modulator, but not as a trigger of the feeding response (Gelperin 1981).

In contrast, dopamine selectively affected buccal motor root activity at lower concentrations (10^{-7} to 3×10^{-6} M) (Figure 11.4B). The interburst interval of the autoactive salivary fast burster cell decreased in a dose-dependent manner, whereas the slow salivary bursters (Copeland and Gelperin 1983) were inhibited; extracellular monitoring revealed no modulation of other units. At

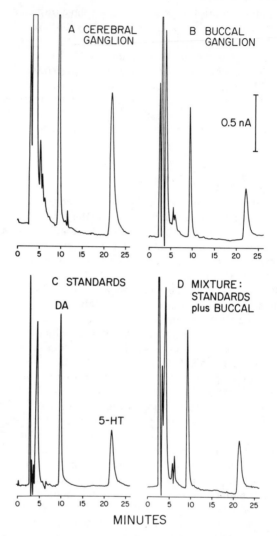

Figure 11.3. Demonstration of monoamines in *Limax* by high-performance liquid chromatography with electrochemical detection. *A*: 20 μl of a 120-μl extract from one pair of cerebral ganglia. *B*: 20 μl of a 50-μl extract from one pair of buccal ganglia. *C*: A mixture of 10.5 pmol dopamine and 5.1 pmol of serotonin standards. *D*: A 1:1 mixture of the same samples shown in parts *B* and *C*.

higher concentrations (3×10^{-6} to 3×10^{-5} M), feeding motor program was triggered in a dose-dependent manner (Figure 11.5). The probability of triggering feeding was 17 percent at 3×10^{-6} M, 33 percent at 10^{-5} M, and 100 percent at 3×10^{-5} M. Dopamine is thus a strong modulator of the feeding system and is a good candidate transmitter for the endogenous trigger of the feeding response.

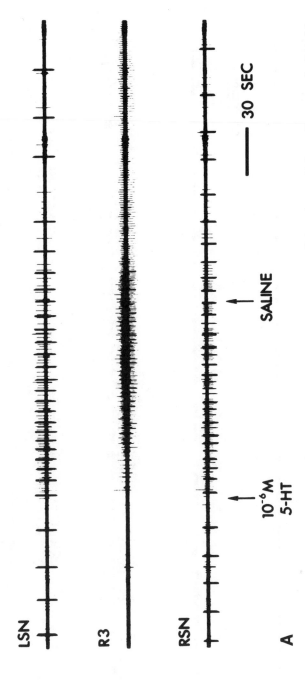

Figure 11.4. Monoamine excitation of the buccal feeding network in *Limax*. *A*: Serotonin-containing saline was substituted for standard saline in a flow-through chamber containing a cerebral ganglion–buccal ganglion system. Extracellular recordings from buccal motor nerve roots showed that all monitored neural activity exhibited excitation. LSN, left salivary nerve; R3, right buccal root 3; RSN, right salivary nerve. Volume of chamber was 200 µl; flowrate 500 µl/min.

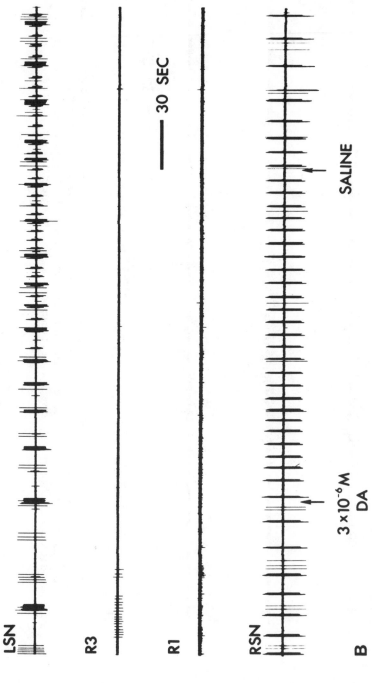

LSN

R3

30 SEC

R1

RSN

3×10^{-6} M
DA

SALINE

B

Figure 11.4 (*cont.*). *B*: Dopamine-containing saline was substituted for standard saline; other conditions as in part *A*. The salivary fast burster cells (thick multispike clusters in LSN and RSN traces) showed excitation as shorter interburst intervals. However, the salivary slow burster cells (thin, resolvable spikes in LSN and RSN) were inhibited; motor roots R1 and R3 showed no excitation.

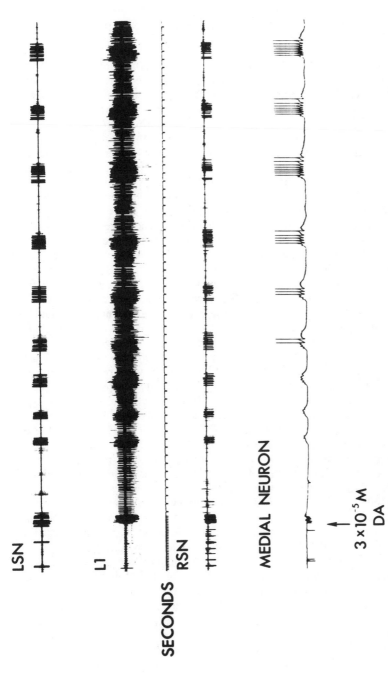

LSN

L1

SECONDS RSN

MEDIAL NEURON

3 × 10⁻⁵ M
DA

Figure 11.5. Induction of feeding motor program by exogenous dopamine: Application of dopamine (3×10^{-5} M) to the nervous system triggered a response that was apparently indistinguishable from lip-stimulated feeding motor program. An intracellular recording from a large buccal motor neuron (B10) shows no apparent direct effect of dopamine on the cell; it was smoothly recruited into the phase-locked feeding in the same pattern seen in chemostimulus-induced feeding.

CHOLINE, CHOLINERGIC SYNAPSES, AND MEMORY
FUNCTION

Stimulated by a growing clinical literature on the possible involvement of cholinergic synapses in human memory function (Bartus et al. 1982; Corkin 1981;
Whitehouse et al. 1982) and the demonstration in several mammalian species
that dietary choline levels can influence whole-brain acetylcholine (ACh) levels
(Cohen and Wurtman, 1976; Haubrich et al., 1976; Wurtman, 1982), we
examined the effects of dietary choline on synaptic function in *Limax*. Using
a very sensitive radioenzymatic assay for choline (Goldberg and McCaman
1974), we found that the concentration of choline in the blood of slugs maintained on a choline-enriched diet averaged 5.5 μM as compared to a value of
3.3 μM choline in the blood of slugs fed a choline-deficient diet (Barry and
Gelperin 1982a). To determine whether changes in free choline availability
within this range can alter synaptic function, a cholinergic neuromuscular
junction was studied in vitro. The presynaptic cholinergic neuron was the salivary fast burster, an autoactive motoneuron innervating the salivary duct
(Beltz and Gelperin 1980a, b; Barry and Gelperin 1982b). Junction potentials
(JPs) recorded from the salivary duct muscle (SD) followed salivary burster
(SB) spikes one-for-one. The SB–SD synapse was isolated in physiological
saline, and stable extracellular recordings of SB spikes and SB-elicited JPs
were made for many hours. The amplitude of SB-elicited JPs fluctuated by
only 10–15 percent during recording periods of 18 hr or longer. In contrast, as
shown in Figure 11.6, an increase of 1.5 μM choline in the saline bathing the
SB–SD synapse increased JP amplitude by more than 100 percent (Barry and
Gelperin 1982a). Elevations in JP size of 150 percent occurred after increasing
exogenous choline from 2.5 to 5.0 μM. These effects took several hours to
become maximal and could be blocked by simultaneous addition of the high-
affinity choline uptake blocker hemicholinium-3 (Barry and Gelperin 1982b).
These and other observations suggest that the increased choline availability
leads to greater choline uptake, ACh synthesis, and ACh release by the presynaptic element, the terminals of the SB neuron. It is clear that alterations in
choline availability within the range produced by dietary manipulation can
alter the efficacy of cholinergic synaptic function in *Limax*.

With these data in hand, it seemed natural to test the memory function of
slugs reared on high- and low-choline diets (Sahley et al. 1981a). Slugs from
the choline-enriched (11.4 g/kg) and choline-deficient (0.3 g/kg) diets were
trained using the associative odor-preference modification task described previously (see Chapter 12) and tested for acquisition 24 hr following training and
for long-term retention at 7 days following training. [Further details can be
obtained from Sahley, Barry, and Gelperin (pers. comm.).] No differences in
learning were apparent between slugs on the choline-deficient and choline-
enriched diets on the initial test at 24 hr. A dramatic difference emerged, however, at the 7-day test. The performance of slugs fed the choline-enriched diet
was as good at 7 days posttraining as it was at 1 day posttraining, whereas the
slugs fed the choline-deficient diet performed at a level close to that of control,
unconditioned slugs when tested 7 days after training. Feeding for 2 weeks on

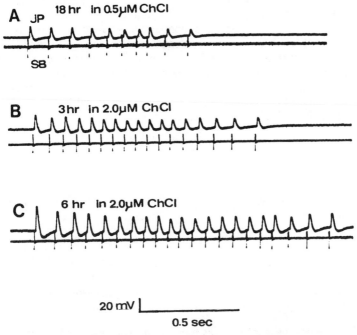

Figure 11.6. Exogenous choline augments a cholinergic synaptic potential. An elevation of only 1.5 μM in exogenous choline concentration produced an increase in amplitude of synaptic potentials at the salivary burster–salivary duct synapse. *A*: The synapse was dissected and bathed in 0.5 μM choline for 18 hr. The choline concentration in the saline was then elevated to 2.0 μM. A 103 percent increase in junctional potential amplitude occurred over the next 6 hr (*B, C*) (SB = salivary burster; JP = junction potential; ChCl = choline chloride).

a choline-enriched diet significantly facilitated memory function. The link between this behavioral observation and the physiological finding of augmented cholinergic synaptic function under conditions of increased choline availability is currently under investigation.

FUTURE DIRECTIONS

Using the several approaches to network analysis discussed above, in combination with multiple intracellular recording and selective ablations of neuron clusters in the cerebral ganglia, we are delimiting the neuroanatomical loci and neurochemical nature of the synapses necessary for expression of associative learning and memory function in the *Limax* isolated CNS preparation. The prominent procerebral lobes of the cerebral ganglia (Van Mol 1967) are prime candidates for such analysis. The cerebral ganglia also contain a number of neurons involved in sexual maturation and reproductive behavior (Van Minnen and Sokolove 1981) so that only a subpopulation of neurons in the cerebral ganglia are involved in the critical, causative synaptic events during learning. Also, it is very clear from work in other systems that a few neurons are invested with very dramatic control over the expression of FMP (Gillette et al. 1982; Rose and Benjamin 1981a, b). It will be very interesting to sample the activity of command neurons for feeding in the *Limax* cerebral ganglion before, during, and after associative learning has occurred.

ACKNOWLEDGMENTS

This work was supported by NSF Grant BNS80-05822, NIH Grant NSMH15698, and NIMH Training Grant T32MH15799.

REFERENCES

Barry, S. R., and Gelperin, A. 1982a. Dietary choline augments blood choline and cholinergic transmission in the terrestrial mollusc, *Limax maximus*. *J. Neurophysiol. 48*:451–457.

———. 1982b. Exogenous choline augments transmission at an identified cholinergic synapse in the terrestrial mollusc. *Limax maximus*. *J. Neurophysiol. 48*:439–450.

Bartus, R. T., Dean, R. L., Beer, B., and Lippa, A. S. 1982. The cholinergic hypothesis of geriatric memory dysfunction. *Science 217*:408–417.

Beltz, B., and Gelperin, A. 1980a. Mechanosensory inputs modulate the activity of salivary and feeding neurons in *Limax maximus*. *J. Neurophysiol. 44*:665–674.

———. 1980b. Mechanisms of peripheral modulation of salivary and feeding neurons in *Limax maximus:* A presumptive sensory-motor neuron. *J. Neurophysiol. 44*:675–686.

Chang, J. J., and Gelperin, A. 1980. Rapid taste-aversion learning by an isolated molluscan central nervous system. *Proc. Nat. Acad. Sci. 77*:6204–6206.

Cohen, E. L., and Wurtman, R. J. 1976. Brain acetylcholine: Control by dietary choline. *Science 105*:1039–1040.

Copeland, J. and Gelperin, A. 1983. Feeding and a serotonergic interneuron activate and identified autoactive salivary neuron in *Limax maximus*. *Comp. Biochem. Physiol. 76A*:21–30.

Corkin, S. 1981. Acetylcholine, aging, and Alzheimer's disease. *Trends Neurosci. 4*:287–290.

Cottrell, G. A., and Osborne, N. N. 1970. Serotonin: Subcellular localization in an identified serotonin-containing neuron. *Nature 225*:470–472.

Culligan, N., and Gelperin, A. 1983. One-trial associative learning by an isolated molluscan CNS: Use of different chemoreceptors for training and testing. *Brain Res. 266*:319–327.

Gelperin, A. 1975. Rapid food-aversion learning by a terrestrial mollusk. *Science 189*:567–570.

———. 1981. Synaptic modulation by identified serotonin neurons. In *Serotonin Neurotransmission and Behavior* (B. Jacobs and A. Gelperin, eds), pp. 288–304. Cambridge, Mass.: MIT Press.

Gelperin, A., Chang, J. J., and Reingold, S. C. 1978. Feeding motor program in *Limax*. I. Neuromuscular correlates and control by chemosensory input. *J. Neurobiol. 9*:285–300.

Gelperin, A., and Culligan, N. In press. *In vitro* expression of *in vivo* learning by an isolated molluscan CNS. *Brain Res.*

Gelperin, A., Culligan, N., and Wieland, S. 1981. Associative learning by the isolated CNS of a terrestrial mollusk, *Limax maximus*, occurs centrally. *Soc. Neurosci. Abstr. 7*:353.

Gelperin, A., and Forsythe, D. 1976. Neuroethological studies of learning in mollusks. In *Simpler Networks: An Approach to Patterned Behaviour and Its Foundations* (J. C. Fentress, ed.), pp. 239–246. Sunderland, Mass.: Sinauer Associates.

Gillette, R., Kovac, M. P., and Davis, W. J. 1982. Control of feeding motor output by paracerebral neurons in brain of *Pleurobranchaea californica*. *J. Neurophysiol 47*:885–908.

Goldberg, A. M., and McCaman, R. E. 1974. An enzymatic method for the determination of picomole amounts of choline and acetylcholine. In *Choline and Acetylcholine: Handbook of Chemical Assay Methods* (Hanin, ed.), pp. 47–61. New York: Raven Press.

Haubrich, D. R., Wang, P. F. L., Chippendale, T., and Proctor, E. 1976. Choline and acetycholine in rats: Effect of dietary choline. *J. Neurochem. 27*:1305–1313.

McClellan, A. D. 1982. Re-examination of presumed feeding motor activity in the isolated nervous system of *Pleurobranchaea*. *J. Exp. Biol. 98*:213–228.

Mefford, I. N. 1981. Application of high performance liquid chromatrography with electrochemical detection to neurochemical analysis: Measurement of catecholamines, serotonin, and metabolites in rat brain. *J. Neurosci. Meth. 3*:207–224.

Osborne, N. N., and Cottrell, G. A. 1971. Distribution of biogenic amines in the slug *Limax maximus*. *Z. Zellforsch. 112*:15–30.

Prior, D., and Gelperin, A. 1977. Autoactive molluscan neuron: Reflex function and synaptic modulation during feeding in the terrestrial slug *Limax maximus*. *J. Comp. Physiol. 114*:217–232.

Rose, R. M., and Benjamin, P. R. 1981a. Interneuronal control of feeding in the pond snail *Lymnaea stagnalis*. I. Initiation of feeding cycles by a single buccal interneurone. *J. Exp. Biol. 92*:187–201.

1981b. Interneuronal control of feeding in the pond snail *Lymnaea stagnalis*. II. The interneuronal mechanism generating feeding cycles. *J. Exp. Biol. 92*:203–228.

Reingold, S. C., and Gelperin, A. 1980. Feeding motor program in *Limax*. II. Modulation by sensory inputs in intact animals and isolated central nervous system. *J. Exp. Biol. 85*:1–19.

Sahley, C. L., Feinstein, S. R., and Gelperin, A. 1981a. Dietary choline increases retention of an associative learning task in the terrestrial mollusc, *Limax maximus*. *Soc. Neurosci. Abstr. 7*:353.

Sahley, C. L., Gelperin, A., and Rudy, J. 1981b. One-trial associative learning modifies food odor preferences of a terrestrial mollusc. *Proc. Nat. Acad. Sci. 78*:640–642.

Sahley, C. L., Rudy, J. W., and Gelperin, A. 1981c. An analysis of associative learning in a terrestrial mollusc: Higher-order conditioning, blocking, and a transient US pre-exposure effect. *J. Comp. Physiol. 144*:1–8.

Van Minnen, J., and Sokolove, P. G. 1981. Neurosecretory cells in the central nervous system of the giant garden slug, *Limax maximus*. *J. Neurobiology 12*:297–301.

Van Mol, J. J. 1967. Etude morphologique et phylogénétique du ganglion cérébroïde des Gastéropodes Pulmonés (Mollusques). *Acad. Roy. Belg. Classe Sci. Mém. 37*(5):1–168.

242

Whitehouse, P. J., Price, D. L., Struble, R. G., Clark, A. W., Coyle, J. T., and DeLong, M. R. 1982. Alzheimer's disease and senile dementia: Loss of neurons in the basal forebrain. *Science 215*:1237–1239.

Wieland, S. J., and Gelperin, A. 1983. Dopamine elicits feeding motor program in *Limax maximus. J. Neurosci. 3*:1735–1745.

Wurtman, R. J. 1982. Nutrients that modify brain function. *Sci. Amer. 246*:50–59.

12 · Associative learning in a mollusk: a comparative analysis

CHRISTIE L. SAHLEY, JERRY W. RUDY, and ALAN GELPERIN

INTRODUCTION

Invertebrates, especially mollusks, are being intensively studied to gain insight into the cellular basis of associative learning. In some laboratories the neural circuits involved in behaviors that can be influenced by associative learning paradigms have been identified, and it is likely that some of the important features of the cellular changes induced by associative learning will be identified shortly (see Chapters 8 and 15).

Against the background of the excitement generated by this prospect is a persistent concern that the emerging principles may not generalize to vertebrates. This is related to the possibility that invertebrates and vertebrates diverged so early in evolutionary history that the cellular mechanisms that underlie their learning abilities may be quite different.

Comparative questions about the properties of neurons that mediate vertebrate and invertebrate learning can only be answered definitively by the application of neurobiological methods to cells known to mediate learned changes in behavior. Unfortunately, such comparisons are presently not available, and consequently there is no basis for comparing vertebrate and invertebrate plasticity at the cellular level. It is possible, however, that comparing vertebrates and invertebrates at another level of analysis, the *behavioral level,* might provide insight into the relation of vertebrate and invertebrate learning processes. In this chapter we will consider this behavioral comparison and examine vertebrate and invertebrate associative learning processes for their *functional similarity.*

THEORETICAL RATIONALE: ASSOCIATIVE LEARNING AND INTEREVENT RELATIONS

Associative learning is a theoretical construct, an inference one makes on the basis of an observed relationship between an organism's behavior and its known past experience. To conclude that associative learning has occurred is to say that the organism's behavior is a product of its experience with a relationship between at least two stimulus events. The burden on the researcher is to demonstrate that it is the organism's experience with the *interevent relationship* per se and not some other variable confounded with that relationship that produced the observed change in behavior. Thus, in the most commonly employed procedure for studying associative learning in animals, the Pavlovian conditioning experiment, the experimenter is required to assess the effects of the conditioned stimulus–unconditioned stimulus (CS–US) pairing operation against the effects of a host of control procedures – for example, CS alone, US alone, random presentations of the CS and US – before concluding that associative learning processes were responsible for the ability of the CS to evoke conditioned responding.

Once a behavioral preparation has been shown to satisfy these general requirements, the researcher attempts to characterize the important operating properties of the associative processes involved by systematically varying the interevent relation that was originally observed to produce the learned behavior changes. The study of associative learning then can be viewed as the experimental analysis of the kinds of interevent relations that influence the organism's behavior.

We have deliberately emphasized the importance of interevent relations in the analysis of associative learning. We think that it is at this relatively abstract level that interesting comparisons of the associative learning of different species can be made. The stimulus events that different species can learn about and the behavioral tendencies influenced by learning reflect unique evolutionary histories and are often quite different. This precludes meaningful comparisons based on the specific content (stimuli and behavior) of the experimental analysis. By focusing on the interevent relations necessary for producing learned behavior changes as the important basis for comparison, however, one can ignore differences in the detailed content of specific experimental preparations.

The interesting comparative question one may then ask is: Do the interevent relations that influence the associative learning processes of vertebrates have a similar influence on invertebrate associative learning processes? Data relevant to this question allow one to compare the associative processes of vertebrates and invertebrates for their *functional similarity*. We can determine if formally similar variations in the interevent relations produce functionally similar effects on the behavior of vertebrates and invertebrates.

The remainder of this chapter will present results obtained by exploiting this strategy. First we will describe some of the more interesting interevent relations known to influence associative learning processes of vertebrates. Then we will describe a research program that has examined the influence of formally similar interevent relations on the conditioning of an invertebrate preparation.

SOME IMPORTANT INTEREVENT RELATIONS FOR VERTEBRATES

The study of associative learning in vertebrate animals is largely identified with the Pavlovian conditioning experiment. The subject is exposed to a relationship between two events, the CS and the US, and associative learning is indexed by a change in the subjects behavior to the CS that can be uniquely attributed to its experience with the CS–US relationship. We will describe some of the interevent relations known to influence associative learning in vertebrates, as revealed by the Pavlovian paradigm.

Temporal relations

Our conception of what kind of interevent relations have an important influence on the associative learning of vertebrates has expanded substantially in the last 15 years. The original conception was that the critical variable governing the outcome of a conditioning experiment was solely the temporal relation of the CS and US events (see Mackintosh 1975). In any behavioral preparation, regardless of species, the time interval separating the CS and US exerts a major influence on the probability that the CS will come to evoke a conditioned response. To be sure, depending on the response measure used to index conditioning, the time interval that will support conditioning varies from milliseconds, as is the case for the rabbit's nictitating membrane response (Gormezano 1972), to hours, as is the case for conditioning a taste aversion in rats (e.g., Revusky and Garcia 1970). Nevertheless, in all cases, as this interval increases, the likelihood that conditioning will occur decreases.

Predictive relations

Contemporary researchers, however, have identified a second interevent relationship that determines the outcome of the Pavlovian conditioning experiment: We will term it the *predictive value of the CS*. Not only must the CS be temporally related to the US, it also must predict US occurrence if it is to acquire the ability to evoke conditioned responding. Several classes of experiments force this conclusion (see Rescorla 1968; Wagner 1969; Kamin 1969). We will only mention two: the contingency experiment of Rescorla (1968) and Kamin's blocking experiment. Rescorla varied the predictive value of the CS by systematically altering the degree to which US occurrence was contingent upon CS occurrence. This was done by giving all animals the same number of CS–US pairings but varying the number of extra US presentations they received. When the US is as likely to occur in the absence of the CS as in its presence, then the CS no longer predicts the US, even though it may have a strong temporal relation with the US. Rescorla found that as the predictive value of the CS decreased, so did its ability to evoke conditioned responding.

Kamin (1969) discovered a dramatic example of the importance of the predictive value of the CS for successful Pavlovian conditioning. In his two-phase experiment animals were first conditioned to one CS, termed S1. Once S1 was capable of evoking strong conditioned responding, another CS, termed S2, was introduced, and the S1–S2 compound stimulus was paired with the US. The important result of Kamin's experiment was that the added CS, S2, failed to acquire the ability to evoke conditioned responding. This was true even though it had been repeatedly paired with the US in a temporal relation that was known to support conditioning. Evidently the prior training to the S1 allowed it to block conditioning to the S2.

On the basis of an experimental analysis of this phenomenon, Kamin concluded that the added stimulus, S2, failed to acquire conditioned properties because the information it provided the subject, with respect to the US occurrence, was redundant with that already provided by S1. He suggested that when the US is well predicted by the presence of a stimulus with which it had been previously paired, the associative processes that produce conditioning will not function.

The blocking effect Kamin originally discovered by studying rats in a conditioned emotional response setting has been observed in a variety of different species and tasks (Cheatle and Rudy 1978; Gillan and Domjan 1977; Holland 1977; Marchant and Moore 1973; Tennant and Bitterman 1975; vom Saal and Jenkins 1970).

Higher-order relations

Another set of interevent relations that influence the associative learning of vertebrates can be termed higher-order relations. The defining feature of this class of interevent relations is that the CS acquires the ability to control conditioned responding without ever being directly paired with the US (see Chapter 2). The most familiar instance of conditioned responding produced by a higher-order relation is the phenomenon of second-order conditioning (see Pavlov 1927). In this two-phase experiment, first-order conditioned responding is established to one conditioned stimulus, S1, by pairing it with the US (S1–US pairings). In phase 2 another CS, S2, is introduced and paired with S1 (S2–S1 pairings). The result of this two-phase training procedure is that S2 acquires the ability to evoke conditioned responding even though it was never paired with the US.

Since Pavlov's (1927) original report, second-order conditioning has been observed in a variety of vertebrate conditioning preparations (Amiro and Bitterman 1980; Cheatle and Rudy 1979; Holland and Rescorla 1975; Rizley and Rescorla 1972).

We have described some of the interesting interevent relations that are known to influence the associative processes mediating vertebrate conditioned responding. We now turn to the comparative issue: Do these interevent relations influence the associative learning processes of invertebrates in a functionally similar manner? In the next section of this chapter we will describe results of work addressed to this question.

ASSOCIATIVE LEARNING IN A TERRESTRIAL MOLLUSK, *Limax maximus*

The behavior of many invertebrates can be influenced by associative learning processes (Carew, et al. 1981; Croll and Chase 1980; Crow and Alkon 1978; Davis and Gillette 1978; Walters et al. 1979). Our research, however, has centered on a terrestrial mollusk, *Limax maximus,* a large slug (Sahley et al. 1981a, b). *Limax* is a generalized herbivore whose feeding behavior is dominated by two classes of stimuli, olfactory and gustatory. It is attracted to potential food sources by their odors, which it detects with olfactory receptors located on its superior tentacles. When a potential food source is located, *Limax* everts its lips so that the gustatory receptors contact the food surface. If a bitter taste is encountered, the food is rejected.

The starting point of our research was an appreciation that the natural odor–taste sequence encountered by the foraging slug provides an opportunity for associative learning processes to contribute adaptively to its food selection behavior. If, for example, *Limax* could associate the independently sensed odor and taste properties of various potential food sources, it might forage more economically by avoiding odors that it had learned signal unacceptable tastes.

The behavioral preparation

This analysis suggested that a laboratory analog of the slug's natural foraging activity might be a useful preparation to study associative learning in *Limax*. The behavioral preparation we developed was simple. Slugs were exposed to normally preferred food odors (either potato or carrot odor) for 2 min and then their lips were placed in contact with a bitter taste, a saturated solution of quinidine sulfate.

After slugs were exposed to an odor–quinidine pairing, their preference for that odor was assessed (usually about 24 hr after training) in a two-choice test procedure. The slug was placed in the test apparatus represented schematically in Figure 12.1, where it was presented two food odors, one (carrot or potato odor) that had been paired with quinidine, and another, generated by laboratory rat chow, that had been its normal food source. From a series of three 60-sec tests we calculated the percentage of total test time (180 sec) that the slug spent over the odor paired with quinidine, and this metric was the dependent variable in all of our experiments. Further details of experimental procedures can be found in Sahley et al. (1981a, b).

A demonstration that associative processes can contribute to the food odor preferences of *Limax* is illustrated by the following experiment (Sahley et al. 1981a). This experiment compared four training conditions. Two experimental groups, PQ and CQ, experienced either a potato odor–quinidine pairing (PQ) or a carrot odor–quinidine pairing (CQ), respectively. Two control groups, PS and CS, experienced potato odor or carrot odor paired with *Limax* saline. All animals were then tested twice. One test assessed their preference for potato odor versus rat chow odor, the other assessed their preference for carrot odor versus rat chow odor.

The results of this experiment, displayed in Figure 12.2, can be easily summarized. Animals that experienced the potato odor–quinidine pairing, group PQ, displayed a reduced preference for potato odor in comparison to all other treatment conditions. They did not, however, display a reduced preference for carrot odor. Animals that experienced the carrot odor–quinidine pairing displayed the symmetrically opposite result, a reduced preference for carrot odor but not for potato odor.

This experiment compels the conclusion that the odor preference behavior of *Limax* can be influenced by associative learning processes. Neither the odor experience nor the quinidine experience alone was sufficient to reduce an odor preference. Slugs only displayed reduced preference for odors that had been *paired* with quinidine. Thus the interevent relation per se was the critical variable that produced the behavior change.

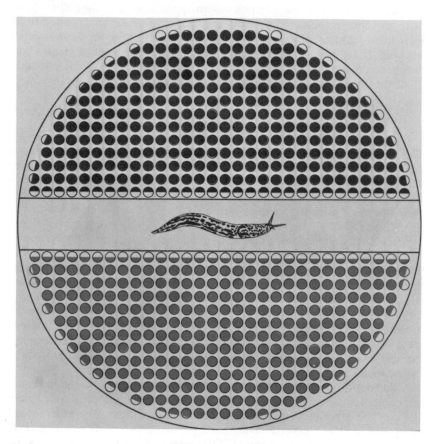

Figure 12.1. Schematic representation of the test apparatus. The slug is in the "neutral zone" that is bordered on each side by two different odor-generating food sources that are directly under the perforated floor of the chamber.

In addition to its value in identifying the involvement of associative learning in the food odor preferences of *Limax*, this experiment revealed a feature of that learning that vastly increases the analytical power of this procedure: The learned changes in *Limax*'s food odor preferences were *selective*. Conditioned reductions in *Limax*'s preference for potato odor did not generalize to carrot odor and *vice versa*. Thus, we have two stimulus events that can serve as conditioned stimuli. This makes it possible to explore some of the more interesting predictive and higher-order interevent relations known to influence vertebrate associative learning (see preceding section "Some Important Interevent Relations for Vertebrates") for their influence on the associative processes of an invertebrate.

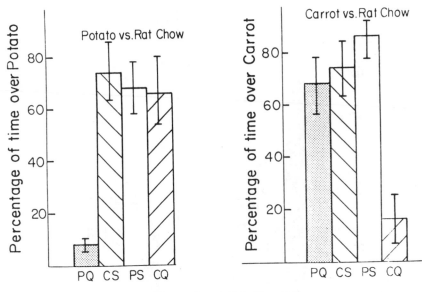

Experimental Condition

Figure 12.2. Mean percentage of time spent over potato odor (left panel) or carrot odor (right panel). P, potato; S, saline; C, carrot; Q, quinidine sulfate. Bars = standard error of the mean.

SOME IMPORTANT INTEREVENT RELATIONS FOR *Limax*

The predictive value of the CS

We have documented the importance of the predictive value of the CS for vertebrate associative learning. In this section we will present evidence of its importance to the associative processes mediating *Limax*'s food odor preferences. This experiment, reported by Sahley et al. (1981b), was modeled after Kamin's (1969) blocking paradigm. Kamin observed that prior conditioning to one element, S1, of a compound stimulus, S1–S2, blocked or prevented conditioning to the second element, S2.

Slugs in the blocking condition, group B, first received three pairings of carrot odor and quinidine (S1–US pairings). In phase 2 they received three pairings of a compound stimulus, composed of carrot and potato odor, and quinidine (S1-S2–US pairings). The empirical question was: Would the prior conditioning to carrot odor (S1) block conditioning to the potato odor (S2)? If this result was observed, we wanted to be able to determine if potato odor (S2) failed to acquire conditioned properties because the prior conditioning to carrot odor (S1) reduced the predictive value of potato odor. This requires one to demonstrate that (1) the phase 1 S1–US pairings, and (2) the presence of S1

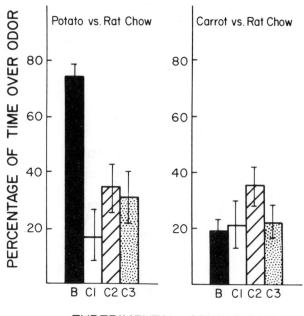

EXPERIMENTAL CONDITIONS

Figure 12.3. Mean percentage of time spent over potato odor (left panel) or carrot odor (right panel). B, blocking group; C1, control group 1; C2, control group 2; C3, control group 3. Slugs in the blocking group (B) failed to condition to the potato odor (S2) component of the compound stimulus as compared to slugs in the three control groups. Slugs in all groups demonstrated conditioning to the carrot odor. Bars = standard error of the mean.

on the phase 2 conditioning trials are *both* necessary conditions for blocking to occur. Thus, we had to include several control procedures in the experiment. To assess the importance of the phase 1 S1–US pairings we included two groups, group C1, that only received the phase 2 compound conditioning trials, and group C2, that also received the phase 2 trials but in phase 1 received backward pairings of quinidine and carrot odor (S1). To assess the importance of the presence of preconditioned S1 during phase 2 trials, slugs in group C3 experienced the phase 1 carrot odor–quinidine pairings of phase 1, but in phase 2 these slugs only experienced potato odor–quinidine pairings; carrot odor (S1) was not presented.

The results of the potato odor (S2) versus rat chow odor preference test are displayed in Figure 12.3. It reveals a powerful blocking effect. Slugs in group B displayed a strong preference for potato odor in comparison to slugs in the three control groups. To appreciate the implications of this finding, the reader should recall that the slugs in group B received the same number of potato odor–quinidine pairings as did the slugs in the control groups. Temporally relating potato odor and quinidine thus was not a sufficient condition for producing associative learning. What was important was the predictive value of potato odor. When its predictive value was reduced by the presence of the pre-conditioned carrot odor, it was not associated with the quinidine. This is the same outcome Kamin reported in rats.

Higher-order conditioning

We have demonstrated that learned changes in the odor preferences of *Limax* are markedly influenced by the predictive value of the CS. In this section we will present evidence that the associative processes of *Limax* can be influenced by the higher-order interevent relations analogous to those that produce sec-ond-order conditioning in vertebrates. In addition, we will analyze the nature of the associations that mediate the second-order conditioned behavioral reac-tions of *Limax* with the methods Rescorla (see Rescorla 1980) has successfully applied to vertebrate second-order conditioning.

To arrange the conditions that produce higher-order conditioning in verte-brates, we (Sahley et al. 1981b) exposed slugs to two phases of training. In phase 1 slugs in group SOC experienced three pairings of carrot odor and quin-idine (S1–US pairings) and in phase 2 they received three pairings of potato odor and carrot odor (S2–S1 pairings). It is important to note that in phase 2 of this experiment, potato and carrot odor were presented to the slug simulta-neously. The empirical question was: Would this training reduce the slug's preference for potato odor (S2), even though it was never directly paired with quinidine? To conclude that such a result is truly second-order conditioning requires one to also demonstrate that it depends on the animal experiencing *both* the phase 1 and phase 2 pairings (e.g., Rizley and Rescorla 1972). This requires that the treatment given group SOC be evaluated against two control groups: one that receives the phase 1 S1–US pairing but not the phase 2 S2–S1 pairing, and the other that receives the phase 2 pairing but not the phase 1 pairing. We included two such control groups in our experiment. Group PU received the phase 1 carrot odor–quinidine pairings but in phase 2 the carrot and potato odor were unpaired. Group UP received unpaired presentations of carrot odor and quinidine in phase 1 and paired presentations of potato and carrot odor in phase 2. Following training all slugs were tested for first-order conditioning to carrot odor (S1) and second-order conditioning to potato odor (S2).

The results of this experiment are presented in Figure 12.4. The test data for first-order conditioning to carrot odor are displayed in the right panel. Note that the two groups that experienced pairings of carrot odor and quinidine, groups SOC and PU, displayed a reduced preference for that odor in compar-ison to slugs that received unpaired presentations of carrot odor and quinidine, group UP.

The important result is presented in the right panel. Note that slugs in group SOC displayed a reduced preference for potato odor in comparison to the slugs in the control groups, groups PU and UP. Thus we have demonstrated higher-order conditioning in an invertebrate. Even though potato odor was never directly paired with the quinidine US, the slugs' preference for it was reduced,

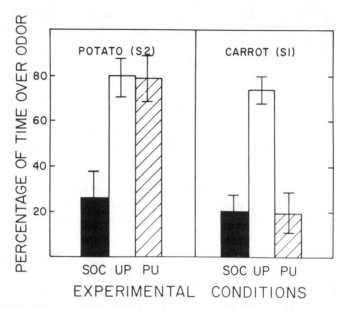

Figure 12.4. Mean percentage of time spent over potato odor (left panel) or carrot odor (right panel). SOC, second-order conditioning group; UP, unpaired–paired group; PU, paired–unpaired group. Slugs receiving SOC training spent significantly less time over potato (S2) odor than slugs in either of the control groups ($p < .01$). Slugs receiving carrot–quinidine pairings in phase 1 (groups SOC and PU) demonstrated first-order conditioning, whereas slugs receiving unpaired presentations of carrot and quinidine showed no first-order conditioning.

and this reduction depended on their experiencing both the phase 1 and phase 2 pairing operations.

ASSOCIATIVE BASIS OF SECOND-ORDER CONDITIONING

Theoretical issues

Rescorla and his colleagues (Holland and Rescorla 1975; Rescorla 1973, 1980; Rizley and Rescorla 1972) have examined the associations mediating second-order conditioning in vertebrates. Rescorla (Chapter 2) discusses the issues involved in considerable detail. During phase 2 of a second-order conditioning experiment (S2–S1 pairings), there are several associations that the subject can learn that could mediate a conditioned reaction to S2. One possibility is an *S2–S1 association.* In this case it is assumed that the subject associates representations of the two CS events. For example, in our study this would imply that the slugs associated representations activated by the potato odor (S2) and carrot odor (S1) experience. According to this view, the second-order conditioned response to potato odor (S2) would be mediated by a two-link chain of associations. First, potato odor (S2) would activate the slugs' representation of carrot odor (S1), which in turn, by virtue of its first-order association with quinidine, would evoke the conditioned response.

A second alternative is the *S2–CR association*. This account assumes that during the phase 2 S2–S1 pairings, S1, by virtue of its previous pairings with the quinidine US, evokes a conditioned response (CR). This allows for the possibility that the subject will associate S2 with the CR activated by the S1. For example, in our experiment, this would mean that the slugs associated potato odor (S2) with the conditioned response that carrot odor (S1) activated. The second-order response to potato odor (S2) during testing would then be directly evoked by potato odor and not indirectly through carrot odor.

Although this analysis is abstract, there is a way of experimentally distinguishing between these two classes of interpretation. The distinguishing experiment makes use of what Rescorla (1975) terms a *postconditioning treatment* strategy. Consider the so-called *S2–S1 association* account. It implies that conditioned responding to S2 is *dependent* upon the ability of S1 to evoke a conditioned response. In contrast, the *S2–CR association* view implies no such dependency. Now consider the effect of a particular postconditioning treatment. Suppose that following the two phases of training necessary to produce second-order conditioning, but prior to testing for second-order conditioning with S2, S1's ability to evoke a first-order conditioned reaction is extinguished (S1 extinction treatment). The *S1–S2 association* view would predict that the S1 extinction treatment would eliminate S2's ability to evoke a second-order conditioned response, because S2's ability depends on the integrity of S1's first-order conditioning properties. In contrast, the *S2–CR association* view predicts that the S1 extinction treatment would not influence S2's ability to evoke a conditioned response, because S2 is assumed to be able to directly evoke conditioned responding.

As Rescorla (1980, 1982; Chapter 2) points out, the implications of these conceptions of the associations involved in second-order conditioning have been widely investigated in vertebrates. Moreover, evidence consistent with both views has been reported. That is, in some instances the S1 extinction procedure has eliminated second-order conditioning (e.g., Cheatle and Rudy 1979; Rizley and Rescorla 1972) and in other instances has not (Rashotte et al. 1977).

Recently some of the variables that determine whether vertebrates learn S2–S1 associations or S2–CR associations have been identified (see Cheatle and Rudy 1978, 1979; Chapter 2). What appears to be especially important is the manner in which S2 and S1 are presented during phase 2. When they are presented *simultaneously* (S2 + S1) the acquisition of S2–S1 associations is favored. When they are presented *sequentially* (S2, S1), acquisition of S2–CR associations is favored.

Experimental analysis

We have recently completed experiments designed to analyze the associations that contribute to the second-order conditioned behavior of *Limax* displays. We sought to determine if the vertebrate generalization – that simultaneous S2 + S1 presentations promote S2–S1 associations, and sequential S2, S1 presentations promote S2–CR associations – also applies to *Limax*.

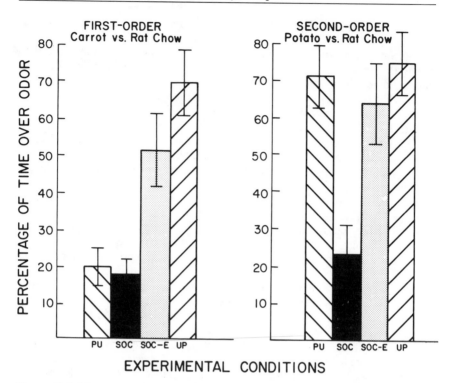

Figure 12.5. Mean percentage of time spent over carrot odor, first-order stimulus, and potato odor, second-odor stimulus (right panel) for slugs experiencing simultaneous presentations of the S1 and S2 during second-order training. Bars = standard error of the mean.

One experiment was designed to ask what associations are learned by slugs exposed to *simultaneous* phase 2 pairings of potato (S2) and carrot (S1) odor. Three of the groups, SOC, PU, and UP, were treated just like the similarly designated groups of the previous experiment. A fourth group, SOC-E, was treated exactly like group SOC, but prior to testing these slugs received a post-conditioning S1 extinction treatment. They were repeatedly exposed to non-reinforced presentations of carrot odor (S1).

The results of this experiment are presented in Figure 12.5. First, it can be seen (left panel) that the S1 extinction treatment administered to slugs in group SOC-E eliminated their first-order conditioned reaction to carrot odor. Note that compared to slugs in both groups PU and SOC, slugs in group SOC-E spent more time over carrot odor. Slugs in group SOC-E, in fact, did not differ from those in group UP that never received the carrot odor–quinidine pairings.

The influence of the S1 extinction treatment on the slugs' second-order reaction to potato odor is displayed in the right panel of Figure 12.5. It can be seen that the S1 extinction treatment also *eliminated* the slugs' second-order reaction. Slugs in group SOC displayed a substantially reduced potato odor (S2) preference in comparison to control slugs in groups PU and UP. In contrast, the potato odor preference of slugs in group SOC-E did not differ from that of control slugs. Both group SOC-E and control slugs preferred potato odor, in comparison to slugs in group SOC.

The second-order reaction to potato odor by slugs exposed to *simultaneous* phase 2 pairings of two odors was *dependent* on the integrity of their first-order reaction to carrot odor. That is to say, extinguishing the first-order reaction also extinguished the second-order reaction. This outcome strongly suggests that the second-order reaction produced by simultaneous S2–S1 pairings was mediated by the slug, in some sense, by associating representations of the two odors, that is, an interodor *S2–S1 association.*

Our second experiment was designed to assess what associations are learned when slugs are exposed to *sequential* pairings of potato (S2) and carrot (S1) odor. It was similar to the previous one, except that during phase 2 the slugs in groups UP, SOC, and SOC-E all received sequential presentations of potato odor (S2) followed immediately by a carrot odor (S1).

The results of this experiment are presented in Figure 12.6. As was the case in the previous experiment, the S1 extinction treatment reduced the first-order reaction to carrot odor (S1). Slugs in group SOC-E spent significantly more time over carrot odor than did slugs in either group SOC or PU that also had received carrot odor–quinidine pairings in phase 1 (see left panel).

The S1 extinction treatment, however, did *not* have a detectable effect on the slugs' second-order reaction to potato odor (S2). Note that slugs in group SOC-E did not differ from slugs in group SOC on the potato preference test (see right panel of Figure 12.6) and that both of these groups displayed reduced preferences for potato odor, in comparison to control slugs in groups PU and UP.

The second-order reaction to potato odor (S2) by slugs exposed to *sequential* pairings of potato and carrot odor evidently was *independent* of their first-order reaction to carrot odor. The S1 extinction procedure eliminated the first-order reaction but had no effect on the second-order reaction. This result provides evidence in support of the so-called *S2–CR association* view. Thus, slugs evidently associated potato odor with some component of the conditioned response complex evoked by carrot odor (S1) during the phase 2 potato–carrot odor pairings.

These two experiments lead to remarkable conclusions: (1) The second-order conditioned change in *Limax*'s odor preference behavior evidently can be mediated by alternative associative structures; and (2) *simultaneous* S2–S1 presentations favor the contribution of an *S2–S1 association,* and *sequential* presentations favor the contribution of an *S2–CR association.* The similarity of these conclusions to those that apply to second-order conditioned reactions of vertebrates is striking.

CONCLUSIONS

The experiments that we have described were designed to probe the associative learning capabilities of an invertebrate using *interevent relations* formally sim-

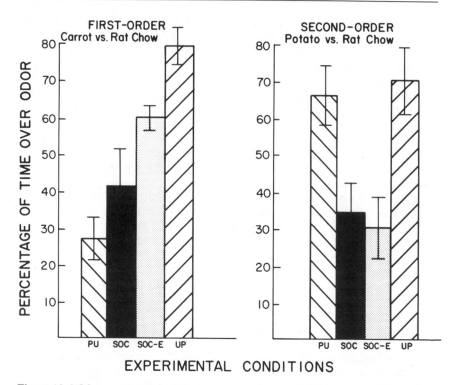

Figure 12.6. Mean percentage of time spent over carrot odor, first-order stimulus (left panel) and potato odor, second-order stimulus (right panel) for slugs experiencing sequential presentations of the S1 and S2 during second-order training. Bars = standard error of the mean.

ilar to those that have a known influence on the associative learning processes of vertebrates. The results of this comparative exercise were instructive. They revealed that the characteristics of the associative processes mediating conditioned changes in the odor preference behavior of *Limax* are *functionally similar* to those that mediate the effects of Pavlovian conditioning on vertebrates.

First, we observed that conditioned changes in the odor preference behavior of *Limax* could be *blocked* by the same interevent relations that Kamin discovered block conditioning in vertebrates. Second, we observed that the higher-order interevent relations that produce second-order conditioning in vertebrates also produce second-order conditioned changes in *Limax*'s odor preference behavior. Third, just as the second-order conditioned behavior of vertebrates can be mediated by alternative associative structures, termed the *S2–S1 association* and *S2–CR association,* so can the second-order conditioned change in *Limax*'s odor preference behavior. Finally, we observed that the interevent relations that favor the acquisition of the so-called *S2–S1 association* and *S2–CR association* by vertebrates, simultaneous and sequential pairings, respectively, also similarly influence what associations contribute to the second-order conditioned behavior of *Limax*.

The interevent relations that we have examined thus influence associative learning of *Limax* in ways that could be predicted by their known influence on vertebrates. At some level, therefore, it would appear that the associative processes of at least one molluscan invertebrate are remarkably similar to those of vertebrates. The learning processes of a mollusk can be influenced by a more interesting range of interevent relations and have a richer potential for association than we previously imagined. Moreover, one cannot help but be impressed that, when the study of associative learning is identified as the experimental analysis of the interevent relations that influence an organism's behavior, a set of learning principles begin to emerge that have considerable generality both within and across species.

Whether these common principles are mediated by homologous or analogous cellular structures and processes, of course, is still an open question. Our data certainly leave open the possibility that there may be important similarities in the cellular basis of vertebrate and invertebrate associative learning processes.

REFERENCES

Amiro, T. W., and Bitterman, M. E. 1980. Second-order conditioning in goldfish. *J. Exp. Psychol. (Anim. Behav.)* 6:41–48.

Carew, T. J., Walters, E. T., and Kandel, E. R. 1981. Classical conditioning in a simple withdrawal reflex in *Aplysia californica*. *J. Neurosci.* 1:1426–1437.

Cheatle, M. D., and Rudy, J. W. 1978. Analysis of second-order odor-aversion conditioning in neonatal rats: Implications for Kamin's blocking effect. *J. Exp. Psychol. (Anim. Behav.)* 4:237–249.

1979. Ontogeny of second-order odor-aversion conditioning in neonatal rats. *J. Exp. Psychol. (Anim. Behav.)* 5:142–151.

Croll, R., and Chase, R. 1980. Plasticity of olfactory orientation to foods in the snail *Achatina fulica. J. Comp. Physiol.* 136:267–277.

Crow, T. J., and Alkon, D. L. 1978. Retention of an associative behavioral change in *Hermissenda. Science* 201:1239–1241.

Davis, W. J., and Gillette, R. 1978, Neural correlates of behavioral plasticity in command neurons of *Pleurobranchaea. Science* 199:801–804.

Gillan, D. J., and Domjan, M. 1977. Taste aversion conditioning with expected versus unexpected drug treatment. *J. Exp. Psychol. (Anim. Behav.)* 3:297–309.

Gormezano, I. 1972. Investigations of defense and reward conditioning in the rabbit. In *Classical Conditioning*, vol. 2 (A. H. Black and W. F. Prokasy, eds.) pp. 151–181. New York: Appleton-Century-Crofts.

Holland, P. C. 1977. Conditioned stimulus as a determinant of the form of the Pavlovian conditioned response. *J. Exp. Psychol. (Anim. Behav.)* 3:77–104.

Holland, P. C., and Rescorla, R. A. 1975. Second-order conditioning with food unconditioned stimulus. *J. Comp. Psychol.* 88:459–467.

Kamin, L. J. 1969. Predictability, surprise, attention, and conditioning. In *Punishment and Aversive Behavior* (R. Church and B. A. Campbell, eds.), pp. 279–296. New York: Appleton-Century-Crofts.

Mackintosh, N. J. 1975. *The Psychology of Animal Learning.* New York: Academic Press.

Marchant, H. G., and Moore, J. W. 1973. Blocking of the rabbit's conditioned nictitating membrane response in Kamin's two-stage paradigm. *J. Exp. Psychol.*
101:155–158.

Pavlov, I. P. 1927. *Conditioned Reflexes.* London: Oxford University Press.

Rashotte, M. E., Griffin, R. W., and Sisk, C. L. 1977. Second-order conditioning of
the pigeon's key peck. *Anim. Learn. Behav. 5*:25–38.

Rescorla, R. A. 1968. Probability of shock in the presence and absence of the CS in
fear conditioning. *J. Comp. Physiol. Psychol. 66*:1–5.

1973. Second-order conditioning: Implications for theories of learning. In *Contem-*
porary Approaches to Conditioning and Learning (F. J. McGuigan and D. B.
Lamsden, eds.), pp. 7–33. Washington, D.C.: Winston.

1980. *Pavlovian Second-Order Conditioning: Studies in Associative Learning.* Hillsdale, N.J.: Erlbaum.

1982. Simultaneous second-order conditioning produces S–S learning in conditioned suppression. *J. Exp. Psychol. (Anim. Behav.) 8*:23–32.

Revusky, S., and Garcia, J. 1970. Learned associations over long delays. In *The Psy-*
chology of Learning and Motivation, vol. 4 (G. H. Bower, ed.), pp. 1–84. New
York: Academic Press.

Rizley, R. C., and Rescorla, R. A. 1972. Associations in second-order conditioning
and sensory preconditioning. *J. Comp. Physiol. Psychol. 81*:1–11.

Sahley, C. L., Gelperin, A., and Rudy, J. W. 1981a. One-trial associative learning in
a terrestrial mollusc. *Proc. Nat. Acad. Sci. 78*:640–642.

Sahley, C. L., Rudy, J. W., and Gelperin, A. 1981b. An analysis of associative learning in a terrestrial mollusc. I. Higher-order conditioning, blocking, and a transient US pre-exposure effect. *J. Comp. Physiol. 144*:1–8.

Tennant, W. A., and Bitterman, M. E. 1975. Blocking and overshadowing in two species of fish. *J. Exp. Psychol. (Anim. Behav.) 104*:22–29.

vom Saal, W., and Jenkins, H. M. 1970. Blocking the development of stimulus control.
Learn. Motiv. 1:52–62.

Wagner, A. R. 1969. Stimulus validity and stimulus selection in associative learning.
In *Fundamental Issues in Associative Learning* (N. J. Mackintosh and W. K.
Honig, eds.), pp. 90–122. Halifax: Dalhousie University Press.

Walters, E. T., Carew, T. J., and Kandel, E. R. 1979. Associative learning in *Aplysia:*
Evidence for conditioned fear in an invertebrate. *Science 211*:504–506.

13 · Short-term memory in bees

RANDOLF MENZEL

Memory as a result of associative learning is not established in its final form immediately after an association has taken place, but needs time to develop. During this time, the hypothetical memory trace changes its properties, and these changes relate to measurable behavioral patterns. For example, the retrievability of the sensory signals involved differs in tests made shortly after learning or later; the effect of newly learned information or the frequency of repetition of the same learning trial changes over time following the initial learning; forgetting has time courses that are thought to reflect different physiological states of the hypothetical memory trace.

The study of the time dependence of memory has been burdened by the controversy whether there is a dichotomy of learning phases (short-term vs. long-term memory) or whether memory formation is a continuous process leading to long-term memory without obvious time phases. I want to stress the point that even if one does not accept the notion of short-term and long-term memory, the undoubtedly different qualities of memory at various times after learning has to be explained. Furthermore, the time dependence of learning and retrieval may be a powerful tool in behavioral studies of learning and may help to characterize the physiological basis of the hypothetical memory trace.

Evidence for a hypothetical short-term processing of information comes from three major sources: human verbal memory studies, studies of experimentally induced amnesia, and chronic disturbances of memory in humans as they appear in psychiatric clinics. It is well documented by now that both for humans and animals the time phases and the properties of the memories for the different paradigms do not fit together at all (e.g., Weiskrantz 1970). Even within one class of experiments as, for example, in studies of induced amnesia, and for just one animal species, the time course of the short-term process may depend on seemingly small differences in procedure. It has been argued, therefore, that the actual time course of short-term processing and storage has very little or nothing to do with basic physiological mechanisms underlying the establishment of a long-term memory trace. Indeed, a hypothesis that correlates the temporal phases of memory directly with different substrates of neuronal activity is much too simplistic for most of the learning tasks used so far to characterize memory phases. Nevertheless, the fact remains that memory formation takes time and has qualitatively different properties at various times after learning. Although there is large variation in experimental results, this seems to be a general feature of associative learning processes, at least those that involve relatively complicated motor responses.

Interestingly, there are very few examples of temporal analysis of the retention of a simple conditioned response. If it is true that the temporal phases of memory formation are the result of retrieval mechanisms reflecting mainly changes in motivation and/or attention but not physiological conditions of the memory trace, one should find less obvious changes and no temporal phases in simple conditioned reflexes. I want to apply an even more radical approach. By selecting an insect, we want to reduce the complexity of behavioral studies and prepare for a neural analysis. In addition, if there is any evidence for a short-term phase of memory, the comparison of its properties with those of short-term memory in mammals should be very interesting. From this point of view we ought to examine especially (1) the time course of retention, (2) sensitivity to interference with newly learned information, (3) storage capacity, (4) modes of transfer to a long-term memory, and (5) localization of a labile, short-term memory in the brain.

BIOLOGICAL BACKGROUND

Honeybees learn quickly and efficiently to use external signals as reference marks on their food-collecting flights. On the first flight from the hive, they learn the location of the colony with respect to the surrounding landmarks and relative to the sun compass. They learn the color, shape, and odor of the hive entrance. On lengthening excursions during the following several days, they learn landmarks further away and use them for relocating the colony and for adjusting their astronomical compass. When a forager bee flies out to collect nectar and pollen from flowers, it learns the direction and distance of the food source, the landmarks that guide it toward the food source, and the signals of the food source itself: color, shape, odor, and distance from immediate surrounding landmarks (see Frisch 1967). These signals of the food source are learned very quickly: natural odors within 1 trial (Lindauer 1970; Koltermann 1973; Lauer and Lindauer 1973), colors within 1–5 trials depending on the color (Menzel 1968), black and white patterns within 5–20 trials (Wehner & Lindauer 1966, Wehner 1981). The honeybee is able to select the learned food source with extreme accuracy, because each bee is continuously informed about the food supply in the surroundings by the dances of its hivemates and the food spreading within the colony. This means that the individual bee does not have continuously to explore to find out if it is still working (i.e., collecting food) in the most economical way, as do other flower-visiting insects. Both under natural conditions and in behavioral tests, bees choose a food source out of two or more alternatives with more than 90 percent accuracy. Pollen-collecting bees, for example, carry a protocol of their choice behavior with them. Less than 10 percent of the pollen load contains more than one kind of pollen grains, proving that bees select one flower out of many with high accuracy.

Besides the rapidity of their learning and the accuracy of their choices, bees offer other advantages for behavioral tests: (1) low genetic variation, because all test animals are sisters within the same colony; (2) test animals are about equal in age and in the same behavioral status (forager); (3) they arrive at the training station in a highly motivated state; (4) new and naive test animals may be recruited by the dances of previously trained test animals; (5) nonreinforcement in the tests has little if any effect on choice behaviors if the tests are not continued too long; (6) one animal can easily make 2000 choices a day or more than 30,000 in its lifetime.

TRAINING PROCEDURES

Karl von Frisch and his coworkers have worked out methods of training individually marked bees to come to a feeding station and of measuring their choice behavior in test situations (see Frisch 1967). In our experiments, the procedure was simplified by (1) working always with a single test bee at a time, and (2) testing the choice behavior of the bee with only two alternatives, one being the learned signal, the other an alternative signal of the same modality. The experiment began with the training of a group of five to eight bees to fly from the hive to the experimental situation by slowly moving a feeding station step by step over the 50- to 120-m distance. The experimental situation is a round table under which there is a light source that projects spectral colors on three ground glass disks on the surface of the table. The ground glass in the center contains sugar water and is visible to the bees during training only; the two others, equally distant from the center, have no sugar solution and are displayed only during tests. The group of marked bees recruit newcomers, each of which is then marked as a test bee and rewarded three times on the unilluminated ground glass. After each reward, the bee returns to its colony. During this period of time, the group of recruiting bees is captured and kept in a box until the experiment with the test bee is finished. During the initial three rewards (pretraining), the test bee becomes familiar with the distance and location of the feeding station and is motivated to search for food in the center of the table and the immediate surroundings. After the pretraining, the test bee views the two colors to be used as discriminative learning signals and is asked about its preference in a test of spontaneous choice. Spectral colors of equal brightness for the bee are used (Menzel 1967). After the 4-min test, the bee is rewarded on one of the colors (λ_+). The alternative color λ_- in the test is the complementary color in all experiments described below. An acquisition function (Figure 13.1) is plotted by inserting a 4-min test period before each of a succession of rewards. Acquisition depends heavily on λ_+, not at all on λ_- if the two colors are discriminable for the bee, and very little on the brightness of the colors if brightness is at least half a log unit above the threshold of the colors (Menzel 1967). Furthermore, acquisition in the first six trials is independent of the amount of sugar reward (Menzel and Erber 1972), and the same acquisition function is found if bees are rewarded several times during one visit, with each landing and sucking counted as a separate reward (Menzel 1968). The acquisition functions mark the genetic boundaries of learning in bees (Menzel et al. 1974). Violet as a food signal is learned fastest, bluish green slowest. Blue (444 nm), the color signal used in most of the experiments described below, is chosen about 75 percent of the time after *one reward*.

A different conditioning procedure was introduced by Kuwabara (1957) and

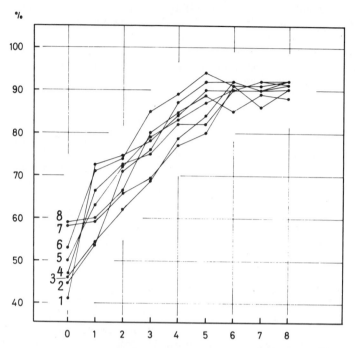

Figure 13.1. Acquisition functions for eight individual bees. The abscissa gives the number of rewards on λ_+, the rewarded color (532 nm). Choice reaction is expressed in percentage of responses to λ_+ relative to the responses to λ_-, the alternative color (413 nm) during the 4-min test situation. The 0 on the abscissa indicates the spontaneous choice test (see text).

has turned out to be extremely useful in studying the physiological basis of learning in bees. Foragers caught at the hive entrance are mounted individually in small tubes (Figure 13.2). Touching the antennae with sugar solution releases a reflex motor program in which both antennae are coordinately directed forward, the mandibles are opened, and the proboscis is extended. This motor pattern can be conditioned to an olfactory stimulus as a conditioned stimulus (CS) (Vareschi 1971; Masuhr and Menzel 1972; Menzel et al. 1974). If a flowery odor is used as the CS, a single conditioning trial changes the response level from about 10 percent (spontaneous response rate) to ≥85 percent. Color signals as CSs are much less effective (Kuwabara 1957; Masuhr and Menzel 1972). Only a third of the bees can be conditioned, and acquisition is very slow.

   The true associative nature of the learning process has been demonstrated in terms of all the usual criteria, including stimulus specificity, the role of repetition, the necessity for temporal association between the stimuli and the motor action involved, and long retention (Menzel et al. 1974; Bitterman et al. 1983). For our purposes here, it is important to bear in mind that a single learning trial produces ≥85 percent correct choice in an odor conditioning

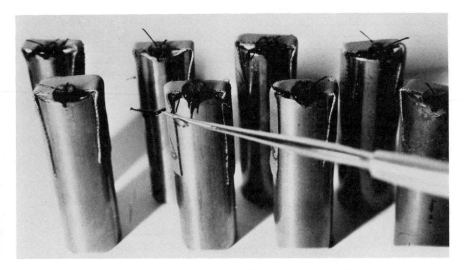

Figure 13.2. Honeybees mounted for the conditioning of the proboscis reflex (Kuwabara preparation). One bee is sucking a sugar solution. Note the positions of the antennae of the sucking bee and of the other bees.

experiment when one of two floral-type odors is rewarded and to about 75 percent in a color conditioning experiment with, for example, blue rewarded and yellow as the alternative. This highly significant change in behavior occurs both in free-flying bees and in bees glued to a stage for conditioning of proboscis reflex with an odor as the conditioned stimulus. In all of the experiments discussed in the following, this *one-trial learning* paradigm was used.

SENSORY MEMORY (STIMULUS TRACE OF CS)

The first experiment demonstrates the existence of a stimulus trace of CS and also includes a control for contingency of CS and unconditioned stimulus/ unconditioned response (US/UR) (Figure 13.3) (Menzel 1968). A free-flying bee approaches a spot where it has been fed on several previous occasions. For the first time, the feeding place is illuminated with blue (444-nm) or yellow (590-nm) light, which is switched off 10 sec or 4 sec or 2 sec before the bee lands and takes the reward (left side of Figure 13.3). In another series of experiments (right side of Figure 13.3) the bee approaches the unilluminated feeding place and the color is switched on at the moment of landing, or 2, 4, or 10 sec later. In all cases, the bee sucks sugar water for 30 sec and then flies back to the hive. When the bee returns from the hive, its choice behavior is measured with the two colors (no reward during the test) that were equally preferred if no training is given. The four curves in Figure 13.3 are averages from many individually tested bees. The results show that color is associated with the sugar

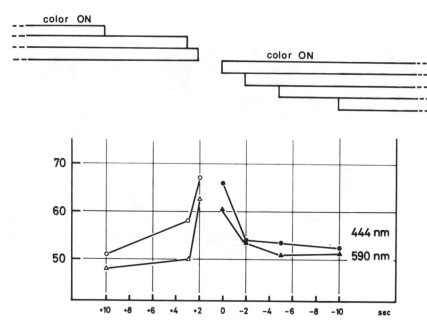

Figure 13.3. Two series of experiments that prove the existence of a stimulus trace of CS. Left side (upper bars and open symbols): The color (blue 444 nm or yellow 590 nm) is seen by the approaching bee and switched off 10, 3, or 2 sec before landing and beginning of sucking (+10, +3, +2 sec). This was achieved by observing an approaching bee and switching off the color at some arbitrary time. The interval between this moment and the landing was measured, and the bees were pooled in three groups (12 to 7 sec for the 10-sec group, 4 to 2.5 sec for the 3-sec group, 2.5 to 1.5 sec for the 2-sec group). The bee was fed sugar water for 30 sec, and its choice behavior was tested when it came back from the hive. Right side: The bees approached an unilluminated disk. The color was switched on immediately at 2, 5, or 10 sec after landing and beginning of sucking. After termination of sucking the bees viewed the color on their flight off to the hive. As in the other tests, the bees were tested after their return from the hive.

water reward even if it is not seen by the bee for 2–3 sec before reward. A color that is shown only after landing and beginning of feeding is not learned. We conclude from this kind of experiment that (1) bees associate a color as a food signal only when they see it during approach and at the very beginning of feeding – the same was found by Opfinger (1931) and Grossmann (1970); and (2) bees have a sensory memory for color in the range of 2–3 sec.

In other experiments it has been shown that even a very brief reward (as little as a fraction of a second) is sufficient to produce a highly significant change in the choice behavior (Menzel 1968; Erber 1975a). This means that the process of association can go on in a few seconds, including a 2–3 sec sensory memory.

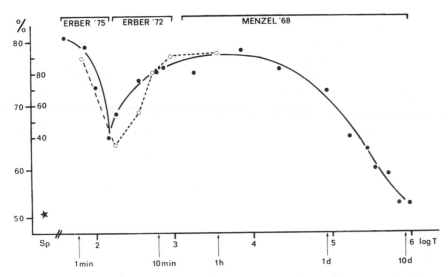

Figure 13.4. Time dependence of retention for free-flying bees trained to a color (long dashes and closed circles; $\lambda_+ = 444$ nm, $\lambda_- = 590$ nm) and fixed bees conditioned to an odor (orange) (short dashes and open circles) (Mercer and Menzel 1982). Note the log scale of the time axis (abscissa). The outer ordinate gives the percentage of correct choices of free-flying bees, the inner ordinate the responses (proboscis extension) of fixed bees after one conditioning trial (see text).

RETENTION

One way to look for transitional periods in the memory trace is to test retention at various times after the learning trial (Figure 13.4). We have done this both for freely flying bees conditioned to colors and fixed bees conditioned to odors. Note that the time is given in a logarithmic scale on the abscissa in Figure 13.4. Retention is best immediately after the one-trial learning, declines in the next 2–3 min, and then slowly increases over the next 10–20 min. After a single trial, retention declines to the spontaneous level of 50 percent within the next 3–5 days, but after as many as three rewards memory is stable for a lifetime of a couple of weeks.

Such a function reminds us of the so-called primacy and recency effects in human verbal learning, where it is found that items encountered first and last are better recalled than the middle items (Weiskrantz 1970). On the basis of such findings it was hypothesized (e.g., Waugh and Norman 1965) that at any instant of time retention will be a joint function of the strength of a short-term and that of a long-term process. If this is the case, the hypothetical short-term memory in the bee should be in the range of few minutes. There is evidence for such an interpretation from experiments with experimentally induced amnesia, but before I describe these experiments I want to report another series

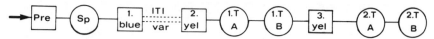

Figure 13.5. Experimental procedure of an experiment that tested the influence of newly learned, controversial information on an initially learned information. Pre, pre-conditioning by three rewards on a grey disk; Sp, spontaneous choice test; 1. blue, first reward on blue; ITI, intertrial interval; 2. yel, second reward on yellow; 1. TA, first session (A) of first test; 1. TB, second session (B) of first test; 3. yel, third reward on yellow; 2. TA, first session (A) of second test; 2. TB, second session (B) of second test.

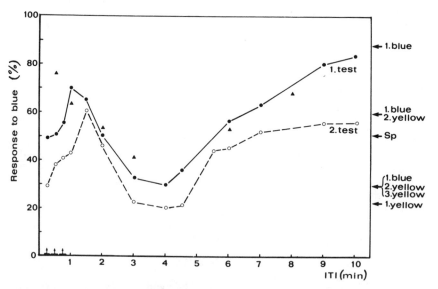

Figure 13.6. The influence of the time interval (ITI) between two learning trials on two different colors (blue and yellow). See procedure in Figure 13.5. The arrows on the right side indicate the response levels after one reward on blue (1. blue); one reward on blue and one reward on yellow (1. blue, 2. yellow); spontaneous choice level of the two colors blue and yellow (Sp); first reward on blue, second and third rewards on yellow (1. blue, 2. yellow, 3. yellow); and one reward on yellow only (1. yellow). 1. and 2. test – see Figure 13.5. Circles are used to illustrate experiments with 5-sec reward durations, triangles represent experiments with 15-sec reward durations.

of experiments in which we examined the effects of new and contradictory information on recently stored information.

INTERFERENCE WITH NEWLY LEARNED INFORMATION

A freely flying bee learns first the color blue (Figure 13.5) in a single trial and then, after a varying interval of 10 sec to 10 min, the color yellow at the same place. In two test sessions, the bee is asked to choose between the blue and the yellow. To look for any recovery effects, the bee is then rewarded once more on yellow and tested again afterward. Figure 13.6 gives the results. Choice behavior is expressed in percentage choices of blue, and the abscissa gives the interval between the two trials. Except for the very first part of the function, intertrial interval (ITI) ≤1 min, the curves have the expected shape if one assumes the resistance of a memory trace to interference from a new learning signal is correlated with the strength of the memory trace: After consolidation, when memory is most stable and most efficient in controlling behavior, it is also most resistant to new and contradictory learned information. Memory seems more labile during the transition from a short-term to a long-term trace. The very early high sensitivity to new information (ITI ≤1 min) is more difficult to interpret, and further experiments are needed to determine whether there are any qualitative differences between this part of the function and the others.

It was well known already to Müller and Pilzecker (1900) that recent memory is more sensitive to interference ("retroactive inhibition") than older memory. It is tempting to speculate that such an "erasing mechanism" may have a neurophysiological basis, as was recently argued by Krasne (1976).

EXPERIMENTALLY INDUCED AMNESIA

A strong retrograde amnestic effect is caused in both freely flying color-trained bees and in fixed bees conditioned to an odor by electroconvulsive shock (ECS), cooling to $+1°C$, and narcosis with CO_2 or N_2 shortly after learning. Details of the experimental procedure can be found in Menzel (1968), Erber (1976), and Menzel et al. (1974). Figure 13.7 gives the results for bees conditioned to blue in one trial. Treatment immediately after the trial has the strongest effect; at an interval of >5 min the choice behavior is not different from a control (B in Figure 13.7), which was sham-treated and then tested more than 15 min after conditioning, as were all the other animals. Another control group (C in Figure 13.7) served to determine whether the ECS acts as a negative reinforcer. Bees were ECS-treated after they had landed on blue but without being conditioned, and their choice behavior was tested afterward. On the average, they chose the blue at the same frequency as untreated animals (50 percent). I want to emphasize that in these experiments freely flying bees came back from the hive when they were motivated to search for food. Therefore, two arguments suggested most frequently against EC experiments do not apply here: The motivational state is not changed, and the treatment does not cause some sort of counterconditioning.

The four treatments used differ in the time courses of their amnestic effects, perhaps because the different speeds with which they effect the bee: ECS acts immediately, causing convulsions from which the bee recovers within the next 5–10 min. Cooling takes 1.5–2.0 min before the bees no longer react to stimulation. Besides these temporal effects, however, there may be additional effects that we have not yet studied in detail.

It is tempting to argue that recent events are in a qualitatively different neuronal state than those more distant events that are untouched by ECS. In a popular view, this early state is called short-term memory (STM), and it is argued that there is a gradual autonomous transfer from short-term storage to permanent storage (long-term memory, LTM). What defines this gradual transfer to long-term storage? Is it simply the time dependency of an internal,

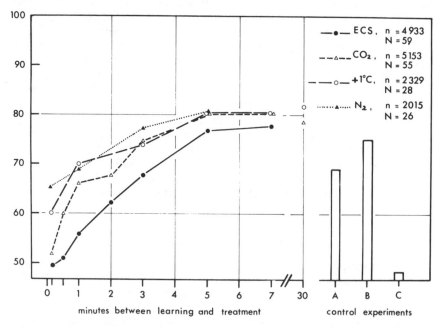

Figure 13.7. Time course of retrograde amnesia produced by four different treatments. The test animals are rewarded once on λ_+ = 444 nm with one reward lasting 20 sec and then treated with ECS, CO_2, N_2, or cooling at different times after onset of reward (abscissa). The curves show the percentage of correct choices (λ_- = 590 nm) in the test. Some control experiments are shown on the right side. In A the experimental animals were handled in the same way as in the ECS experiments but not shocked. In B the animals were restrained in a box for 30 min following the reward. In C the animals were tested for the spontaneous choice behavior and then shocked while sitting on a colored disk but without reward. In the following test the two alternative colors were chosen in the same way as spontaneously. N, number of bees tested; n, number of choices.

autonomous process? Is the transfer a gradual progression from STM to LTM? These are basic questions of memory research that can be addressed experimentally in bees. There is ample knowledge from studies of humans and of laboratory mammals that the serial progression assumption is very unlikely and that time dependency alone does not explain the various experimental data (Weiskrantz 1970). But how about such a "primitive" creature as the bee with its little brain, less than a millionth of the size of the human brain?

Time dependency may be a more crucial factor in the bee because we find very similar time courses in the two experimental paradigms we have used so far, the freely flying color-learning bee and the fixed odor-conditioned bee. A closer analysis reveals, however, that in the bee, too, time dependency is just one factor; parallel processing and immediate access to long-term memory are others.

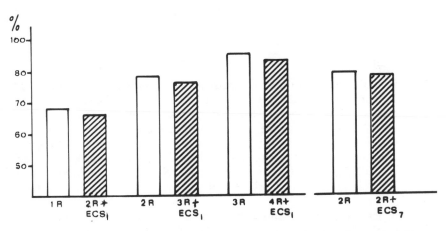

Figure 13.8. The effect of ECS on massed training trials. The first three pairs of bars give the choice reaction after one, two, and three training trials (1R, 2R, 3R) in comparison with always one more training trial plus immediate application of ECS (2R + ECS$_i$, 3R + ECS$_i$, 4R + ECS$_i$). Choice reaction is equal in experiments with x number of training trials and $x - 1$ numbers of training trials plus immediate ECS. If ECS is applied 7 min after the training trials (see right pair of bars, 2R + ECS$_7$), no effect of ECS is seen.

Erber (1975a, b) has studied these questions by combining massed trials with ECS treatment. The results in Figure 13.8 are twofold. First it is shown that a 2-sec reward causes the same change of behavior as a 30-sec reward. Then it is found that an ECS treatment immediately applied after two, three, or five short (2-sec) rewards within less than 30 sec eliminates only the learning effect of only one learning trial: Choice behavior is similar in bees trained with n_L rewards plus immediate ECS and in bees trained by $n_L - 1$ rewards without ECS. Figure 13.8 shows another control: if ECS is applied 7 min after the series rewards, choice behavior is similar to that of bees without ECS treatment.

As Erber (1975b) pointed out, there are two possible explanations to his findings: Either (1) there is direct access to LTM if the hypothetically limited STM capacity is occupied; or (2) there is a speed–up of transfer from short- to long-term memory in massed trials. To test these alternatives, a preliminary experiment was carried out by conditioning fixed bees to two different odors shortly after each other, one trial with each odor. Immediately afterward, the bees were cooled to $+5°C$. The results so far are in agreement with the second interpretation, indicating a limited storage capacity of STM. It is tempting to speculate, as has been done for verbal items in humans (Müller and Pilzecker 1900; Miller 1956; Peterson and Peterson 1959; Weiskrantz 1970) and for various conditioning treatments of laboratory mammals, that the transfer from a limited-capacity short-term store to a long-term store is dependent both on time and information flow. The result is that under certain conditions long-term storage is reached nearly immediately, giving the impression of parallel input to both STM and LTM (see for comparison Shallice and Warrington 1970).

NEURAL STRUCTURES INVOLVED IN STM–LTM TRANSFER

Experiments of the kind I have described so far aim at a more direct analysis of the brain mechanisms underlying learning and memory. The goal is to relate brain events with behavior, in particular the change of behavior as a result of experience. We wish to find the neural substrate of the change in behavior, as opposed to the substrate of the behavior per se. The paradigms used must permit one to distinguish between neurophysiological substrates of learning and performance. There must be changes that develop within the brain system involved in learning and memory. How do we search for the location of these brain systems?

Brain lesions are inappropriate methods, as we know from many studies (e.g., Hebb 1949), but if one could locally and temporally bind the transient destructive effect on the short-term store, one might get information on the participation of certain brain areas in the establishment of long-term memory. We have used local cooling immediately after one-trial learning of an odor signal to probe various structures in the bee brain (Masuhr and Menzel 1972; Menzel et al. 1974).

As you may suspect, the bee brain is relatively small, so the cooled pencil has to be sharpened before use. Actually, the bee brain, more accurately the supraesophageal ganglion, has a volume of about 1 mm^3 and contains about 900,000 neurons. About 80 percent of the neurons are in the two large optic lobes. The midbrain is dominated by a paired neuropil of densely packed neurons, the corpora pedunculata or mushroom bodies. These mushroom bodies have been suspected to be the center of intelligence in bees for a long time (von Alten 1910). This suggestion is supported by the structural regularity of the globula fibers, the input from olfactory and visual centers, and the extreme high density of fibers and synaptic connections (Schürmann 1970, 1972; Mobbs 1982). Each mushroom body is divided in two caplike structures (the calyces) connected by two short fused stalks (pedunculus) and two lobes (α- and β-lobe). The input regions are the calyces, the output regions, the pedunculus, and the α- and β-lobes. Figure 13.9 shows in addition the outline of the pair of lobulae, the third visual neuropil, and the pair of antennal lobes, the sensory neuropil of the antennae. Primary sensory projections from the antennae reach the antennal lobes, the subesophageal ganglion, and a region caudal to the antennal lobes. Secondary olfactory neurons connect strictly ipsilaterally the antennal lobes with the calyces and two regions in the lateral protocerebrum. Note the circular arrangements and length of the secondary olfactory neurons. For local cooling, we have used one or two needles with a tip diameter of 150 μm cooled to $+1\,^\circ$C. The main results of these experiments are given in Figure 13.9. The shaded circles show the areas cooled for a short period of time (2–5 sec). In all three experimental series considered here, both antennae received the odor signal and both paired structures were cooled.

The cooling effect is limited to a period of a few minutes after the one-trial learning. Cooling the antennal lobes later than 2 min, the α-lobes later than 3–

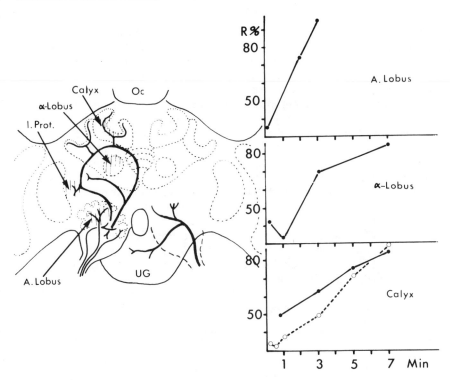

Figure 13.9. The effect of local cooling of small brain regions during the transition from short- to long-term memory. Left: Outlines of some brain structures. A. Lobus, antennal lobe (sensory neuropil of the antennae); 1. Prot., lateral protocerebrum; α-Lobus and Calyx, two parts of the mushroom bodies (corpora medunculata); Oc, ocelli; UG, subesophageal ganglion. The thick lines indicate the primary (into antennal lobe and subesophageal ganglion) and secondary projections from the antennal nerve (see text). Right: Time courses for the cooling effect in three brain areas. In all three cases both structures in the left and right hemisphere of the brain were cooled for a few seconds. The dashed line in the bottom figure gives the time course for cooling the whole animal. (For details see text and Erber et al. 1980.)

4 min, and the calyces later than 5 min after learning cause no reduction of response in a test 15 min later. Blockage of the transfer to a permanent store, therefore, is caused only if the sensory integration centers (antennal lobes) are cooled shortly after the learning trial. For both compartments of the mushroom body, the susceptibility to cooling is considerably longer, and the time course for the calyces is very similar to that of cooling the whole animal (dashed line), although only the two frontal calyces of those on either side have been cooled. No reduced responsiveness was found when the lateral protocerebrum close to the lobula area was cooled.

We conclude from these experiments, which have been performed with various permutations of parameters and a number of controls: (1) A circulating neural activity between antennal lobes, mushroom bodies, and lateral proto-cerebrum is not the substrate for STM. Vowles (1961, 1964) formulated such a hypothesis on the basis of the striking circular feature of the second-order olfactory neurons and lesion experiments; (2) STM is not localized in a limited area of synaptic interaction – instead, several widespread neuropiles take part, although with different time courses.

More generally, with these experiments we have found another surprising parallel relationship to memory in mammals, namely the participation of several widespread neural structures in the initial memory process. Our data do not allow us to separate between the possibility of several memory traces, as Hebb (1949) argued, or of one trace changing its property over time differently in the participating structures. What our data exclude definitely, however, is the possibility of a unique synaptic focus for the changes that occur when the bee learns to associate a specific odor with food reward.

In summary, short-term memory in bees is characterized by (1) a maximal duration of a few minutes, as seen in retrograde amnesia experiments; (2) its precise control of behavior; (3) its initially high sensitivity to interference from new learning; (4) its limited capacity; (5) its rapid transfer to long-term memory under conditions of high information flow; and (6) its widespread representation in the brain. Furthermore, (7) STM is preceded by a sensory store. Considering the small brain of the bee, the different structural basis of its neuronal wiring, and the different evolutionary adaptations of its learning system, the parallels with mammalian learning are striking.

ACKNOWLEDGMENTS

I want to thank Dr. Erber for providing Figure 13.8 and part of Figure 13.4, and Dr. Bitterman for his most helpful comments on the manuscript. This work was supported by grant DFG Me375.

REFERENCES

Alten, H. von. 1910. Zur Phylogenie des Hymenopterengehirns. *Jena Z. Naturwiss.* 46:511–590.
Bitterman, M. E., Menzel, R., Fietz, A., and Schäfer, S. 1983. Classical conditioning of proboscis-extension in honeybees. *J. Comp. Physiol. Psychol.* 97:107–119.
Erber, J. 1975a. The dynamics of learning in the honeybee *(Apis mellifica carnica)*. I. The time dependence of the choice reaction. *J. Comp. Physiol.* 99:231–242.
   1975b. The dynamics of learning in the honeybee *(Apis mellifica carnica)*. II. Principles of information processing. *J. Comp. Physiol.* 99:243–255.
   1976. Retrograde amnesia in honeybees *(Apis mellifica carnica)*. *J. Comp. Physiol. Psychol.* 90:41–46.
Frisch, K. von, 1967. *The Dance Language and Orientation of Bees.* Cambridge: Cambridge University Press.

Grossmann, K. E. 1970. Erlernen von Farbreizen an der Futterstelle durch Honigbienen während des Anflugs und während des Saugens. *Z. Tierpsychol. 27:*553–562.

Hebb, D. O. 1949. *The Organization of Behavior.* New York: Wiley.

Kolterman, R. 1973. Rassen- bzw. artspezifische Duftbewertung bei der Honigbiene und ökologische Adaptation. *J. Comp. Physiol. 85:*327–360.

Krasne, E. B. 1976. Invertebrate systems as a means of gaining insight into the nature of learning and memory. In *Neural Mechanisms of Learning and Memory* (M. R. Rosenzweig and E. L. Bennett, eds.), pp. 401–429. Cambridge: MIT Press.

Kuwabara, M. 1957. Bildung des bedingten Reflexes vom Pavlov Typus bei der Honigbiene *(Apis mellifica). J. Fac. Sci. Hokkaido Univ.* series 6, *Zoology 13:*458–464.

Lauer, J., and Lindauer, M, 1973. Die Beteiligung von Lernprozessen bei der Orientierung. *Fortschr. Zool. 21*(2,3):349–370.

Lindauer, M. 1970. Lernen und Gedächtnis – Versuche an der Honigbiene. *Naturwiss. 57:*463–467.

Masuhr, Th., and Menzel, R. 1972. Interhemispheric transfer of learning performance in the honeybee. In *Processing of Information in the Visual System of Arthropods* (R. Wehner, ed.), pp. 315–322, New York: Springer-Verlag.

Menzel, R. 1967. Untersuchungen zum Erlernen von Spektralfarben durch die Honigbiene *(Apis mellifica). Z. Vergl. Physiol. 56:*22–62.

1968. Das Gedächtnis der Honigbiene für Spektralfarben. I. Kurzzeitiges und langzeitiges Behalten. *Z. Vergl. Physiol. 60:*82–102.

Menzel, R. and Erber, J. 1972. Influence of the quantity of reward on the learning performance in honeybees. *Behaviour 41:*27–42.

Menzel, R., Erber, J., and Masuhr, Th. 1974. Learning and memory in the honeybee. In *Experimental Analysis of Insect Behavior* (L. Barton-Browne, ed.), pp. 195–217. New York: Springer-Verlag.

Mercer, A., and Menzel, R. 1982. The effects of biogenic amines on conditioned and unconditioned responses to olfactory stimuli in the honeybee *Apis mellifera. J. Comp. Physiol. 145:*363–368.

Miller, G. A. 1956. The magic number seven plus or minus two: Some limits on our capacity for processing information. *Psychol. Rev. 63:*81–97.

Mobbs, P. G. 1982. The brain of the honeybee, *Apis mellifera.* I. The connections and spatial organization of the mushroom bodies. *Phil. Trans. Roy. Soc B298:*309–354.

Müller, G. E., and Pilzecker, A. 1900. Experimentelle Beiträge zur Lehre vom Gedächtnis. *Z. Psychol. 1:*1–288.

Opfinger, E. 1931. Ueber die Orientierung der Biene an der Futterstelle. *Z. vergl Physiol. 15:*431–487.

Peterson, I. R., and Peterson, M. J. 1959. Short-term retention of individual verbal items. *J. Exp. Psychol. 58:*193–198.

Schürmann, F. 1970. Ueber die Struktur der Pilzkörper des Insektengehirns. I. Synapsen im pedunculus. *Z. Zellforsch. Mikrosk. Anat. 103:*365–381.

1972. Ueber die Struktur der Pilzkörper des Insektengehirns. II. Synaptische Schaltungen im -Lobus des Heimchens Acheta domestica. *Z. Zellforsch. Mikrosk. Anat. 127:*240–257.

Shallice, T., and Warrington, E. K. 1970. The independent functioning of the verbal memory stores: A neurophysiological study. *Q. J. Exp. Psychol. 22:*73–97.

Vareschi, E. 1971. Duftunterscheidung bei der Honigbiene: Einzelzell-Ableitungen und Verhaltensreaktionen. *Z. Vergl. Physiol. 75:*143–173.

Vowles, D. M. 1961. Neural mechanisms in insect behavior. In *Current Problems in Animal Behavior* (W. H. Thorpe and O. L. Zangwill, eds.), pp. 5–29. London: Cambridge University Press.

1964. Olfactory learning and brain lesions in the wood ant, *Formica rufa. J. Comp. Physiol. 58*:105–111.

Waugh, N. C., and Norman, D. A. 1965. Primary memory. *Psychol. Rev. 72*:89–104.

Wehner, R. 1981. Spatial vision in arthropods. In *Handbook of Sensory Physiology,* vol. 7/6C (H.-J. Autrum, ed.), pp. 287–616. New York: Springer-Verlag.

Wehner, R., and Lindauer, M. 1966. Zur Physiologie des Formensehens bei der Honigbiene. *Z. Vergl. Physiol. 53*:290–324.

Weiskrantz, L. 1970. A long-term view of short-term memory in psychology. In *Short-Term Changes in Neural Activity and Behaviour* (G. Horn and R. A. Hinde, eds.), pp. 63–74. London: Cambridge University Press.

14 · Response changes of single neurons during learning in the honeybee

JOACHIM ERBER

It is obvious why bees are well suited for an analysis of neural events underlying complex forms of learning. They can learn very rapidly a variety of different sensory signals, and their brain structure is relatively simple compared to that of vertebrates. A free-flying bee needs one reward to learn an odor and about three rewards for a color. Even if the reward duration is as short as 100 msec, a color signal is learned by the bee (Erber 1975). Memory formation can be disrupted by different treatments such as narcosis, electroconvulsive shock, and cooling (Erber 1976). Many features of the learning process and the formation of memory in the bee are similar to that of vertebrates (Menzel and Erber 1978), which also makes studies of learning in this insect highly interesting for comparative reasons.

Obviously, an electrophysiological study of the neural events accompanying and perhaps underlying learning in the bee has to be based on behaviors that are apparent in the laboratory situation. Fortunately, bees can learn even under very restricted circumstances. For electrophysiological recordings the bees are mounted in tubes, the head capsule is opened, and only the antennae, mandibles, and tongue are allowed to move freely. In this situation bees can be conditioned to odors and to moving stripe patterns. In contrast to free-flying bees, sensitization of different responses is an important feature of behavioral change in the restricted laboratory situation. The basic characteristics of learning and memory formation are very similar in laboratory and free-flying bees (Erber et al. 1980). Although restrained bees do not learn a large variety of sensory signals, the similarities with the free-flying situation make bees suitable subjects for an electrophysiological analysis of learning.

It is the major goal of this study to identify the neurons that display response changes during learning, to characterize the physiological properties of these neural elements, and to establish, if possible, causal relationships between neural substrate and behavioral response.

THE BEE BRAIN

The brain of the worker bee consists of approximately 850,000 neurons (Witthöft 1967). Many of these cells, especially in the sensory neuropils, are organized in a columnar array of parallel elements. The limiting factor for an elec-

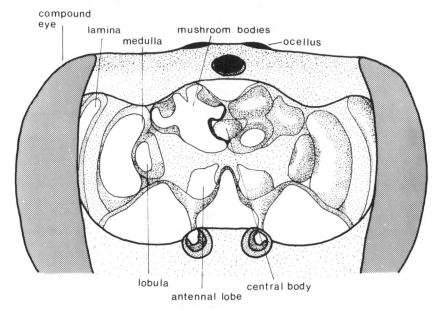

Figure 14.1. Schematic drawing of the bee brain. The head capsule is opened and the main neuropil areas are indicated.

trophysiological analysis of learning is the size and density of the neurons in the bee brain. Only about a dozen of them have axon diameters of 10 μm or more; the vast majority of cells have axon diameters of less than 2 μm. As the somata of central neurons in the insect brain are located far from the electrically active parts of the cells, it is not feasible, in most cases, to record neural activity from the cell bodies. The small size of neurons and their enormous density in the bee brain makes it necessary to record the responses of single cells intracellularly from the axons or the dendritic regons.

The main neuropil regions of the bee brain are illustrated in the schematic drawing of Figure 14.1. Information from the compound eyes is processed in the lamina, medulla, and lobula. In the lamina axons of the photoreceptors converge onto monopolar cells. Neurons of the medulla reveal complex receptive fields and a variety of intensity response functions (Hertel 1980). The receptive fields of lobula cells are larger than those of medulla neurons. A number of movement sensitive neurons in the lobula are direction-specific (Hertel 1980; Erber and Gronenberg 1981). Integrating neurons with large dendritic fields in the lobula project into the central parts of the brain and into the optic ganglia of the contralateral hemisphere.

The axons of antennal sensilla converge onto first-order interneurons in the antennal lobes. In the worker bee approximately 65,000 sensory neurons in each antenna that respond to olfactory, mechanical, various chemical, temperature, and gustatory stimuli converge onto 7000 neurons in the antennal lobes (Esslen and Kaissling 1976). Several hundred interneurons connect the antennal lobes with the mushroom bodies. This highly ordered neuropil region also receives input from the optic system. The mushroom bodies in the bee are made up of over 150,000 intrinsic neurons with small axon diameters and extremely dense packing. In the fly, a density of 1.6×10^7 neurons/mm^3 has been estimated for the intrinsic mushroom body neurons (Strausfeld 1976). The highest density of neural packing for vertebrates was found in the cerebellum, with 7×10^6 neurons/mm^3 (Braitenberg and Atwood 1958).

This summary of the important features of the bee brain demonstrates that this brain is certainly not a "simple system" for electrophysiological studies of learning. It is clear that the type of electrophysiological analyses routinely carried out on mollusk neurons, which sometimes have somata as big as the entire bee brain, are not feasible in the bee. On the other hand, complex behavior is only possible with a complex nervous system, and at present we know very little about the basic rules of information processing used by complex nervous systems in organizing behavior and behavioral adaptivity. Since intracellular recordings in the bee brain can be made during learning, it is possible to characterize and identify the neural elements involved in behavioral change using this animal.

SINGLE-CELL RECORDINGS DURING ODOR LEARNING

The proboscis extension reflex is used for olfactory conditioning. An odor is offered to the bee, and one antenna of the animal is stimulated with a drop of sugar water. In response to the gustatory stimulus the bee extends its proboscis. Following proboscis extension, the bee is rewarded by allowing it to drink a small drop of sugar water. When the odor is presented again, about a minute later, the proboscis is now extended in response to the conditioned stimulus (Erber 1980, 1981). One-trial learning is apparent in up to 80 percent of the bees. The level of learning depends on the season and probably also on the function of the experimental bee in the hive. One-trial odor learning is so fast that neural activity can be recorded intracellularly in single cells while the animal is learning.

Where in the nervous system could this type of learning take place? Olfactory information is processed in the antennal lobes, in the mushroom bodies, and in the central part of the brain. The antennal lobes and the mushroom bodies play an important role during the formation of olfactory memory. When these neural structures are locally cooled after one-trial learning, the conditioned response is reduced in later tests. The disruption of memory formation is limited to the first minutes after one-trial learning (Erber et al. 1980; Chapter 13).

So far there is no electrophysiological evidence for response changes during learning in first-order interneurons projecting from the antennal lobes to the calyces of the mushroom bodies (Homberg 1981, pers. comm.). In the first-order interneurons of the olfactory pathways there is already a remarkable convergence of different sensory modalities. Signal processing in the mushroom bodies has been analyzed by measuring field potentials (Kaulen et al., pers. comm.) and recently by recording the activity of single intrinsic cells (Erber, pers. observ.). As yet, there is no evidence of response changes in intrinsic neu-

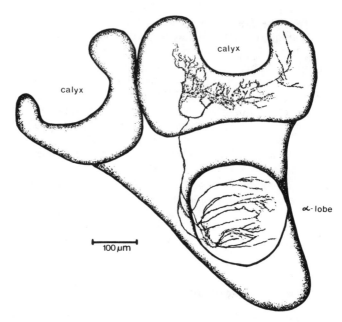

Figure 14.2. Neuroanatomical structure of a neuron connecting the α-lobe with the median calyx of the mushroom body. These neurons could serve as feedback loops between the output and input of the mushroom body system.

rons of the mushroom bodies during odor learning. Responses of intrinsic neurons are characterized by aftereffects sometimes lasting more than 30 sec. The aftereffects in the mushroom bodies could be due to multiple feedback loops connecting the output of the system (α-lobe) with the input region (calyx); Figure 14.2 shows such a neuron. Based on our present electrophysiological data, we conclude that different sensory modalities converge in the mushroom bodies and that neurons of this neuropil region are characterized by maintaining activity following sensory stimuli. This maintained activity could be a necessary prerequisite for olfactory memory formation. Interruption of neural processing by cooling may interfere with olfactory memory by disrupting the maintained activity of mushroom body neurons.

Higher-order olfactory interneurons of the central part of the bee brain can display response changes during olfactory learning. Neurons of this area are characterized by multimodal convergence. So far, changes of responses in single cells during olfactory conditioning have only been found in multimodal neurons that respond to both odor *and* to sugar water stimuli (Erber 1978, 1980, 1981). All such cells show selective changes of the response for odor during conditioning. It is difficult to assess the significance of these neural changes in single cells for olfactory learning because very little is known about the physiology of olfactory signal processing in higher-order neurons of the insect brain. Most of the cells that display response changes during learning respond to a variety of different odors. We have not observed response changes that are specific for the conditioned odors.

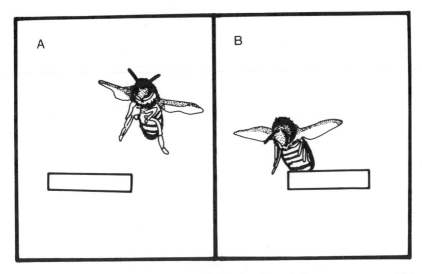

Figure 14.3. Antennal positions of a bee during landing. *A*: The antennae are still in flight position. *B*: Shortly before landing, they are moved downward.

SENSITIZATION OF A VISUAL REFLEX

In contrast to the olfactory system, we know a lot more about the neurophysiology of visual signal processing in the bee brain. Recently it has been possible to demonstrate associative and nonassociative visual learning for bees in the laboratory situation (Erber and Schildberger 1980). Based on electrophysiological analyses of the visual system it is possible to formulate hypotheses on the neural basis of behavioral change and to test them with single-cell recordings during learning.

When a moving stripe pattern is presented to a fixed bee, the animal moves its antennae in the direction opposite the movement of the stimulus. When the pattern moves upward, the antennae move downward, and vice versa. The same antennal reaction can be observed in free-flying bees as they land. Figure 14.3 shows two pictures of a landing bee. Shortly before landing, when surrounding visual cues move upward relative to the bee, the antennae move downward (Figure 14.3*B*). The direction-specific antennal response in the laboratory can be measured with an optical device using photodetectors (Erber and Schildberger 1980). This device counts how often the antennae move downward for a given direction of stripe movement. Differences in the frequency of antennal movements across the photodetectors for different directions of stripe motion can be used as a measure of direction-specific antennal responses. In the context of learning it is important that the antennal reflex is modified after the bee is stimulated with sugar water. The response for the upward-moving pattern increases, whereas the response for downward motion does not change. This type of direction-specific sensitization is apparent after some sugar water stimuli. Figure 14.4 shows a summary of this experiment for a group of bees. The antennal reflex is direction specific at the beginning of the experiment as indi-

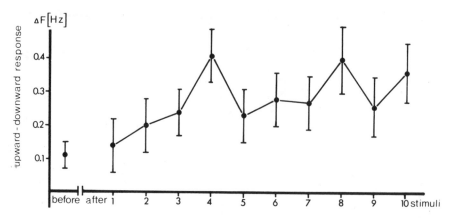

Figure 14.4. Direction-specific sensitization of the antennal reflex. The ordinate indicates the frequency difference for antennal passages across the photodetectors for upward- and downward-moving stripe pattern. The abscissa shows the response before and after sugar water stimuli applied to the antennae and proboscis. For details, see text.

cated by positive values of ΔF. If the bees are stimulated with sugar water at the antennae and proboscis and retested 1 min later, the direction specificity of the reflex changes, reaching a maximum after about 5 stimulations.

The most effective stimulus for a modification of the antennal reflex is sugar water presented to both the antennae and the proboscis. The direction specificity of the reflex is also modified when only the antennae are stimulated with sugar water, but not when only the proboscis is stimulated. Figure 14.5 shows a summary of sensitization experiments. Input from gustatory receptors on the antennae is necessary to sensitize the antennal reflex. A single sugar water stimulus to antenna and proboscis changes the reflex, although the response change takes several minutes to build up. Sensitization is side-specific: When the contralateral antenna is stimulated and the ipsilateral compound eye is tested with the stripe pattern, there is no change of the reflex response. Mechanical stimuli applied to the antennae do not modify the reflex.

This specific form of sensitization is of great interest for analyses of associative learning in the bee since gustatory stimuli have been found to modulate the response for another stimulus modality.

RESPONSE CHANGES OF VISUAL NEURONS DURING SENSITIZATION

There are many different possibilities as to where the neural events underlying behavioral change could be located in the bee brain. In principle there are three different levels in the nervous system where these changes could take place. The modulation of the antennal response could be a function of the motor sys-

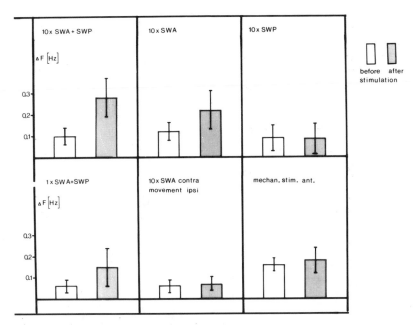

Figure 14.5. Summary of sensitization experiments. The ordinates show the response difference for upward- and downward-moving stripe pattern, as described in Figure 14.4 and the text. The responses were tested before and after various stimulations. The upper row indicates stimulations with sugar water applied to the antennae (SWA) and/or proboscis (SWP); the bees were stimulated 10 times and tested 1 min after each stimulation. The lower row indicates experiments with one sugar water stimulation (1 × SWA + SWP); the bees were tested 10 times. Stimulation at the contralateral side caused no differences in the ipsilateral response (middle panel, lower row). Antennae stimulation also caused no differences in response (right panel, lower row). In each group 30 bees were tested.

tem, a feature of central neurons of the unstructured neuropil, or a characteristic of visual interneurons; or it could be a result of neural changes at all three levels. Changes at the cellular level of motoneurons during motor learning are well documented in insects (Woolacott and Hoyle 1977), and the electrophysiological recordings during odor learning in the bee demonstrate that central, multimodal neurons can change their response properties. There is no experimental evidence, so far, that neural changes during odor learning in the bee occur at the level of the first sensory interneurons.

In the visual pathway of the bee, movement-sensitive neurons with relatively small visual fields were recorded and identified in the medulla (Hertel 1980). Some of these neurons were movement-sensitive but did not reveal direction specificity. There is little multisensory convergence in medullary neurons. The visual cells of the lobula, however, have larger visual fields than those of the medulla, and a number of these neurons are movement-sensitive and direction-specific (Hertel 1980; Erber and Gronenberg 1981). The degree of multimodal

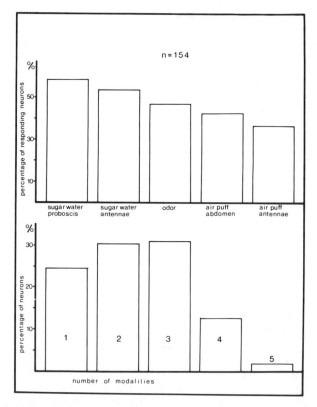

Figure 14.6. Multimodality of movement-sensitive neurons in the bee brain. In the upper graph, the ordinate indicates the percentage of movement-sensitive neurons responding also to various other modalities. In the lower graph the percentage of neurons is shown that respond to the indicated number of modalities.

convergence in the lobula is higher than in the medulla. The analysis of neural events underlying the changes of the antennal reflex during sensitization concentrated on movement-sensitive cells in the lobula for two reasons: With axon diameters of up to 5 μm, the activity of these cells can be recorded intracellularly over several minutes without major difficulties; and secondly, direction sensitivity is a property of movement-sensitive lobula neurons.

Firstly, we recorded from movement-sensitive neurons in the lobula to determine the degree of multisensory convergence in these cells. Figure 14.6 shows a summary of this investigation. Over 75 percent of the movement-sensitive neurons respond to more than one modality. A large proportion of these cells respond to sugar water stimuli applied to the proboscis or the antennae. Many cells receive input from olfactory or mechanical receptors. The high degree of multisensory convergence in movement-sensitive neurons is remarkable (Erber and Gronenberg 1981).

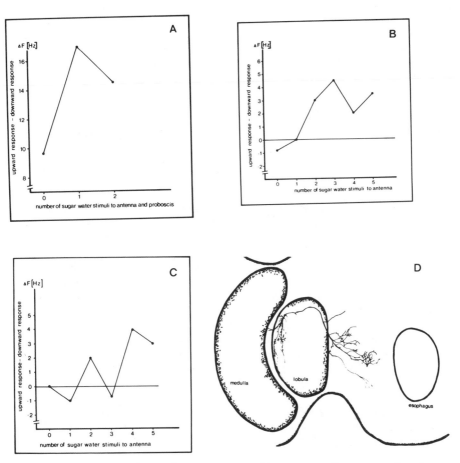

Figure 14.7. Response changes in movement-sensitive neurons after sugar water stimulation. *A, B, C*: Three examples of neuronal responses. The ordinates indicate the response specificity as in Figures 14.4 and 14.5; the abscissae show the stimulations after which the movement response was tested. *D*: Anatomical pattern of the neuron in *C;* the other neurons had similar neuroanatomical structure.

When the bee is stimulated with sugar water, the direction-specific responses of such multimodal, movement-sensitive neurons can change drastically. The observed neural changes correlate directly with the behavioral modifications (Erber 1980, 1981). The recorded cells can be identified with dye injection (Lucifer yellow). Figure 14.7 shows three examples of cells that display changes of the direction specificity after sugar water stimulation. These neurons have projections in the lobula and in the central part of the brain. All neurons with such properties were at least bimodal, responding to visual stimuli and to sugar water taste. Similar features can be found in central, multimodal neurons that display movement sensitivities. We conclude that plasticity of the response is a property of visual interneurons of the lobula. It remains an open question whether neurons in the medulla, which precede those in the lobula, also show similar properties. Nondirectionality of the movement responses and the low degree of multisensory convergence at this level of optic signal processing make this possibility unlikely.

CONCLUSION AND PERSPECTIVES

Response changes in single, identifiable neurons of the bee brain can be recorded during associative and nonassociative learning. There is a general rule that only those cells that receive input from the two sensory modalities that are involved in learning show response changes. In the insect brain convergence of different modalities is already apparent at the level of first-order interneurons. For the visual system of the bee, response plasticity is a property of higher-order sensory interneurons. The electrophysiological experiments indicate that there are two types of higher-order visual neurons: a group with stable responses to the sensory stimuli that give a fixed reference of the visual input, and a second group of neurons with adaptive response properties that depend on the combination and succession of certain multimodal stimuli. It is now necessary to identify all the neurons of the two groups in order to quantify their anatomical characteristics. For the visual system we expect to find only a small number of interneurons between movement-sensitive cells and the motoneurons of the antennae. The problem of establishing causal relationship between neural and behavioral change for the antennal reflex, therefore, may be solved.

The studies of the neural basis of learning in the bee provide important information on the basic rules that a complex nervous system uses in organizing adaptive behavior. By limiting the investigations to specific forms of learning that are based on identifiable networks of neurons, these studies can help to bridge the wide gap between "simple" invertebrate preparations and the more complex forms of vertebrate learning.

ACKNOWLEDGMENTS

I want to thank Allison Mercer for valuable suggestions concerning the manuscript. This work is supported by the Deutsche Forschungsgemeinschaft.

REFERENCES

Braitenberg, V., and Atwood, R. P. 1958. Morphological observations on the cerebellar cortex. *J. Comp. Neurol. 109*:2–27.
Erber, J. 1975. The dynamics of learning in the honey bee *(Apis mellifica carnica)*. I. The time dependence of the choice reaction. *J. Comp. Physiol. 99*:231–242.
      1976. Retrograde amnesia in honey bees *(Apis mellifera carnica)*. *J. Comp. Physiol. Psychol. 90*:41–46.

1978. Response characteristics and after effects of multimodal neurons in the mushroom body area of the honey bee. *Physiol. Entomol. 3*:77–89.

1980. Neural correlates of non-associative and associative learning in the honey bee. *Verh. Deut. Zool. Gesell. 73*:250–261.

1981. Neural correlates of learning in the honey bee. *Trends Neurosci. 4*:270–273.

Erber, J., and Gronenberg, W. 1981. Multimodality and plasticity of visual interneurons in the bee. *Verh. Deut. Zool. Gesell.* (Abstract) *74*:177.

Erber, J., Masuhr, Th., and Menzel, R. 1980. Localization of short-term memory in the brain of the bee, *Apis mellifera. Physiol. Entomol. 5*:343–358.

Erber, J., and Schildberger, K. 1980. Conditioning of an antennal reflex to visual stimuli in bees *(Apis mellifera L.). J. Comp. Physiol. 135*:217–225.

Esslen, J., and Kaissling, K. E. 1976. Zahl und Verteilung antennaler Sensillen bei der Honigbiene *(Apis mellifera L.). Zoomorphology 83*:227–251.

Hertel, H. 1980. Chromatic properties of identified interneurons in the optic lobes of the bee. *J. Comp. Physiol. 137*:215–231.

Homberg, U. 1981. Recordings and Lucifer yellow stainings of neurons from the tractus olfacto-globularis in the bee brain. *Verh. Deut. Zool. Gesell.* (Abstract) *74*:176.

Menzel, R., and Erber, J. 1978. Learning and memory in bees. *Sci. Am. 239*:102–110.

Strausfeld, N. J. 1976. *Atlas of an Insect Brain.* New York: Springer-Verlag.

Witthöft, W. 1967. Absolute Anzahl und Verteilung der Zellen im Hirn der Honigbiene. *Z. Morph. Tiere 61*:160–184.

Woolacott, M., and Hoyle, G. 1977. Neural events underlying learning in insects: Changes in pacemaker. *Proc. Roy. Soc. 195*:395–415.

BIOPHYSICS AND BIOCHEMISTRY

Introduction to Part III

DANIEL L. ALKON

In asking how neural systems store learned information with membrane and subcellular physiologic mechanisms, investigators have taken a variety of approaches. Some of these approaches, but by no means all, are represented by contributions in this part. One approach is to study biophysical and/or biochemical steps important for transmitting, responding to, and storing information in general without directly linking such processes to behavioral change. Elegant examples of such an approach included here are provided by the biochemical studies of Howard Rasmussen and the biophysical work of John Connor. The promise of this approach with regard to learning is that parallels to physiology of cellular changes during learned behavior will suggest themselves and thus motivate predictions on a basic level about such physiology. This is illustrated by the biophysical and biochemical steps described here for *Hermissenda* associative learning. An accumulation of intracellular Ca^{2+} during acquisition was implicated as causing increased activation of Ca^{2+}-calmodulin–dependent protein kinase (see Chapter 15), which in turn causes long-term changes of phosphorylation (see Chapter 16) responsible for long-term reduction in the number of open voltage-dependent K^+ channels.

Another approach involves analysis of membrane physiology involved in prolonged functional changes within the nervous system, again without directly relating these changes to behavior. This is well exemplified by some of the past work on synaptic posttetanic potentiation (see Erulkar and Rahamimoff 1978) and synaptic facilitation (see Zucker 1982). Still other experimental strategies attempt to link biochemical and/or functional changes more directly to behavior. Hyden, Flexner, Barondes, and others observed differences of protein synthesis metabolism as a function of training, but localizing these differences to particular cells or cell groups was difficult. Furthermore, establishment of the causal role for such changes proved elusive. Ladislav Tauc, together with his colleagues Bruner, Hughes, and Kandel, measured synaptic changes of *Aplysia* using a nonassociative learning paradigm known as habituation. Eric Kandel (see Chapter 9) and his colleagues then established a parallelism between behavioral and synaptic changes during habituation and determined some of the relevant neural circuitry. Because the *Aplysia* and crayfish (see Zucker 1972a,b) synapses at which habituation has been analyzed are not readily accessible, it has not been possible thus far to directly identify specific presynaptic membrane currents, as has been accomplished for the squid synapse.

Indirect attempts have been made to formulate hypotheses about presynaptic currents by measuring soma currents during habituation (see Chapter 9).

Other investigators have attempted to relate membrane changes to associative learning behavior. Woollacott and Hoyle (1977) obtained evidence that changes (lasting for some minutes) of impulse activity within an identified locust motorneuron occurred following an associative learning paradigm. The relevant neural systems of the gastropods *Pleurobranchaea, Limax,* and *Aplysia,* however, have not been sufficiently well characterized to specifically correlate changes of identified neurons or even identified groups of neurons directly with associative learning behavior. Close correlations have been obtained between changes of impulse activity in identified neural aggregates and associative learning of vertebrate preparations (see Chapters 4, 5, and 6). Once such correlations are found, an investigation of causal biophysical changes can begin. These, when obtained, might then be compared to the persistent (for days) biophysical changes of specific ionic channels in identified neurons that have now been shown to be sufficient to cause associative learning of *Hermissenda* (see Chapters 10 and 15, and Introduction to Part II).

The diversity of experimental approaches and preparations in this part as well as others of this volume reflects the primitiveness of our understanding of learning and its underlying cellular mechanisms. We can agree on obvious classifications of learning behavior – for example, associative and nonassociative – but we don't yet know really what associative learning of snails will teach us about our own ability to learn. At this point we only know that there is ample evidence for conservaton during evolution of the means for accomplishing biological functions. We know that there also is an economy of cell biological processes. Part of the biochemical sequence involved in release of thyroid-stimulating hormone may also be useful for causing long-lasting but reversible membrane changes of learning. Some of the transformations of ionic channels and neuronal structure that occur during development of organisms may also be involved in permanent learning within our nervous systems. We continue then to be in a position of having to collect information with these diverse approaches using a full range of scientific disciplines and resources with an open mind, constantly testing for generality of principles across species and discarding no clues that together will ultimately reveal to us the subject of our interest.

REFERENCES

Erulkar, S. D., and Rahamimoff, R. 1978. The role of calcium ions in tetanic and post-tetanic increase of miniature end-plate potential frequency. *J. Physiol.* (London) *278*:501–511.

Woollacott, M. H., and Hoyle, G. 1977. Neural events underlying learning in insects: Changes in pacemaker. *Proc. Roy. Soc. B 195*:395–415.

Zucker, R. S. 1972a. Crayfish escape behavior and central synapses. I. Neural circuit exciting lateral giant fiber. *J. Neurophysiol. 35*:599–620.

Zucker, R. S. 1972b. Crayfish escape behavior and central synapses. II. Physiological mechanisms underlying behavioral habituation. *J. Neurophysiol. 35*:621–637.

Zucker, R. S. 1982. Processes underlying one form of synaptic plasticity: Facilitation. In *Conditioning* (C. D. Woody, ed.), pp. 249–264. New York: Plenum Press.

15 · Persistent calcium-mediated changes of identified membrane currents as a cause of associative learning

DANIEL L. ALKON

INTRODUCTION

In considering general classes of behavioral responses, reflexes can be regarded as having evolved when only one type of response, such as approach or avoidance, has net survival value in most environmental situations that are encountered by a species. It is, for example, almost always adaptive for us as well as a gastropod to withdraw from and thus avoid noxious stimuli. Learning, particularly associative learning, can be considered as having evolved when the most adaptive type of response can vary drastically depending on the environmental situations that are encountered by individual animals. The bee in one setting may have a neutral response to the color blue, but an approach response to red because this color had been, in prior experience, temporally associated with food (see Chapter 13). In another milieu, with different-colored flowers, the bee might have learned the opposite responses.

ASSOCIATIVE LEARNING OF *Hermissenda*

Such a potential for learning different behavior responses to the same stimulus, light, has been demonstrated with the nudibranch mollusk *Hermissenda crassicornis* (Figure 15.1). During the daylight portion of its light cycle the animal will move quickly, although often with a somewhat circuitous route, toward a light source (Alkon 1974a; Lederhendler et al. 1980). The positive phototactic behavior during the daytime is also manifest with another measure. The animal will avoid darkness (Figure 15.2) by turning at light–dark edges or borders (Alkon and Fuortes 1972; Lederhendler, pers. comm.). The variability characteristic of *Hermissenda*'s response to light is in contrast to the stereotypic quality of reflexive behavior and suggests potential for different stimuli when paired with light to produce different behavioral outcomes. In fact, after light is repeatedly temporally associated with rotation when the animal's head is oriented *toward* the center of rotation, on subsequent days the animal no longer approaches light with its former rapidity (Alkon 1974; Crow and Alkon 1978; Farley and Alkon 1982). On the other hand, after light is repeatedly temporally associated with rotation when the animal's head is oriented *away* from the center of rotation, it approaches light more rapidly than prior to training (Farley and Alkon 1980).

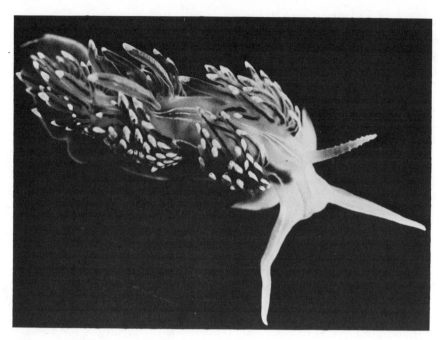

Figure 15.1. The nudibranch mollusk *Hermissenda crassicornis*. Note the small black spot (the right eye) at the base of the lower rhinophore. Length of the animal is ～4 cm. (From Alkon 1980a)

These behavioral changes have been shown to have the defining features of associative learning observed with many vertebrate species. Paired light and rotation (i.e., the onset of maximal rotation followed the onset of light by ～1.0 sec) produce the change. Randomized or explicity unpaired light and rotation do not produce the behavioral change. The change increases as a function of practice (i.e., shows acquisition), shows savings, and persists for many days (i.e., shows retention). The learned behavior is also stimulus-specific. After training with paired light and rotation, the animal's response to light is changed but not its response to darkness, a gravitational force, or food (Crow and Offenbach 1979; Alkon 1980a; Farley and Alkon 1980, 1982).

For the associative learning to occur there is also a requirement that there be a contingent relationship rather than simply a contiguous relationship between the associated stimuli. That is, the conditioned stimulus (in this case, light) must predict the occurrence of the unconditioned stimulus (in this case, rotation). This was demonstrated by Joseph Farley when he showed that the learning behavior was degraded by adding light or rotation stimuli that were not paired with rotation during the period of training with paired stimuli. The associative learning behavior could also be extinguished by the presentation of light stimuli following the training period (Farley et al. 1982). Finally, the light stimulus, which in this paradigm can be regarded as a conditioned stimulus

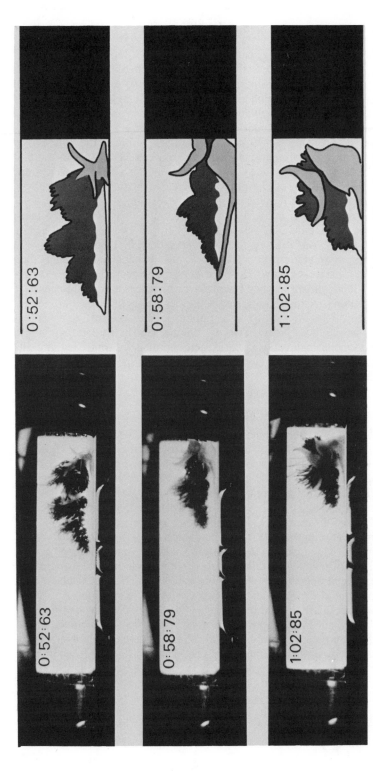

Figure 15.2. Sequence of timed photographs of the response by the individual *Hermissenda* to a shadow. The animal is approaching the stimulus in the top frame. The eyes, located under the dorsal tentacles, are about to encounter the dark in the central frame. In the bottom frame the animal has completed the response. The line drawings are tracings of the photographic images. The light gray areas emphasize the animal's foot and buccal region that are in contact with the inner surface of the tube.

(CS), after associative training, can, to some extent, be regarded as taking on the quality of the unconditioned stimulus (UCS) rotation. This is manifest by the findings that the UCS ordinarily causes the animal to reduce its overall level of motor activity by clinging to surfaces and that light (the CS) also elicits reduced overall initial movement only for the conditioned animals (see Chapter 10; Farley and Alkon 1982; Lederhendler, pers. comm.).

How does nature construct a machine that has within it the potential to respond differently in different environmental contexts? By virtue of the biophysical characteristics and the synaptic interactions of a number of neurons within the neural systems of *Hermissenda,* we can explain how cellular physiology accounts for salient features of associative learning – for example, associativity or temporal specificity, contingency, stimulus specificity, persistence or long duration, acquisition, and extinction. It will be seen that it *is the response of a system of neurons to the training stimuli,* as expressed by the effects of these systemic responses on specific ionic conductances of individual neurons, that ultimately endows *Hermissenda* with associative learning capacity. Such systemic effects as they interact with specific ionic conductances of individual neurons are to be contrasted with phenomena that occur at individual synapses involving such processes as fatigue, facilitation, posttetanic potentiation, and habituation. Neural system-mediated associative learning is also in contrast to global motivational states, neurohormonally controlled and diffusely affecting neuron aggregates such as flight, fight or generalized arousal, and sensitization.

NEURAL SYSTEMS

The organization of vertebrate neural systems provides for both serial and parallel processing of information within patterns of stimuli arising from the environment. Serial processing is exemplified by progressive extraction of features of the visual world as the signals pass from the retina to the lateral geniculate and then the cortex of the mammalian visual system (see Hubel and Wiesel 1978). Parallel processing has been indicated by the findings, among others, of Hubel and Wiesel in their work on visual cortex (Hubel and Wiesel 1972), of Llinas (Llinas and Simpson 1981; Pellionisz and Llinas 1979) in studies of the cerebellum, and Mountcastle (1976) in his studies of somatosensory cortex. Both serial and distributed (parallel) neural organization properties are important in the ultimate generation of our conscious experience, particularly as it depends on complex arrays of learned associations.

When we seek to understand biophysical and biochemical mechanisms that store associatively learned information, it seems of great importance not to oversimplify the problem although general cellular principles should have economy and simplicity in their formulation. The behavioral phenomenon of associative learning as exhibited by vertebrates has been characterized explicitly as meeting certain criteria, which have been met by the associative learning of *Hermissenda* (see above and Alkon 1974; Crow and Alkon 1978; Chapter 10, this volume), and, in part, by the learning behavior of *Pleurobranchaea* (Mpitsos and Collins 1975) and other gastropods (e.g., Chapter 12). It is equally necessary to adequately define the gastropod neural systems, in all their complexity, to arrive at mechanisms that may have analogy for more evolved species. It is not possible to identify a few types of synaptic interaction, for instance, between central interneurons and motoneurons, and from postulated repetition of such interactions to arrive at conclusions about how the system functions during stereotypic, let alone learning behavior. Similarly, it is not sufficent to characterize with electrophysiologic and morphologic techniques several of the large, somewhat recognizable neurons, but not the small, less identifiable neurons, in these "simple system" preparations if a functional understanding of these systems is to be achieved. Unfortunately, most of the gastropod preparations have hosts of smaller neurons, many of which are responsible for important integrative processing. Furthermore, with the relatively small number of neurons in these systems, the specificity of the function of single neurons (many with diameters ~ 15 μm) is astounding (see below and Alkon 1980a).

In determining the neural systems of *Hermissenda* we attempted to arrive at a complete enough understanding to follow sensory stimuli as they affect, serially, receptor cells, interneurons, motoneurons, and ultimately muscle movement, and, in addition, to follow such stimuli as they affect in a distributed manner parallel neuronal pathways. It follows that to understand this distributed information processing it was essential to study convergence of sensory information not only within the same modality but also between distinct modalities. By repeatedly making intracellular recordings from two, three, and four neurons simultaneously (Alkon and Fuortes 1972; Alkon 1973, 1974; Tabata and Alkon 1982; Goh and Alkon 1982) as well as marking pre- and postsynaptic terminal endings with intracellular horseradish peroxidase (Crow et al. 1979) we arrived at a working knowledge of the animal's visual pathway as well as the interaction of this pathway with the statocyst and tentacular or chemosensory pathways. A brief description of *Hermissenda* neural systems follows.

SENSORY STRUCTURES

Eyes. There are five photoreceptors in each eye, two type A and three type B. Each one of these cells can be recognized by the morphologic, electrophysiologic, and synaptic characteristics that were established by recording simultaneously from hundreds of pre- and postsynaptic cell pairs. For example, the medial type B photoreceptor (impulse amplitude ~ 15 mV) has mutually inhibitory synaptic relations with the lateral and intermediate type B cells and inhibits as well the medial and lateral type A cells (impulse amplitude ~ 45 mV). Because the rhabdomeric (light-sensitive) membranes are precisely oriented around the lens of each eye and because the pigment cells isolate these rhabdomes from stray light not transmitted by the lens, the five cells can detect (as enhanced by their "lateral" synaptic inhibition) the movement of lights or shadows across the visual field of the eye (Alkon and Fuortes 1972). All five photoreceptors in each eye depolarize with increased number of superimposed impulses in response to illumination (Figure 15.3).

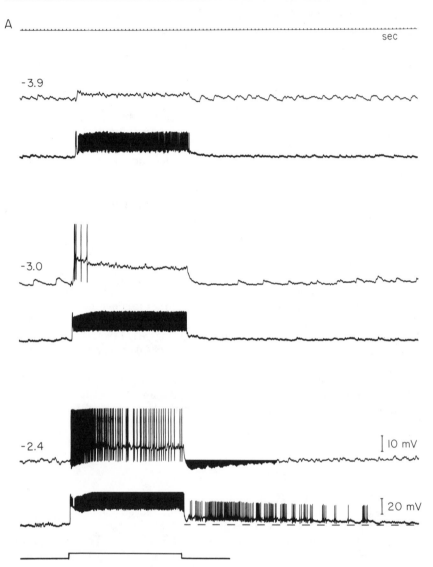

Figure 15.3. Responses to light steps of simultaneously impaled photoreceptors. Intact type A and type B photoreceptors are exposed to steps (indicated by bottom trace) of increasing intensity (expressed in −log units on left of traces). With sufficient light intensity a long-lasting depolarization (LLD) appears in the response of type B photoreceptor (lower trace in each pair of records). For this same intensity and with the same time course, a long-lasting hyperpolarization (LLH) appears in the response of a simultaneously impaled type A cell. For illustration purposes the difference of the type A membrane potential from the resting level during the LLH is represented by the darkened area. (From Alkon and Grossman 1978)

Statocysts. There are 13 hair cells in each statocyst. Hair cells that are dia-metrically opposed are mutually inhibitory (Detwiler and Alkon 1973). Hair cells that are located ~90° from each other on a statocyst equator show uni-directional inhibition. There are also inhibitory synaptic interactions (as well as electrical synapses) between hair cells of the right and left statocysts. The "lateral" synaptic inhibition between the cells enhances contrast between responses of the cells to differences of spatial orientation. The neurophysiologic effects of orientation differences can also be elicited by rotation of the statocyst. Rotation, like gravity, causes $CaCO_3$ crystals known as statoconia that are pro-pelled by the inherently motile intraluminal hairs of the cells to press against the hairs of those cells in front of the centrifugal force vector (Alkon 1975). The force of accelerated statoconia is transmitted by the hairs to cause con-ductance changes of the hair cell membrane in the vicinity of the hair's inser-tion points, that is, the area surrounding the basal bodies (Grossman et al. 1979; Stommel et al. 1980; Alkon 1982a,b). These conductance changes cause depolarization and increase of impulse activity of the hair cell (see Fig. 15.12).

Tentacles. These are chemosensory organs containing very elaborate neural networks. Information concerning chemosensory stimuli is transduced by tufted receptor cells on the tentacular suface (Kuzirian, in press) and trans-mitted to the intratentacular network, which then results in changes of impulse activity within the tentacular nerve that controls membrane potential of inter-neurons within the cerebropleural ganglia.

INTERNEURONS

The three paired sensory structures, the eyes, statocysts, and tentacle, synapse on groups of interneurons within the optic ganglion and cerebropleural gan-glion (Figure 15.4). A distributed or parallel organization is suggested for these sensory postsynaptic cells. The best understood are those cells that receive visual information. These neurons fall into at least four separate categories: (1) other sensory cells, namely the hair cells; (2) optic ganglion cells located in close proximity to the eye; (3) a few interneurons located in the vicinity of the photoreceptor axon's point of entry into the cerebropleural ganglion; and (4) a cluster of central neurons on the dorsal surface of the cerebropleural ganglia.

Visual information, then, is *distributed* onto at least four separate cell types, some of which themselves synaptically interact and which process this infor-mation in different contexts, that is, converging with other synaptic inputs that are characteristic for each of the four cell types. Chemosensory information and information concerning spatial orientation from the statocyst is distributed in a similar manner. Statocyst hair cells synapse on other sensory receptors, the photoreceptors (Figure 15.4), optic ganglion cells, and cerebropleural cen-tral neurons. Tentacular input synaptically affects hair cells and photorecep-tors, optic ganglion cells, and central ganglial neurons (Figure 15.4). If this distribution of sensory information seems nonspecific and somewhat arbitrary, it is neither. Careful intracellular recording from these neurons in thousands of preparations reveals an incredibly reproducible specificity. For example, impulses of type B photoreceptors but not type A are followed in one-for-one

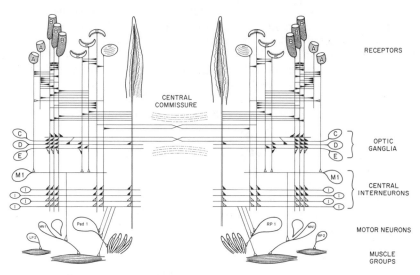

Figure 15.4. Summary of synaptic interactions within and between the visual, stato-cyst, and chemosensory pathways. This is a partial but not complete representation of all the known interactions established by simultaneous pre- and postsynaptic intracel-lular recordings and by morphologic criteria of apposed processes stained with horse-radish peroxidase. Filled endings represent inhibitory, open circles excitatory, synaptic interactions. A few of the many known ipsilateral–contralateral synaptic interactions are also included. Ped 1, LP 2, RP 1, and RP 2 are identified central neurons in the pedal ganglion of *Hermissenda*.

fashion by inhibitory synaptic potentials recorded from ipsilateral optic gan-glion cells. Cephalic (but not caudal) hair cells are inhibited by type B photo-receptors. Caudal hair cells inhibit both ipsilateral and contralateral photore-ceptors. Interactions between interneurons and from sensory receptors onto interneurons are equally reproducible. The "E" optic ganglion cell (Tabata and Alkon 1982), presumed to be electrically coupled to the "S" cell, causes excit-atory postsynaptic potentials (EPSPs) recorded from all ipsilateral type B (but not type A) photoreceptors, and causes inhibitory postsynaptic potentials (IPSPs) recorded from the ipsilateral caudal hair cell (Figures 15.1 and 15.5). All other cells within the same optic ganglion have no synaptic interactions, but "C" optic ganglion cells inhibit contralateral "D" optic ganglion cells, thereby enhancing contrast between these cells' responses to illumination of each of the two eyes.

Perhaps most important for the ability of *Hermissenda* to move away from shadows and toward areas of maximal light intensity, a complete input–output pathway has been determined for the animal's ability to turn in response to light (Goh and Alkon 1982). The medial type A receptor impulses are followed in one-for-one fashion by EPSPs recorded from a specific cerebropleural inter-neuron whose impulses in turn are followed in a one-for-one fashion by EPSPs recorded from a specific motoneuron (MN1), as in Figures 15.6 and 15.7,

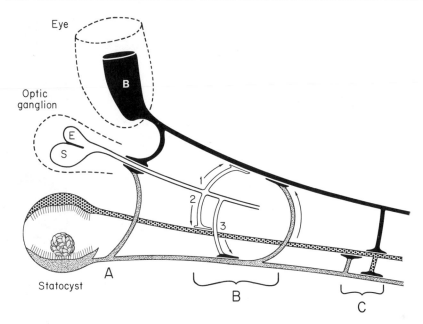

Figure 15.5. Intersensory integration by the *Hermissenda* nervous system. *A*: Convergence of synaptic inhibition from type B and caudal hair cells on S–E cell. *B*: Positive synaptic feedback onto type B photoreceptor. 1, Direct synaptic excitation. 2, Indirect excitation: E–S excites cephalic hair cell that inhibits caudal hair cell and thus disinhibits type B cell. 3, Indirect excitation: E–S inhibits caudal hair cell and thus disinhibits type B cell. 4, Indirect excitation: B-cell inhibits C-cell, and thus disinhibits E-cell; C-cell effects are not illustrated. *C*: Intra- and intersensory inhibition. Cephalic and caudal hair cells are mutually inhibitory. Type B cell inhibits mainly the cephalic hair cell. All filled endings indicate inhibitory synapses; open endings indicate excitatory synapses. (From Tabata and Alkon 1982)

responsible for turning movements (see below). Other cerebropleural central neurons in an aggregate on the nervous system's dorsal surface receive, in some cases, EPSPs and in others IPSPs, both of which follow in a one-for-one fashion impulses of ipsilateral type B photoreceptors.

Although our knowledge of the *Hermissenda* visual system and its relationship to systems of other sensory modalities is sufficiently comprehensive to provide a causal sequence for the behavioral expression of biophysical changes that store the learned association of stimuli (see below), it is by no means a complete knowledge. Just as the behavioral response of the animal to light is comprised of several components (e.g., arousal, discrimination of intensity differences), so are visual stimulus effects distributed over at least several neuronal classes that probably subserve different behavioral functions. One function already mentioned concerns the animal's discrimination of illumination intensity differences. This is provided for by the separation of the rhabdomes of individual photoreceptors as they abut in specific locations around an approximately spherical lens. Discrimination of spatial differences of illumination is also aided

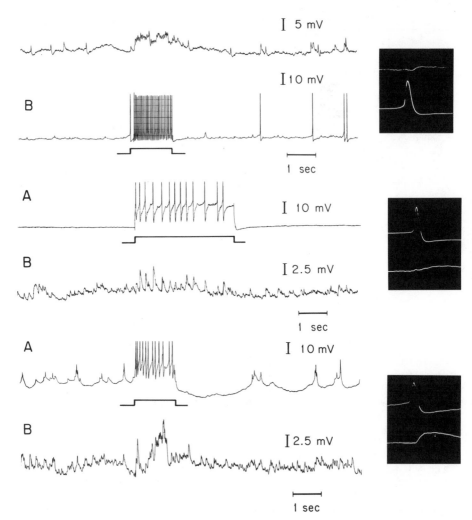

Figure 15.6. *Top:* Interneuron effect on MN1 cell. *A*: MN1 cell, hyperpolarized with steady current (-0.9 nA). *B*: Interneuron, hyperpolarized with steady current (-0.15 nA) except during positive current pulse. When the interneuron is excited by positive current injection (0.15 nA) EPSPs in MN1 cell follow, one-for-one, impulses of the interneuron. *Inset:* Photographic record of a single interneuron impulse followed by unitary EPSP recorded from MN1 cell. *Center:* Type A photoreceptor input to interneuron. *A*: Medial type A photoreceptor. *B*: Interneuron. Note the EPSPs recorded from the interneuron follow type A photoreceptor impulses in a one-for-one fashion. Interneuron is hyperpolarized with a steady current injection (-0.4 nA). Type A cell is hyperpolarized with steady current (-0.8 nA) before and after the depolarizing current pulse ($+0.8$ nA). *Inset:* photographic record of type A photoreceptor impulse followed by unitary EPSP recorded from interneuron. *Bottom:* Hair cell input to interneuron (in this case, ipsilateral caudal hair cell). Note hair cell impulses are followed by EPSPs recorded from the interneuron, in a one-for-one fashion. Hair cell is hyperpolarized with steady current (-0.3 nA) before and after positive current pulse ($+0.3$ nA). Interneuron is hyperpolarized with a steady current (-0.4 nA). *Inset:* photographic record of hair cell impulse followed by unitary EPSP recorded from interneuron. (From Goh and Alkon, in press)

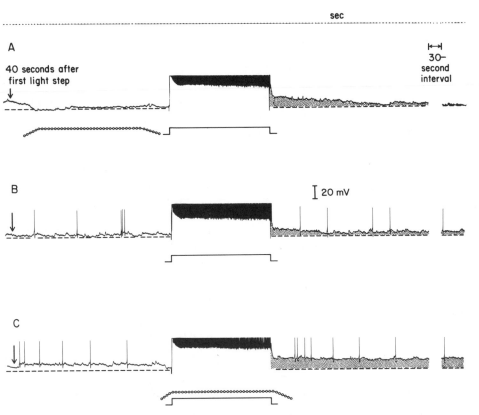

Figure 15.7. Intracellular voltage recordings of *Hermissenda* neurons during and after light and rotation stimuli. Responses of a type B photoreceptor to the second of two succeeding 30-sec light steps (with a 90-sec interval intervening). The cell's initial resting potential, preceding the first of the two light steps in *A*, *B*, and *C*, is indicated by the dashed lines. Depolarization above the resting level after the second of the two light steps is indicated by shaded areas. *A*: Light steps ($\sim 10^4$ erg cm^{-2} sec^{-1}) alternating with rotation (caudal orientation) generating ~ 1.0 g. The end of the rotation stimulus preceded each light step by 10 sec. *B*: Light steps alone. *C*: Light steps paired with rotation. By 60 sec after the first and second light steps, paired stimuli cause the greatest depolarization and unpaired stimuli the least. The minimal depolarization was in part attributable to the hyperpolarizing effect of rotation. Depolarization after the second presentation of paired stimuli was greater than that after the first. (From Alkon 1980b)

by synaptic inhibition between photoreceptors and inhibition between certain visual interneurons of the right and left optic ganglion (Tabata and Alkon 1982).

The interactions between the visual system and neurons mediating stimuli of other modalities facilitate choice behavior. When, for example, the animal is sufficiently hungry, its responsiveness to chemosensory stimuli will be most sensitive and cause inhibition of both the visual and statocyst pathways. The animal will therefore "choose" (Alkon et al. 1978) to move toward a food substance rather than areas of maximal light intensity. Other functions, served, for instance, by the rhythmic output of the optic ganglion, could include circadian variation of activity and states of arousal.

A detailed "blueprint" of the animal's nervous sytem, although not complete, was sufficiently comprehensive to enable one to ask a number of questions concerning the animal's ability to learn an association between a light stimulus and the aversive stimulus rotation. What biophysical, biochemical, and anatomic changes within the neuronal pathways known to underlie the animal's behavioral responses to light, rotation, and combinations thereof could be correlated with the acquisition and retention of the associative learning behavior? If such correlates could be obtained, was it possible to construct a causal sequence for the production of these changes within the nervous system, the retention of the changes, and ultimately the behavioral expression of these changes? Furthermore, could such changes be differentiated as to which were primary and which were secondary or simply consequential to the primary changes? Finally, it could be asked, can we actually produce the associatively learned behavior by causing the biophysical changes in single identified neural elements that had been implicated as sites of primary change? What follows is essentially a summary of progress toward answers to these questions.

Most broadly speaking, we knew that the visual and statocyst pathways mediate the training stimuli to produce the associative learning. When the eyes are removed or obscured, the animals are no longer able to discriminate light – dark differences. When the statocysts cannot transduce gravitational stimuli, it is not possible to associatively train the animals. This was accomplished (Crow and Harrigan 1979; Harrigan et al. pers. comm.) by using laboratory-reared animals with abnormally few statoconia or crystals that load the hairs of statocyst cells in response to a gravitational stimuli. Thus we have observed that statocyst hair cells that are themselves normal respond with much less depolarization and fewer superimposed impulses when 1–2 instead of 150–200 statoconia press on their intraluminal hairs. With such statocysts in otherwise normal nervous systems, the *Hermissenda* do not learn from paired presentations of light and rotation. To further implicate neurons within these pathways, we identified electrophysiologic changes that were closely correlated with the associative learning.

CORRELATED ELECTROPHYSIOLOGIC CHANGES

Immediately following three 1-hr associative training (but not control-procedures) periods on successive days, a number of changes were recorded from neurons at the input and output stages of the *Hermissenda* visual pathway. Type B photoreceptors were more depolarized in darkness, showed elevated input resistance, and responded to light with larger and more prolonged depolarizing generator potentials following conditioning but not control procedures (Crow and Alkon 1980; Farley and Alkon 1981, 1982). Type A photoreceptors were more hyperpolarized, received more inhibitory postsynaptic potentials (IPSPs), and responded to light with fewer impulses superimposed on depolar-

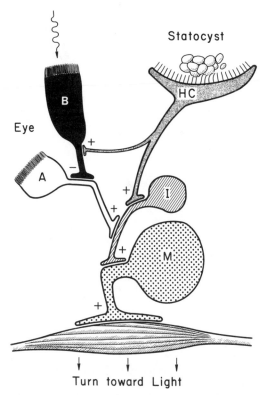

Figure 15.8. Schematic diagram of visual pathway and its convergence with the statocyst pathway. The type B photoreceptor (B) causes monosynaptic inhibition of the medial type A photoreceptor (A). The medial type A photoreceptor causes monosynaptic excitation of ipsilateral interneurons (I), which are also excited by ipsilateral hair cells (HC). Ipsilateral hair cell impulses cause a transient inhibition (not shown here) and a long-lasting synaptic excitation of the type B photoreceptor. This excitation consists of disinhibition from hair cell impulses and an increased number of excitatory synaptic potentials from the S–E optic ganglion cell. The ipsilateral interneurons (I) in turn cause monosynaptic excitation at specific motoneurons (M). (From West et al. 1982)

izing generator potentials (Alkon 1976; Farley and Alkon 1981, 1982). Putative motoneurons (Fig. 15.8) responded to light with less depolarization during and following light steps (Takeda and Alkon 1982; Goh and Alkon 1982; Lederhendler et al. 1982; Chapter 10). All of these changes were recorded immediately after training and thus are most reasonably related to the process of acquisition. This inference is supported by other observations during the actual presentation of the training stimuli. Intracellular recordings from type B cells *during* repeated presentations of paired light and rotation reveal progressive and prolonged depolarization (Alkon 1980b), and progressive enhancement of the light response (Figure 15.7).

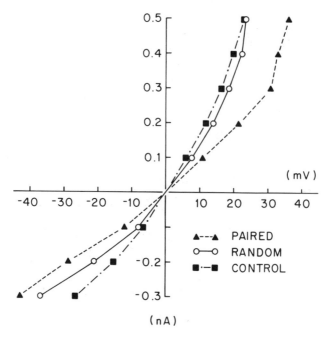

Figure 15.9. Current–voltage relations of type B photoreceptors. Postregimen values were obtained by measuring the steady-state voltage changes produced by current pulses through an intracellular microelectrode. The paired cells showed significantly greater voltage changes, particularly for positive pulses. This difference of paired cells (as compared to random and control cells) was greater for the post- than for the pre-regimen values. (From West et al. 1982)

On retention days following associative training (i.e., 1, 2, and 3 days *after* training) additional electrophysiologic changes were recorded. In darkness, type B cells were no longer depolarized (West et al. 1981; Farley and Alkon 1981) but did show a substantial increase of input resistance and enhancement of the sustained depolarizing response during and after a light step (Figures 15.9 and 15.10). Putative motoneurons also showed no differences in darkness but did show decreased depolarization during and particularly following a light step (Takeda and Alkon 1982; Goh et al. 1982, pers. comm.; Lederhendler et al. 1982; Chapter 10).

Given these electrophysiologic differences that were closely correlated with the associative learning of *Hermissenda,* is it possible to construct a causal sequence for the associative learning within the visual pathway? For a number of reasons we believe this to be the case, and the sequence begins with the type B photoreceptor. That is, changes of the type B photoreceptor are a primary cause of the associatively learned behavioral changes. What is the evidence for this?

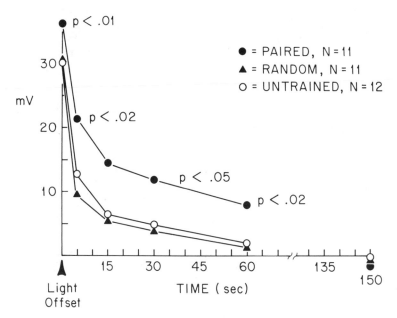

Figure 15.10. LLD responses of type B photoreceptors. Values taken from actual voltage recordings at prechosen time points (0, 5, 15, 30, 60 sec) following the first light step. Note that the paired LLD values are significantly greater than random and control, using a two-tail Mann–Whitney U-test. The significance levels indicated refer to comparison of paired with respect to random values. (From West et al. 1982)

THE CAUSAL ROLE OF TYPE B PHOTORECEPTOR CHANGES

The close correlation of type B electrophysiologic change with the ability of individual animals to move toward a light source following associative training was itself some indication of a causal role (Crow and Alkon 1980). Consideration of the *Hermissenda* visual system synaptic interactions gives further support for such a role. Type B impulses cause one-for-one IPSPs in type A photoreceptors. The medial type A photoreceptor impulses in turn cause through an interneuron, excitatory postsynaptic potentials (EPSPs) on ipsilateral motoneurons that are responsible for turning the animals in response to light. Increased frequency of type B impulses during and following a light step following associative training will therefore cause sequentially fewer type A impulses, fewer interneuron impulses, fewer EPSPs on ipsilateral motoneurons, fewer motoneuron impulses, and reduced turning away from darkness (Figure 15.7). If indeed this sequence of changes causes associative learning, beginning with the first or primary change within the type B photoreceptor this latter change should be intrinsic to the type B cell. That is, when the type B cell is physically and electrically isolated from all other *Hermissenda* neurons, it should retain changes after associative training but not other control procedures. That type B photoreceptors after isolation by axotomy do show changes only after associative learning is perhaps the most important correlative evidence for the causal role of these changes. Type B photoreceptors isolated on retention days show longer and more prolonged depolarization during and following a light step (Crow and Alkon 1980; West et al. 1980). Type B input resistance in the dark was increased on retention days only for conditioned animals (West et al. 1980). More recently, increases of type B input resistance were found to predict the decreased depolarizing response of the MN1 cell to light (using a blind experimental procedure) for conditioned animals. Naive and explicitly unpaired control animals showed neither enhanced type B input resistance or changes of the MN1 response to light (Lederhendler et al. 1982). Furthermore, even in naive controls, the magnitude of the MN1 depolarizing response to light was correlated with the rate with which the animals moved toward a light source as well as turn at light – dark borders (Lederhendler et al. 1982). To summarize these data, learning-specific changes intrinsic to the type B membrane could explain the enhanced response of this cell to light and, by close correlation and by virtue of the known neuronal pathways, the decrease responses of the MN1 cell and the decreased movement of the animal toward light.

Another experimental approach was directed to the issue of causality. If, as we proposed, the type B membrane changes arose out of temporal and stimulus-specific responses of the visual–statocyst neural systems, it should be possible by impaling and critically stimulating crucial neural elements within these systems to produce the same type B cell membrane changes that occur during associative learning. This in fact was possible in experiments conducted by Joseph Farley.

We then asked whether it was possible to more directly demonstrate the causal role of this type B membrane change in producing the learned behavior. Also in experiments conducted by Joseph Farley we impaled single identified type B cells of intact animals and simulated the effects of associative training (Farley et al. 1983). This was accomplished by pairing (or, for controls, explicitly unpairing) injection of a depolarizing current step with a light step. This "pairing" (the current step immediately followed the light step) simulated the effects produced by sensory stimulus pairing (a light step with rotation) on the neural system mediating the convergence of the visual and statocyst pathways (Figure 15.7) and thus on the type B photoreceptor itself (Tabata and Alkon 1982). Following the training simulation of effects on the single type B photoreceptor, the electrode was withdrawn and the animal's response to light after recovery was tested on subsequent days. There was in fact a pairing-specific decrease in the rate of movement toward a light source. Thus, membrane changes in this single neuron were sufficient to *cause* the same behavioral change previously observed to be produced as a result of associative learning. We might ask: How do these membrane changes that are intrinsic to the type B cell body (since they were present after axotomy) arise as a result of paired light and rotation? What happens to the soma membrane to produce the change? Put in another way, how is the information regarding the association of stimuli actually stored within the type B cell? First of all, what happens during acquisition of the learned behavior?

To understand the biophysical basis for the enhanced light response during acquisition and retention, it was first necessary to characterize the ionic currents that flow across the type B membrane during darkness as in response to light.

TYPE B MEMBRANE CURRENTS

In addition to the outward K^+ current ("leak current") that helps determine the resting level of membrane potential but is not voltage-dependent, there are two potassium currents that flow across the membrane of the type B photoreceptor (from the inside to the outside of the neuron) when the membrane potential becomes more positive. That is, during depolarization of the type B cell in darkness, two outward potassium currents are turned on: an early, large transient current I_A and a late, smaller sustained current I_B. These currents are activated in darkness, that is, entirely independent of any phototransduction functions of the cell, and they are not unique to the type B cell, to photoreceptors, or to molluscan or other invertebrate neurons. In short, they are encountered in a variety of neurons in a variety of species. These two outward currents have very different and very characteristic properties. I_B is blocked by intracellular tetraethylammonium (TEA) and by superfusion with Ba^{2+}–seawater solution (Alkon et al. 1982). I_A is blocked by 4-aminopyridine, but is unaffected by Ba^{2+}. I_A, unlike I_B, is also substantially inactivated for many seconds following prolonged membrane depolarizations (≥ 5 mV). That is, making the membrane potential more positive by 30 mV for 10 sec is not only enough to activate I_A, but is enough to cause its inactivation for 30–50 sec later. Making the membrane potential more negative by 30 mV for 10 sec causes prolonged activation of I_A. By elevating extracellular divalent cations and blocking the two voltage-dependent potassium currents (with 4-aminopyridine and TEA), another voltage-dependent current was revealed. Intrinsic to the type B cell body is an inward voltage-dependent calcium current. The magnitude of this current varies with the concentration of extracellular calcium and is greatly reduced or eliminated in the presence of extracellular cadmium. The increase of intracellular calcium associated with the activation of the voltage-dependent calcium current causes still another current. This is a calcium-activated outward potassium current. Because this current depends on the level of intracellular calcium, it follows with some delay the onset of the voltage-dependent calcium current.

Voltage clamp analysis of the type B photoreceptor *during* a light step revealed two important light-induced currents: an early, rapidly inactivating inward Na^+ current and a rapidly inactivating outward Ca^{2+}-dependent K^+ current. The Na^+ current is not voltage dependent. Out of voltage clamp at the resting potential for the type B cell when light first causes an inward Na^+ current, there is a small I_A but no Ca^{2+} current. As sodium rushes into the type B cell it causes the membrane to depolarize and thus to activate I_A and the Ca^{2+} current. The Ca^{2+} current decays so slowly that it persists for many seconds after the cessation of a light step and thus contributes significantly to the prolonged depolarizing response (LLD) of the type B cell following a light step (Figure 15.11).

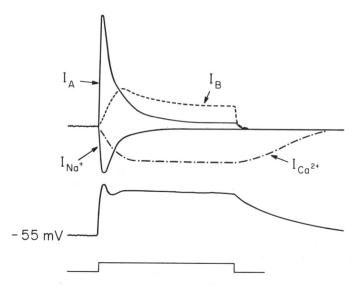

Figure 15.11. Summary of major membrane currents (not drawn to represent relative magnitude) of type B photoreceptor. The lowest recording (-55 mV) indicates voltage response of the type B cell during and after a light step ($\sim 10^4$ erg cm^{-2} sec^{-1}). The lowest trace indicates duration of the light step. There is a two light-induced inward current, I_{Na+}, a large transient early sodium current. I_{Ca2+}, a smaller, sustained late calcium current, is voltage-dependent. Note that I_{Ca2+} persists long after the cessation of the light step. There are two dark outward currents, I_A, a large transient early K$^+$ current, and I_B, a smaller late K$^+$ current, both of which are voltage-dependent. Another, light- and voltage-induced Ca^{2+}–K$^+$ current is not shown.

ACQUISITION: ITS BIOPHYSICAL BASIS

Intracellular recordings from the type B cell (Figure 15.7) demonstrated an enhanced depolarization following light paired with rotation (for a caudal orientation) when compared to depolarization following light alone, light alternating with rotation, or paired stimuli for a cephalic orientation (i.e., head pointed away from the center of rotation). These recordings were made during and following successive presentations of these light and rotation stimuli. We found that enhanced depolarization of the type B cell following a single stimulus pair is pairing-specific and stimulus-specific. This pairing and stimulus specificity is a precise function of and was predicted by the synaptic organization within and between the visual and statocyst pathways (Figure 15.12).

Caption of Figure 15.12 (*cont.*)
(caudal hair cell, upper record) after light paired with rotation. The LLD after stimulus pairing is greater than that after light alone (line of long dashes). The line of short dashes indicates level of resting membrane potential. The lowest trace indicates light duration; top trace, angular velocity of turntable (effecting 1.2 g). (From Alkon 1979)

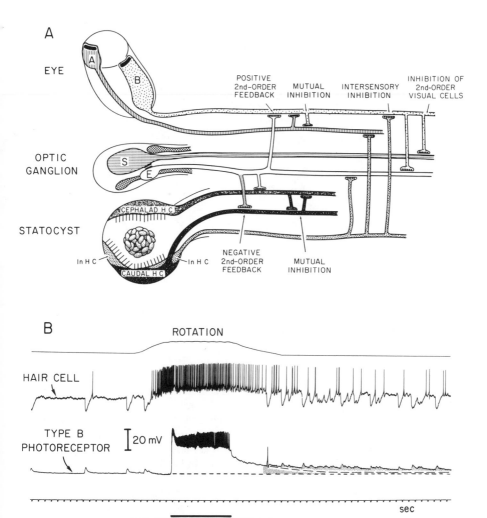

Figure 15.12. Neural responses to stimulus pairing. Maximal rotation onset usually follows (not as depicted here) light onset for training of intact animals as well as isolated nervous systems. *A*: Neural system (schematic and partial diagram) responsive to light and rotation. Each eye has two type A and three type B photoreceptors; each optic ganglion has 13 second-order visual neurons; each statocyst has 12 hair cells. The neural interactions (intersection of vertical and horizontal processes) identified to be reproducible from preparation to preparation are based on intracellular recordings from hundreds of pre- and postsynaptic neuron pairs. Abbreviations: In HC, hair cell ~45° lateral to caudal north–south equatorial pole of statocyst; S, silent optic ganglion cell, electrically coupled to E-cell; E, optic ganglion cell, presynaptic source of EPSPs in type B photoreceptors. The E second-order visual neuron causes EPSPs in type B photoreceptors and cephalad hair cells and simultaneous IPSPs in caudal hair cells. *B*: Intracellular recordings (simultaneous) from caudal hair cell and type B photoreceptor show increase of EPSPs (type B cell, lower record) and simultaneous IPSPs

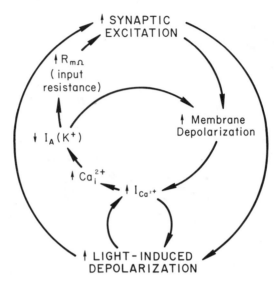

Figure 15.13. Regenerative changes of voltage-dependent Ca^{2+} and K^+ conductances during stimulus pairing. The net effect of the current changes is a mutual enhancement of light-induced and synaptic excitation.

Through the synaptic interactions of the type B cell with other elements of these pathways, it receives, following a light step, at least two types of synaptic excitation (Tabata and Alkon 1979, 1982): increased numbers of EPSPs from the ipsilateral optic ganglion and decreased numbers of IPSPs from ipsilateral hair cells (Figure 15.12). When light is paired with rotation-induced stimulation of the caudal hair cells (i.e., when the animal's head points toward the center of rotation) this synaptic excitation is increased by frequency changes of EPSPs and IPSPs. Synaptic excitation following stimulus pairing also increases by means of a positive feedback process, which is as follows. Enhanced synaptic excitation following each stimulus pair due to changes of synaptic potential frequency causes a net steady-state shift of membrane potential in the positive direction. This net depolarization causes a voltage-dependent Ca^{2+} current (Alkon et al. 1983) to increase. Associated with the inward Ca^{2+} current is a prolonged rise of intracellular Ca^{2+} as measured by differential absorption spectrophotometry of Arsenazo III–injected type B cells (Connor and Alkon 1982). Elevated intracellular Ca^{2+} produced by stimulus pairing or by direct injection under voltage clamp conditions has now also been shown (Alkon et al. 1982b) to cause prolonged reduction of an early voltage-dependent K^+ current that contributes significantly to the type B cell's resting membrane conductance (and thus input resistance). Thus, elevated intracellular Ca^{2+}, by decreasing a specific membrane current, causes increased input resistance and increased depolarization (Figure 15.13). The integrated effect of enhanced synaptic excitation, then, is to increase the long-lasting after-depolarization (LLD) and to increase membrane resistance following a light step. By Ohm's law, any current flowing across a larger membrane resistance will cause a larger voltage change. Therefore, the currents associated with synaptic excitation will cause a larger voltage change. Thus sequentially, enhanced synaptic excitation following each stimulus pair causes a dc depolarization that causes a larger LLD. A larger LLD will cause an enhanced synaptic excitation by increasing total membrane resistance (Alkon 1979). This then is the positive feedback process that occurs during each pairing of light and rotation (Figure 15.13). Synaptic excitation enhances the LLD, which in turn enhances synaptic excitation, and so on. For a single stimulus pair this positive feedback cycle or oscillation can be expected to progressively diminish; that is, the oscillation would be damped.

Because, however, the long-lasting depolarization following a stimulus pair decays very slowly, some remains when the next stimulus pair occurs. The residual depolarization potentiates the depolarizing effect of the next stimulus pair still further and in this way the depolarization accumulates (Alkon 1979, 1980b). Because of its cumulative nature, this depolarization provides a biophysical translation at the cellular level of the behavioral phenomenon of acquisition.

To summarize, stimulus and pairing specificity for the associative learning of *Hermissenda* are expressed by enhanced depolarization that arises out of a positive feedback cycle between synaptic and light-induced depolarization. Acquisition of the associative learning is expressed by cumulative depolarization that results during repeated stimulus pairings and persists for some hours (Crow and Alkon 1980; Alkon 1980b; Farley and Alkon 1982). What changes persist for days? Again, how is the information stored during the retention period of the associatively learned behavior? As already mentioned, there is no residual depolarization on retention days, yet the type B response to light (particularly its LLD) is enhanced (West et al. 1981; Alkon 1980b).

RETENTION: ITS BIOPHYSICAL BASIS

If an engineer were looking for a device to act as a switch between alternative responses, such as approach and avoidance, I_A, the early voltage-dependent outward K^+ current, which is sensitive to the magnitude and duration of prior depolarization such as occurs during acquisition, would be an excellent candidate. If an aversive stimulus, when paired with a CS, by effects on neural systems caused membrane depolarization, I_A would subsequently remain inactivated whereas a positively reinforcing stimulus when paired with a CS might produce via the neural systems membrane hyperpolarization that would cause I_A to subsequently remain activated.

Recently, together with my colleagues Drs. Izja Lederhendler and Jon Shoukimas, I completed a blind voltage clamp study of type B photoreceptors isolated from *Hermissenda* that had been conditioned with paired light and rotation or from control animals that were either naive or trained with randomized light and rotation. One and two days following training, the "paired" cells had a significantly smaller I_A than the "random" and "naive" cells (Figure 15.14, Table 15.1). This then is a biophysical change that provides a mechanism for *retention* of the learned behavior. How can a smaller I_A, which is a transient dark current and is not involved in the phototransduction function of the type

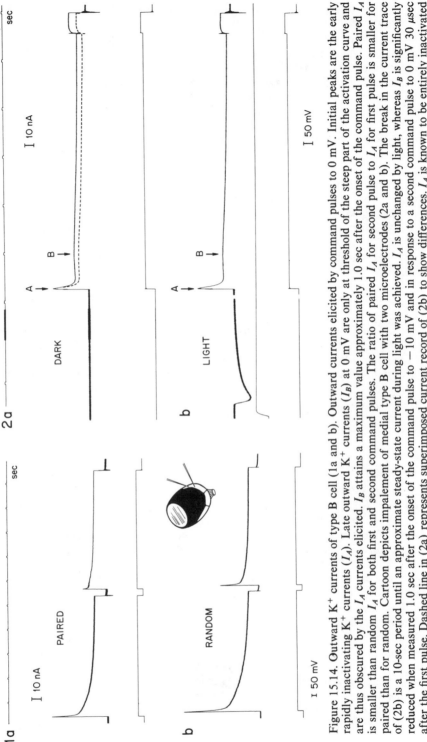

Figure 15.14. Outward K⁺ currents of type B cell (1a and b). Outward currents elicited by command pulses to 0 mV. Initial peaks are the early rapidly inactivating K⁺ currents (I_A). Late outward K⁺ currents (I_B) at 0 mV are only at threshold of the steep part of the activation curve and are thus obscured by the I_A currents elicited. I_B attains a maximum value approximately 1.0 sec after the onset of the command pulse. Paired I_A is smaller than random I_A for both first and second command pulses. The ratio of paired I_A for second pulse to I_A for first pulse is smaller for paired than for random. Cartoon depicts impalement of medial type B cell with two microelectrodes (2a and b). The break in the current trace of (2b) is a 10-sec period until an approximate steady-state current during light was achieved. I_A is unchanged by light, whereas I_B is significantly reduced when measured 1.0 sec after the onset of the command pulse to −10 mV and in response to a second command pulse to 0 mV 30 μsec after the first pulse. Dashed line in (2a) represents superimposed current record of (2b) to show differences. I_A is known to be entirely inactivated during the second command pulse. (From Alkon et al. 1982a)

Table 15.1. *Voltage-dependent current characteristics of medial type B cells*

	Paired (P)	N	Random (R)	N	"Naive" control (C)	N	Comparison	P <
I_A 60D (nA) ± SD	39.10 ± 10.70	10	55.13 ± 19.11	8	56.07 ± 15.27	7	P < R	.025
							P < C	
							R = C	
I_A 50D (nA) ± SD	16.82 ± 5.51	11	24.20 ± 9.53	7	24.92 ± 8.50	6	P < R	.05
							P < C	.025
							R = C	N.S.
I_{A2}/I_{A1} ± SD	0.348 ± 0.099	10	0.523 ± 0.110	8	0.439 ± 0.067	7	P < R	.005
							P < C	.001
							R = C	N.S.
t_{pp} (msec) ± SD	211.00 ± 18.17	5	250.00 ± 54.31	6	233 ± 13.51	5	P = R	N.S.
							P < C	.05
							R = C	N.S.

Key: N = number of medial cells, one per eye. I_A 60D = I_A elicited by command step to 0 mV. I_A 50D = I_A elicited by command step to −10 mV. I_{A2}/I_{A1} = Ratio of maximum I_A value at 0 mV for second to maximum value for first of twin command pulses. t_{pp} = half-time of decay. N.S. = not significant. Comparison: by Student *t*-tests.

B cell, explain an enhanced depolarizing response during and following a light step? (This enhanced type B response, particularly following the light step, was found as already mentioned, only for "paired" cells during the retention period and not controls.) To understand the role of I_A it is crucial to recognize that the membrane potential of any cell is determined by the summed effects of all the currents flowing across the membrane at any given time. *In the dark,* quantitative analysis (Shoukimas pers. comm.) revealed that although I_A is greatly inactivated at the type B cell's resting membrane potential, enough I_A remains activated during the steady state that a substantial proportion (~40 percent) of the cell's total resting membrane conductance (and thus its input resistance) is due to the conductance for I_A. This analysis received support from an additional observation (Acosta-Urquidi pers. comm.) that superfusion of the type B cell with 10 mM 4-aminopyridine (which blocks I_A conductance) causes a marked rise of input resistance in the dark. Taken together, these data motivate a prediction that was confirmed by our voltage and current clamp studies of conditioned and control animals. Since I_A is reduced only for the conditioned animals (and not for random and naive controls) type B input resistance measured in darkness should be elevated only for the conditioned animals. This prediction has now been confirmed by a number of independent studies of type B cells within intact circumesophageal nervous systems (Crow and Alkon 1980; Farley and Alkon 1982) and of type B cells isolated (after training) by axotomy from all synaptic interactions and excitable axonal membrane (i.e., membrane capable of generating and propagating action potentials) (West et al. 1981, 1982).

Further quantitative analysis of the currents that flow across the type B membrane during and following presentation of a light (Shoukimas pers. comm.) indicate that during the initial transient peak depolarization, the transient inward light-induced Na^+ current predominates, whereas I_A (because of the balance of its activation and inactivation characteristics) plays a more minor role. The reduction of I_A specific to conditioning, therefore, should only reduce the peak depolarizing voltage response by 3–5 mV. This prediction was in fact consistent with the voltage responses recorded from type B cells isolated after training (West et al. 1982).

During the steady-state phase of the light response, I_{Na+} is very small, and I_B is abolished, not because of its inactivation properties but due to the effect of light most likely mediated by cyclic AMP–dependent protein kinase (Alkon et al. 1983). Since light itself has no direct effect on I_A (Figure 15.14), and because of its steady-state activation and inactivation characteristics, some I_A current persists during the steady-state light response. The voltage-dependent Ca^{2+} current, although small, also persists. The membrane potential of the type B cell during and following the steady-state phase of its light response should be determined largely by these two opposing currents: I_A, which causes the membrane to hyperpolarize, and the sustained Ca^{2+} current, which would cause the membrane to depolarize. Thus, a reduced I_A for conditioned cells would be predicted to cause more depolarization and thus an elevated response during and following a light step. This prediction again was confirmed by a number of independent studies (Crow and Alkon 1980; West et al. 1981, 1982; Farley and Alkon 1982). Although I_A is to a considerable degree transient, it can control the magnitude of the sustained light response through its effect in opposing the voltage-dependent Ca^{2+} current. When conditioning reduces I_A and thus enhances the type B light response, it can, then, via the neuronal chain described above, be expected to decrease turning behavior in response to light. Type B impulses inhibit the medial type A photoreceptor that excites an ipsilateral interneuron and thereby an ipsilateral motoneuron, the MN1 cell (Goh and Alkon 1982, and in press). More type B impulses during a light step will cause less type A excitation of the MN1 cell that causes turning toward a light source. Furthermore, if, as the animal moves, it goes from an illuminated to a darkened area, it will have less ability to resolve the transition from light to dark when the type B after-depolarization is enhanced. That is, because the type B cell of conditioned animals, due to a reduced I_A, remains depolarized longer *after* the light goes off, the animal will perceive the light as still on for a longer time when it encounters a dark border. We have shown in our laboratory that the animal's ability to discriminate light–dark differences is reduced on retention days when I_A is reduced by training (Lederhendler pers. comm.).

What of other membrane current changes during retention of the associative conditioning? We know from past studies that the non–voltage-dependent K^+ current (the "leak" current) and the delayed voltage-dependent K^+ current I_B are not changed during the retention period, whereas I_A, as described above, is reduced and more rapidly inactivates. That I_{Na+}, the light-induced sodium current, is not changed was indicated by the observations of West et al. (1982) that the peak depolarizing transient (which largely results from I_{Na+}) is not significantly changed by conditioning. Furthermore, between-group differences for the peak, steady-state, and LLD responses to light were eliminated by hyperpolarizing the type B cell with steady negative current injection. Such hyperpolarization would be expected to enhance between-group differences due to changes of I_{Na+} since I_{Na+} for all cells is larger (due to an increased driving force) with hyperpolarization.

Additional observations, however, clearly raised the possibility that not only the I_A current was modified by conditioning. When the type B cell was depolarized with steady positive current injection, between-group differences in the response to light were enhanced (West et al. 1982). I_A should be significantly inactivated during such prior depolarization and thus might be expected to cause less of a conditioning-induced change in the type B light response. Since present evidence indicates that I_B, I_{Na+}, and the "leak" current do not change, likely candidates, in addition to I_A, responsible for the light response differences during steady depolarization are the voltage-dependent calcium current and/or the calcium-dependent potassium current (I_C). This interpretation is consistent with the results of an additional experiment designed to uncover conditioning-induced current changes other than I_A (Farley and Alkon 1983). Positive current steps were injected into type B cells immediately prior to the onset of a light step. Such depolarization should also significantly inactivate I_A currents. Type B cells from conditioned animals following such depolarization still showed an enhanced steady-state and LLD response to light steps. A changed voltage-dependent calcium current and/or calcium-dependent potassium current might account for these remaining differences. However, additional experiments examining type B responses when I_A and I_B are totally blocked (respectively by external 4-aminopyridine and TEA ions) are now being conducted to rule out alternative explanations. Also, the voltage-dependent calcium currents and calcium-dependent potassium currents are now being measured directly for conditioned and control animals.

In summary, a voltage-dependent potassium current I_A has been found to be reduced only in conditioned animals. This reduction of the I_A current has been shown to play a causal role in transforming the type B cell and via known pathways the muscular response of *Hermissenda* to light. There may be, however, additional changes of membrane currents that also contribute to encoding and expression of the conditioned response. Furthermore, other cells, such as the medial type A photoreceptor, also can serve as a primary locus of membrane changes responsible for the conditioned behavior. Ongoing studies that implicate the type A cell as another site for storing the learned information (Farley pers. comm.) suggest that cells that have synaptic interactions (e.g. the type B and type A photoreceptors) may undergo complimentary membrane changes. These complimentary changes, then, may have the effect of biasing a neural system (Alkon 1983) so that changes of elicited neuronal responses are amplified by the integrated effect of a neural system's (i.e., synaptically linked neurons) response to a conditioned stimulus following conditioning.

LONG-TERM CHANGES OF MEMBRANE CURRENTS

The persistence of I_A reduction produced by the conditioning of *Hermissenda* is within a temporal domain of days. Because *Hermissenda* can be conditioned for 3 or more weeks (Harrigan pers. comm.), I_A reduction may last much longer. In any case, no other identified membrane current has been shown to undergo stimulus-induced transformations for more than many minutes. Rahamimoff and others have studied, at the neuromuscular junction, posttetanic potentiation lasting for many minutes. Atwood and his colleagues have observed synaptic facilitation lasting for many minutes, and occasionally hours. Carew and coworkers also have evidence of persistent synaptic habituation (see Chapter 8). These synaptic phenomena, which have no demonstrated relation to mechanisms of associative learning, have only in some cases been attributable to alterations of recognizably distinct membrane channels. In the case of posttetanic potentiation and synaptic facilitation, ionic currents (Na^+ and Ca^{2+}) have been implicated for a time course of many minutes (see Rahamimoff et al. 1976; Zucker 1982). In the case of habituation at an *Aplysia* synapse, it has not been possible to identify membrane currents, because of the inaccessibility of this synapse either to voltage clamp analysis or focal extracellular recordings. Investigators have speculated about the relevant synaptic currents by attempting to identify the soma currents of the same cell (see Chapter 9). There is, however, abundant evidence that membrane currents cannot be expected a priori to be uniform with a neuron's geometry. This is true for *Hermissenda* neurons, for example, some of whose axonal membranes can generate impulses when soma membranes cannot. It is true for the squid, where synaptic currents predominate with entirely different ratios than within the axon. It is true of cerebellar Purkinje cells where soma membrane supports sodium impulses and dendritic membranes are the sites of origin for Ca^{2+} impulses (Llinas 1981).

Although the conditioning-induced reduction of I_A lasting for days cannot yet be compared to other known stimulus- or training-induced change of membrane conductance, since no others have yet been measured, specific conductance changes that do persist have been observed during early development (Cowan 1981). As we determine the transformation of I_A (and possibly other currents) for *Hermissenda* that have been conditioned for many weeks, the relationship of this transformation to developmental membrane changes could become closer or at least clearer.

BIOCHEMICAL ANALYSIS OF LEARNING-INDUCED CHANGES
OF IONIC CHANNELS

To the biophysicist, a central question has concerned the biochemical processes that actually determine the opening and closing of channels (or related structures) that allow ions to flow across biological membranes. The very fast kinetics with which such channels open and close has made biochemical analysis exceptionally difficult. Although physical measurements of currents associated with single channels are now becoming possible (see Neher et al. 1978), biochemical analysis of these currents remains elusive. Not only are the kinetics unfavorable, but because the number of open channels at any given time is relatively small, biochemical states correlated with channel "openness" are not easily defined. The persistent reduction of I_A channels within the type B cell membrane, therefore, offers a new opportunity for biochemical questions of membrane physiology as well as of mechanisms of learning. With a substantial *steady-state* decrease of I_A channels in the open state (which decrease causes associative learning) we can ask what are the observable correlated biochemical differences. Within the neural systems of *Hermissenda* we found many correlated neural changes, and it then became necessary to determine those changes that were primary, which actually stored the learned information and cause the learned behavior. Similarly in our laboratory, we are now surveying the correlated biochemical differences in those cells that cause the associative learning (see Chapter 16) and searching among these correlated differences for the primary or causal biochemical changes. We have found, for example, that for conditioned animals, but not random and naive controls, incorporation of radioactive ^{32}P into a specific protein band ($\sim 20,000$ MW) was changed. Because this suggested that persistent states of protein phosphorylation might regulate the flow of ionic currents, we injected protein kinases into isolated type B cell bodies. The cyclic AMP–dependent (cAMP-dependent) catalytic subunit of protein kinase reduced the late voltage-dependent K^+ current (Figure 15.15) I_B to a greater extent than the early current, I_A, which was implicated in causing the associative learning (Alkon et al. 1983). As might be expected from other experiments (see above) that implicated progressive elevation of intracellular Ca^{2+} and its reduction of I_A in acquisition of the associative learning, the injection of small quantities of Ca^{2+}–calmodulin–dependent protein kinase very specifically affected I_A but not I_B (Acosta-Urquidi et al. 1982). Whereas the cAMP-dependent protein kinase caused an immediate effect on I_A and I_B, the Ca^{2+}–calmodulin–dependent protein kinase reduction of I_A only followed Ca^{2+} influx into the type B cell. Thus, the specificity of these biochemical effects suggests which are important for the associative learning. Since learning did not affect I_B but did affect I_A, cAMP-dependent phosphorylation is less likely to be involved than Ca^{2+}–calmodulin–dependent phosphorylation in producing the persistent change of I_A channels observed for the conditioned but not the control animals. However, it should be noted that elevated intracellular cAMP has been shown to reduce the light- and voltage-dependent Ca^{2+} current (Alkon 1979). A complex interaction of Ca^{2+} and cAMP-sensitive reactions may therefore be involved.

How direct is the involvement of Ca^{2+}–calmodulin–dependent phosphorylation in modifying the I_A channel? Future research should help establish what intervening biochemical steps, if there are such, link this phosphorylation to the persistent changes of membrane conductance observed. In the process, which step or steps are actually modified in a rate-limiting manner may become clear. For example, it has been suggested (Howard Rasmussen pers. comm.) that the Ca^{2+}–calmodulin–dependent protein kinase may phosphorylate itself in a fairly irreversible manner. Such phosphorylation may in turn regulate the activity of other enzymes more directly responsible for the membrane conductances.

Changes of phosphorylation may be critical for learning changes lasting for days. Other types of changes – for example, those involving protein synthesis

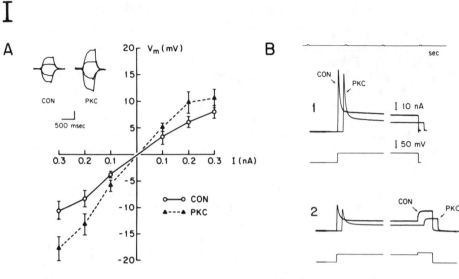

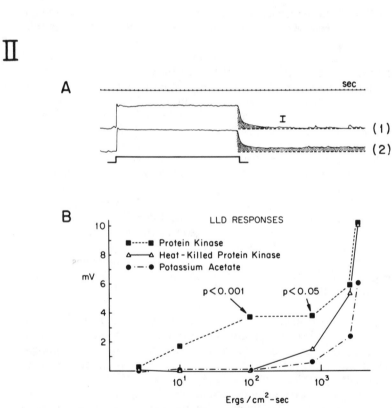

Figure 15.15. *IA:* Effects of iontophoretic PKC injection on dark membrane effective input resistance in type B photoreceptors. Values for open circles – control injection of carrier solution (for each point, $N = 12 \pm SEM$). Inset (top) shows representative

– are now being considered for *Hermissenda* learning that lasts for many weeks. Ultimately, a sequence of biochemical changes may provide for different temporal domains of membrane and thus behavior change. The longest-lasting biochemical changes may then affect neuronal geometry in ways that bear some relationship to developmental transformations.

Based on our current data, associative learning of *Hermissenda* can be understood to occur as follows: Temporally associated stimuli specifically perturb genetically constrained neural systems to cause in the type B cells, membrane depolarization and elevation of intracellular Ca^{2+}. During repetitive temporal association – that is, during conditioning – this depolarization and intracellular Ca^{2+} accumulate. When intracellular Ca^{2+} is sufficiently elevated, it activates the Ca^{2+}–calmodulin–dependent protein kinase. This activated enzyme causes increased fairly irreversible phosphorylation of itself,

Caption to Figure 15.15 (*cont.*)

voltage responses to current injection (± 0.1, 0.2 nA). Vertical calibration bar is 5 mV. *IB:* Outward K^+ currents of type B cell. 1, Outward currents elicited by command pulses to 0 mV from a V_H of -60 mV (lower trace indicates $+60$-mV command). I_S currents are the early peak transients. Late outward K^+ currents (I_B) at 0 mV are only at threshold of the steep part of the activation curve and are thus masked by the predominant I_A currents elicited. Note that I_A (peak value) following protein kinase injection (PKC) is only somewhat smaller than I_A before injection (CON), whereas I_B (PKC) is substantially reduced when measured 1.0 sec after the onset of the command pulse. Note that the full recordings are not presented for a 10-sec interval indicated by the interruption of traces. 2, Outward currents elicited by command pulses to -10 mV from a V_H of -60 mV (prolonged lower trace indicates $+50$-mV command). I_B attains a maximum value approximately 1.0 sec after the onset of the command pulse. Note that I_B is significantly reduced following protein kinase injection (PKC) when measured 1.0 sec after the onset of the command pulse to -10 mV and in response to a second command pulse to 0 mV, 30 μsec after the first pulse. *II:* Effects of protein kinase injection on type B photoreceptor LLD responses following offset of light steps. PKC injection was accomplished by causing the voltage across the microelectrode to undergo an electrical voltage oscillation ("ringing") of approximately 30 V peak to peak for approximately 1 sec. Since the PKC is negatively charged at pH 8.6, the 15-V negative phases of the oscillation were probably directly responsible for the actual enzyme injection. Responses of type B photoreceptors impaled with microelectrodes filled with PKC solution were recorded (after "ringing") during and after light steps of increasing intensity following 10 min of dark adapation. Control responses were also recorded (after "ringing") from type B photoreceptors impaled with microelectrodes filled with heat-denatured (in boiling H_2O for 1 min) PKC solutions or potassium acetate (KAc) alone. *A:* Responses to a 30-sec light step (intensity 7.5×10^2 erg cm^{-2} sec^{-1}, lower trace) recorded intracellularly with K^+ acetate (upper trace 1) and protein kinase solution (lower trace 2) in the microelectrodes. Recordings with KAc electrodes were not appreciably different from those obtained with heat-denatured protein kinase solutions. Lower trace monitors light step. *B:* Effects of PKC injection on the amplitude of long-lasting depolarization (LLD) measured at 30 sec following light offset as a function of different light intensities. Values for each are the means of five or more cells from different animals. Statistical differences were tested by Mann–Whitney U-comparison test. Energy values indicated intensity (unfiltered) of light stimulus, provided by a tungsten 6-V, 15-W bulb (Philips, Netherlands). Calibration bar indicates 10 mV. (From Alkon et al. 1983) which in turn controls membrane phosphorylation responsible for persistent closure of the I_A (and possibly $I_{Ca^{2+}-K^+}$) potassium channels. On days when the associative learning is retained, membrane polarization and intracellular Ca^{2+} returns to pretraining or normal levels, while the Ca^{2+}–calmodulin–dependent phosphorylation and K^+ channel reduction persists.

IMPLICATIONS

What cellular principles of associative learning are suggested by these findings?

1. The convergence of distinct sensory pathways responsive to distinct stimuli must precede the learning of an association between these stimuli. There were, within the nervous system, several convergence points – that is, sites of synaptic interaction between the visual and statocyst pathways – that were demonstrated with intracellular recordings in every adult animal examined. These convergences were crucial to the production of the primary neural changes that were found. Since preexisting convergence preceded the association, growth and development of synapses and/or axonal branches were not necessary for *Hermissenda* learning lasting for days and may not be necessary for vertebrates as well. The developmental stage of a nervous system would determine, however, the number of convergence points and thus constrain the number and kind of associations that can be learned. This may be one means by which critical periods for learning are expressed.

The critical result of these convergences is enhanced synaptic excitation of a light-induced depolarization of the type B cell when light and rotation occur together in time. This enhanced depolarization of the type B cell illustrates a second cellular principle of associative learning.

2. Temporal association of distinct sensory stimuli must result in unique neuronal responses. This uniqueness of encoding paired sensory stimuli arises out of the preexisting neural systems' synaptic organization. Just as paired sensory stimuli are precisely encoded, repetition of these pairings is also encoded by biophysical changes within neuronal membranes – which is a third principle, namely the following:

3. Repetition of paired sensory stimuli must result in unique long-lasting neuronal changes. In our gastropod model the long-lasting inactivation of I_A was measured at the cell body – separated from axon and synaptic interactions. For decades it has been assumed that learning must occur at synapses and possibly terminal branches. Our model suggests that synaptic organization is critically important for the specificity of learned associations because the association of stimuli elicits responses that are endowed with unique characteristics by virtue of the synapses. But the site of primary change, the locus of the actual storage of information, need not be at the synapses.

4. A final cellular principle concerns a paradox. From Karl Lashley's experiments and those of others we know that learning is a diffuse process: It occurs simultaneously in many areas of the mammalian brain. From our own personal learning experiences we are aware of a related diffuseness: Within the context of what we can perceive, any one stimulus can be associated with any other. Aside from the enormous number of convergences that must exist, there must

be an incredible number of neurons that have common biophysical characteristics. The long-lasting depolarization that we have found with our learning model is in no way unique; it has been observed in vertebrate preparations for many years. The I_A and its ready inactivation by depolarization is also not unique – these have been identified for a wide variety of neurons. Thus, a final cellular principle suggested by our results is that the necessary conditions for associative neuronal changes are widespread in neural systems. I would expect, therefore, that the necessary biophysical steps we have identified for *Hermissenda* for associative learning also are widespread in vertebrate brains and therefore serve some memory function in these as well. How important a function, of course, remains unclear. Thus far, there are no other *known* biophysical mechanisms for associative learning for vertebrates or invertebrates. Other conductance changes with different biochemical bases may very likely serve such functions. But as an example of how such biophysical steps in sequence can produce associative learning, our experience with *Hermissenda* will hopefully be of some value.

The more detailed is our understanding of the biochemistry of the persistent membrane conductance changes that cause *Hermissenda* associative learning, the more precise predictions can be formulated to test for generality of cellular principles of learning. As mentioned above, iontophoretic injection of Ca^{2+}–calmodulin–dependent protein kinase in the type B cell caused long-term reduction of the I_A current (which changed with learning) after intracellular Ca^{2+} was elevated by a voltage-dependent Ca^{2+} flux (see above). This observation motivated a prediction. The same enzyme injection into vertebrate neurons (of awake cats) that had been implicated as sites of learning-induced change (see Chapter 5) should cause changes of input resistance (presumeably due to I_A reduction) in a Ca^{2+}-dependent manner. Although the presence of an I_A current and a voltage-dependent Ca^{2+} current were not demonstrated for these cells, it was hypothesized that, should they be present, the enzyme iontophoresis followed by a depolarizing current step would produce the predicted input resistance change. That this prediction has now been experimentally confirmed by Charles Woody, Bruce Hay, and myself, raises the possibility that the biophysical mechanisms established for *Hermissenda* associative learning apply to a wide range of species.

Whatever the biophysical mechanisms are revealed to be for vertebrate species, their behavioral manifestations will probably be far more complex than those for a less evolved animal such as a nudibranch mollusk. For *Hermissenda,* long-lasting differences of specific ion channel function have been shown to actually cause associatively learned behavior. For *Hermissenda* a discrete stimulus, light, has been associated with another well-defined stimulus, rotation, to result in changes of channels and thus behavior. For vertebrate species such as cats, discreteness of the associated stimuli is probably only approximated in elemental learning paradigms used by Thompson (see Chapter 4) and Woody (see Chapter 5) and their colleagues. It seems more likely that even during the association of a glabellar tap with a click a variety of contextual cues are also being registered or associated by the animal's nervous system during training. Associative learning, therefore, probably results in many associations that are diffusely represented at neuronal convergence sites distributed across the animal's brain. Some of these sites undoubtedly are more important than others in actually producing a manifest behavioral change. For vertebrates, however, it seems reasonable that there are many sites for storing associatively learned information that do not have direct behavioral expresson. The mechanisms of storage, however – for instance, persistent changes of ionic channels – may be the same. Modification of membrane currents intrinsic to hippocampal neurons, therefore, will be important to define even if the hippocampus need not be present to observe an associatively learned behavior. In conclusion, although a distributed set of neural changes in a mammalian brain will not bear the same direct causal relationship to behavioral outcomes as has been shown for *Hermissenda,* the biophysical processes by which these changes arise and the biochemistry of their storage may be in many respects similar. Provided we don't oversimplify the similarities or exaggerate the differences, common principles will hopefully emerge.

REFERENCES

Acosta-Urquidi, J., Neary, J. T. and Alkon, D. L. 1982. Ca^{2+}-dependent protein kinase regulation of $K^+(V)$ currents: a possible biochemical step in associative learning of *Hermissenda. Soc. Neurosci. Abstr. 8*:825.

Alkon, D. L. 1973. Intersensory interactions in *Hermissenda. J. Gen. Physiol. 62*:185–202.

1974. Sensory interactions in the nudibranch mollusc *Hermissenda crassicornis. Fed. Proc. 33*:1083–1090.

1975. Neural correlates of associative training in *Hermissenda. J. Gen. Physiol. 65*:46–56.

1976. Neural modification by paired sensory stimuli. *J. Gen. Physiol. 68*:341–358.

1979. Voltage-dependent calcium and potassium ion conductances: A contingency mechanism for an associative learning model. *Science 205*:810–816.

1980a. Cellular analyis of a gastropod *(Hermissenda crassicornis)* model of associative learning. *Biol. Bull. 159*:505–560.

1980b. Membrane depolarization accumulates during acquisition of an associative behavioral change. *Science 210*:1375–1376.

Alkon, D. L., Acosta-Urquidi, J., Olds, J., Kuzma, G., and Neary, J. 1983. Protein kinase injection reduces voltage-dependent K^+-currents. *Science 219*:303–306.

Alkon, D. L., Akaike, T., and Harrigan, J. F. 1978. Interaction of chemosensory, visual, and statocyst pathways in *Hermissenda. J. Gen. Physiol. 71*:177–194.

Alkon, D. L., and Fuortes, M. G. F. 1972. Responses of photoreceptors in *Hermissenda. J. Gen. Physiol. 60*:631–649.

Alkon, D. L., and Grossman, Y. 1978. Long-lasting depolarization and hyperpolarization in eye of *Hermissenda. J. Neurophysiol. 41*:1328–1342.

Alkon, D. L., Lederhendler, I., and Shoukimas, J. J. 1982a. Primary changes of membrane currents during retention of associative learning. *Science 215*:693–695.

Alkon, D. L., Shoukimas, J., and Heldman, E. 1982b. Calcium-mediated decrease of a voltage-dependent potassium current. *Biophys. J. 40*:245–250.

Connor, J. A., and D. L. Alkon. In press. Light-induced changes of intracellular Ca^{++} in *Hermissenda* photoreceptors measured with Arsenazo III. *J. Neurophysiol.*

Cowan, W. M. 1981. *Studies in Developmental Neurobiology: Essays in Honor of Viktor Hamburger.* New York: Oxford University Press.

Crow, T. J., and Alkon, D. L. 1978. Retention of an associative behavioral change in *Hermissenda. Science 201*: 1239–1241.

1980. Associative behavioral modification in *Hermissenda:* Cellular correlates. *Science 209*:412–414.

Crow, T., and Harrigan, J. F. 1979. Reduced behavioral variability in laboratory-reared *Hermissenda crassicornis* (Eschscholtz, 1831) (Opisthobranchia: nudibranchia). *Brain Res. 173*:179–184.

Crow, T., Heldman, E., Hacopian, V., Enos, R., and Alkon, D. L. 1979. Ultrastructure of photoreceptors in the eye of *Hermissenda* labeled with intracellular injections of horseradish peroxidase. *J. Neurocytol. 8*:181–195.

Crow, T., and Offenbach, N. 1979. Response specificity following behavioral training in the nudibranch mollusc *Hermissenda crassicornis. Biol. Bull. 157*:364.

Detwiler, P. B., and Alkon, D. L. 1973. Hair cell interactions in the statocyst of *Hermissenda. J. Gen. Physiol. 62*:618–642.

Farley, J., and Alkon, D. L. 1980. Neural organization predicts stimulus specificity for a retained associative behavioral change. *Science 210*:1373–1375.

1981. Associative neural and behavioral change in *Hermissenda:* Consequences of nervous system orientation for light- and pairing-specificity. *Soc. Neurosi. Abstr. 7*:352.

1982. Associative neural and behavioral change in *Hermissenda:* Consequences of nervous system orientation for light- and pairing-specificity. *J. Neurophysiol. 48*:785–808.

1983. Changes in *Hermissenda* Type B photoreceptors involving a voltage-dependent Ca^{++} current and a Ca^{++}-dependent K^+ current during retention of associative learning. *Soc. Neurosci. Abstr. 9*:167.

Farley, J., Richards, W. G., and Alkon, D. L. 1982. Extinction of associative learning in *Hermissenda:* Behavior and neural correlates. *Psychonom. Soc. Abstr. 20*(3):144.

Farley, J., Richards, W. G., Ling, L. J., Liman, E., and Alkon, D. L. 1983. Membrane changes in a single photoreceptor cause associative learning in *Hermissenda. Science 221*:1201–1203.

Goh, Y., and Alkon, D. L. 1982. Convergence of visual and statocyst inputs on interneurons and motoneurons of *Hermissenda:* A network design for associative conditioning. *Soc. Neurosci. Abstr. 8*:824.

In press. Sensory, interneuronal and motor interactions within the *Hermissenda* visual pathway. *J. Neurophysiol.*

Grossman, Y., Alkon, D. L., and Heldman, E. 1979. A common origin of voltage noise and generator potentials in statocyst hair cells. *J. Gen. Physiol. 73*:23–48.

Hubel, D. H., and Wiesel, T. N. 1972. Laminar and columnar distribution of geniculocortical fibers in the macaque monkey. *J. Comp. Neurol. 146*:421–450.

Kuzirian, A. M. In press. Neuroanatomy of *Hermissenda:* A model system for the study of associative learning. *J. Moll. Stud., Suppl. 12A.*

Lederhendler, I. I., Barnes, E. S., and Alkon, D. L. 1980. Complex responses to light of the nudibranch *Hermissenda crassicornis. Behav. Neural Biol. 28*:218–230.

Lederhendler, I. I., Goh, Y., and Alkon, D. L. 1982. Type B photoreceptor changes predict modification of motorneuron responses to light during retention of *Hermissenda* associative conditioning. *Soc. Neurosci. Abstr. 8*:824.

Llinas, R. R. 1981. Electrophysiology of the cerebellar networks. In *Handbook of Physiology,* Section 1, *The Nervous System,* Volume II, Part 2, *Motor Control* (V. B. Brooks, ed.), pp. 831–76. Bethesda, Md.: American Physiological Society.

Llinas, R. R., and Simpson, J. I. 1981. Cerebellar control of movement. In *Handbook of Behavioral Neurobiology: Motor Coordination,* vol. 5, (A. L. Towe and E. S. Luschei, eds.), pp. 231–302. New York: Plenum Press.

Mountcastle, V. B. 1976. The world around us: Neural command functions for selective attention. *Neurosci. Res. Prog. Suppl. 14*:1–47.

Mpitsos, G. J., and Collins, S. D. 1975. Learning: Rapid aversive conditioning in the gastropod mollusk *Pleurobranchaea. Science 188*:954–956.

Neher, E., Sackmann, B., and Steinbach, I. H. 1978. The extracellular patch clamp: A method for resolving currents through individual open channels in biological membranes. *Pflugers Arch. 375*:219–228.

Pellionisz, A., and Llinas, R. R. 1979. Brain modeling by tensor network theory and computer simulation. The cerebellum: Distributed processor for predictive coordination. *Neuroscience 4*:232–348.

Rahamimoff, R., Erulkar, S. D., Alnase, E., Meiri, H., Rotshenker, S., and Rahamimoff, H. 1976. Modulation of transmitter release by calcium ions and nerve impulses. *Cold Spring Harb. Symp. Quant. Biol. 40*:107–116.

Stommel, E. W., Stephens, R. E., and Alkon D. L. 1980. Motile statocyst cilia transmit rather than directly transduce mechanical stimuli. *J. Cell Biol. 87*:652–662.

Tabata, M., and Alkon, D. L. 1979. Control of synaptic feedback in a molluscan visual system. *Invest. Ophthal. Vis. Sci. Suppl. 77*.

   1982. Positive synaptic feedback in visual system of nudibranch mollusk *Hermissenda crassicornis. J. Neurophysiol. 48*:174–191.

Takeda, T., and Alkon, D. L. 1982. Correlated receptor and motorneuron changes during retention of associative learning of *Hermissenda crassicornis. Comp. Biochem. Physiol. 73A*:151–157.

West, A., Barnes, E., and Alkon, D. L. 1981. Primary neuronal changes are retained after associative learning. *Biophys. J. 33*:93a.

   1982. Primary changes of voltage responses during retention of associative learning. *J. Neurophysiol. 48*:1243–1255.

Zucker, R. S. 1982. Processes underlying one form of synaptic plasticity: Facilitation. In, *Conditioning,* C. D. Woody, Ed. New York: Plenum Press, pp. 249–264.

16 · Biochemical correlates of associative learning: protein phosphorylation in *Hermissenda crassicornis,* a nudibranch mollusk

JOSEPH T. NEARY

Recent neurophysiological studies have advanced our understanding of the cellular basis of learning and behavior (for reviews, see Refs. 1–5). However, very little is known about the biochemical processes underlying behavioral modification, especially at the cellular level. Previous work in behavioral biochemistry often utilized complex nervous systems containing many cells and cell types, thus making it difficult to determine specific sites of biochemical changes. There are now available, however, experimental preparations as well as biochemical and immunological techniques that make it possible to avoid many of the problems and criticisms of the earlier studies in behavioral biochemistry. Thus, recent advances in biochemical and immunological techniques, such as two-dimensional gel electrophoresis, single-cell analyses, and monoclonal antibodies, and the availability of simplified, neuronal systems with well-defined pathways, such as invertebrate preparations and hippocampal tissue slices, make this an exciting and hopefully fruitful period to investigate biochemical processes that play a role in learning and behavior.

POSSIBLE BIOCHEMICAL PROCESSES RELATED TO LEARNING AND BEHAVIOR

It is unlikely that one biochemical process can explain the diversity of learning and behavior exhibited by organisms. All neurons contain nucleic acids, proteins, lipids, carbohydrates, metabolites, and cofactors as well as combinations of substances in these groups such as glycoproteins, lipoproteins, and glycolipids. Specific molecules in some or all of these categories may be utilized by cells in order to process and store information needed to initiate and retain a behavioral modification. The purpose of biochemical studies in this area of research is to discover what molecular changes occur during learning and what biochemical mechanisms regulate these changes.

In order to focus attention on specific biochemical processes that might be related to learning and behavior, the following outline is proposed as a possible biochemical sequence that might be utilized by cells in a specific pathway that mediates stimulation and response. The sequence includes a fast, reversible event and a slow, more permanent change. In the vicinity of the sensory receptor on the plasma membrane, the structure and function of the membrane may be altered by enzyme-catalyzed protein and lipid modification reactions or by

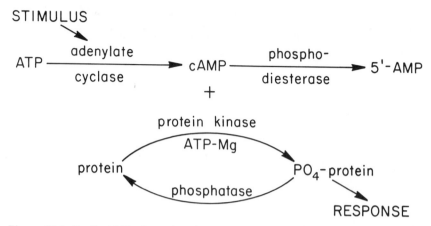

Figure 16.1. Cyclic AMP–dependent protein phosphorylation in a stimulus–response pathway.

insertion of new components into the membrane. This could give rise to the formation of an intracellular messenger that would activate a different biochemical system, such as gene expression, to provide for retention of the information. Gene products or their nucleic acid templates could then be available for use in a manner analogous to antibody synthesis or for use in membrane or cytoskeleton specialization or compartmentalization.

This chapter will focus on posttranslational, covalent protein modification reactions with an emphasis on protein phosphorylation in the nudibranch mollusk *Hermissenda crassicornis*. Protein covalent modification reactions offer several important features for the processing of sensory and synaptic information: (1) Since the reactions are catalyzed by enzymes, they can occur rapidly; (2) several of the reactions can be reversed by another enzymatic process, thereby providing a mechanism by which a substrate molecule can exist in active and inactive forms; (3) enzymes that catalyze the reactions can exhibit a high degree of substrate specificity and can be regulated by effector molecules; and (4) an extracellular stimulus could be amplified by cascade-type enzyme-catalyzed reactions.

Some examples of reversible posttranslational protein modification reactions include phosphorylation, methylation, acetylation, adenylation, and ribosylation. Protein phosphorylation is probably the most extensively studied covalent protein modification reaction in the nervous system (for review, see Ref. 6). Phosphorylation is catalyzed by protein kinases and dephosphorylation by protein phosphatases. Four types of protein kinases have been found and classified on the basis of the molecules that activate them: cAMP, cGMP, Ca^{2+}–calmodulin, and Ca^{2+}–lipids. Of these, cAMP-dependent protein kinase is the most extensively studied. By comparison, very little is known about the regulation of the protein phosphatases. A simplified scheme for the cAMP activation of protein phosphorylation is shown in Figure 16.1. An extracellular stimulus such as a neurotransmitter or neurohormone activates membrane-bound

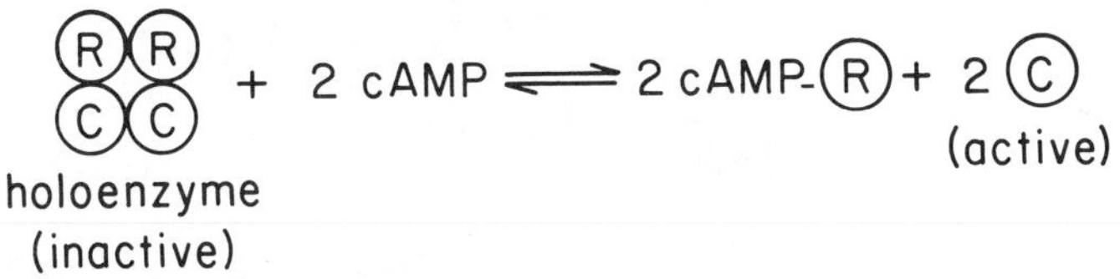

$$\begin{array}{c} \text{\textcircled{R}\textcircled{R}} \\ \text{\textcircled{C}\textcircled{C}} \end{array} + \ 2 \ cAMP \ \rightleftharpoons \ 2 \ cAMP\text{-}\text{\textcircled{R}} + \ 2 \ \text{\textcircled{C}}$$

holoenzyme
(inactive)

(active)

Figure 16.2. Dissociation of catalytic (C) and regulatory (R) subunits of cAMP-dependent protein kinase by cAMP.

adenylate cyclase by binding to a membrane receptor, thereby leading to the production of cAMP. (The level of cAMP is also affected by the activity of phosphodiesterase which degrades cAMP to 5′-AMP.) cAMP activates cAMP-dependent protein kinase by binding to the regulatory subunits of the holoenzyme and releasing the catalytic subunits (Figure 16.2), which in the presence of a Mg–ATP complex catalyze the phosphorylation of specific proteins. The terminal phosphate group of ATP is transferred to serine, threonine, or tyrosine residues at the hydroxyl position, thereby giving the protein a more negative charge. A change in the charge on a protein could affect physiological functions such as neurotransmitter release or the kinetics of ionic permeability. In this manner, protein phosphorylation is one type of biochemical process that can provide for the conversion of a molecule from an active to an inactive state, or vice versa, thereby leading to a new or altered response that was activated by an extracellular stimulus.

Greengard has discussed the possible roles of phosphorylated proteins in neuronal functions and has speculated that protein phosphorylation–dephosphorylation may be one of the biochemical mechanisms by which short-term and long-term information storage and retrieval are regulated [7]. Recently, it has been shown that changes in brain phosphoprotein bands occur following electrical stimulation of nerve pathways in hippocampal slices known to produce long-term changes in synaptic activity [8,9]. In addition, transient changes in the phosphorylation of cortical membrane proteins have been reported following electroconvulsive shock [10]. For marine invertebrates, the presence of neuronal protein kinases has been reported in *Aplysia* [11–15], *Loligo* [16], *Hermissenda* [17], and lobster [18]. Electrical stimulation [13,16], cyclic nucleotides and their analogs [11–15], and neurotransmitters [11,14,18] have been shown to alter levels of neuronal phosphoproteins and protein kinase. Serotonin and cAMP can mimic presynaptic facilitation of postsynaptic potentials in *Aplysia* motor neurons that underlie sensitization of the gill withdrawal reflex [19]. These agents can also stimulate phosphorylation of a membrane phosphoprotein band in abdominal ganglia [14], thereby suggesting a role for protein phosphorylation in short-term sensitization.

PROTEIN PHOSPHORYLATION AND ASSOCIATIVE LEARNING
IN *Hermissenda*

For a biochemical analysis of learning, invetebrate nervous systems offer several advantages in that they typically contain a relatively small number of neu-

AUTORADIOGRAMS

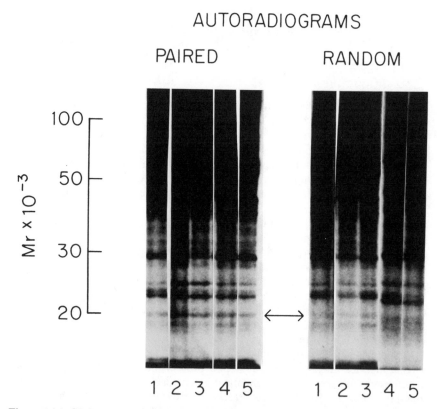

Figure 16.3. SDS electrophoresis gels comparing endogenous protein phosphorylation in eyes of *Hermissenda* presented with paired or random light and rotation. Each lane represents an eye sample from one animal. Time of exposure is 3 days using intensification techniques [22,25]. Data for the more heavily labeled phosphoproteins (MW > 40,000) were obtained from autoradiograms of shorter duration and are presented in Table 16.1. See Ref. 22 for additional experimental details.

rons, some of which have large, identifiable somata. A model of associative learning has recently been developed in the nudibranch mollusk *Hermissenda crassicornis*. Drs. T. Crow and D. L. Alkon have shown that the phototactic behavior of *Hermissenda* can be modified by an associative conditioning procedure [20] and that cellular neurophysiological changes in the type B photoreceptors of the eye are correlated with the behavioral modification [21]. In an effort to elucidate the biochemical processes underlying associative learning and its neurophysiological correlates, we have investigated protein phosphorylation in the eyes of trained and control animals. We have observed an increase in the level of incorporation of ^{32}P into a specific phosphoprotein band [22]. Since the eye of *Hermissenda* is a relatively simple structure containing only five photoreceptors, a lens, and a few pigment and epithelial cells, a biochemical change related to associated learning has now been localized to a few cells within a nervous system.

The training and testing procedures [20] involve (1) measuring latencies of individual animals to enter an illuminated area; (2) training with simultaneous (paired) light and rotation, or random, unpaired, or no light and rotation for control groups; and (3) retesting to measure the effects of training on the animals' latencies to enter the illuminated area. Training and testing are conducted in an automated apparatus [23]. Animals trained with paired light and rotation exhibit significantly longer latencies than random or unpaired control groups [20]. For biochemical analysis, the circumesophageal nervous system, consisting of eyes, statocysts, and ganglia, was dissected from paired and control groups following testing and incubated for 2 hr at 15°C in artificial seawater containing glucose and ^{32}P-inorganic phosphate (for details of experimental procedures, see Ref. 22). The nervous systems were then rinsed in an ice-cold isotonic solution and the eyes and a small amount of surrounding connective tissue were dissected on a cold plate and lysed. Aliquots of eye samples from individual animals were analyzed for specific phosphoprotein bands by SDS-polyacrylamide gel electrophoresis [24] and autoradiography [25].

Autoradiograms of eyes from animals receiving paired or random light and rotation are shown in Figure 16.3. An increase was observed in the incorporation of ^{32}P into a 20,000-MW phosphoprotein band in the eyes of animals whose phototactic behavior was altered by training with paired light and rotation. This increase can also be seen in the densitometric scans of the relevant portions of the autoradiograms, as shown in Figure 16.4. An analysis of variance revealed an overall significant difference ($F_{2,30} = 6.49$, $P < .01$) among the groups in the 20,000-MW phosphoprotein band (Table 16.1). Planned comparisons (two-tailed t-tests) revealed that the paired group was significantly different ($P < .01$) from the random and unpaired control groups whereas the two controls were not different from each other. No significant differences were observed in 10 other phosphoprotein bands that were detectable in eyes from all groups studied (Table 16.1).

The 20,000-MW phosphoprotein band constitutes about 1 percent of the total phosphoproteins in the eyes of untreated animals. The level of incorporation in the 20,000-MW phosphoprotein band from eyes of animals receiving paired stimuli increases about twofold over that in eyes from untreated animals and 80 percent over eyes from animals receiving random or unpaired stimuli.

In some cases ($n = 5$), a second minor phosphoprotein band ($\sim$27,000 MW) was observed in eyes from animals receiving paired stimuli; although this band has not been consistently present, we have never observed significant levels of this phosphoprotein band in eyes from random, unpaired, or untreated animals. We have also observed the 27,000-MW phosphoprotein band in eyes from isolated nervous systems that have been stimulated with paired light and rotation (personal observations).

The change in the 20,000-MW phosphoprotein band following acquisition is highly specific since (1) 10 other phosphoprotein bands in the same sensory structure were not affected and (2) the alteration was observed in eyes from animals presented with paired, but not random, stimuli. There are several possibilities that could explain the change in the 20,000-MW phosphoprotein band. During training, pairing of stimuli could activate or induce a protein kinase that catalyzes the phosphorylation of the 20,000-MW protein. Alternatively, a phosphatase could be activated or induced during acquisition,

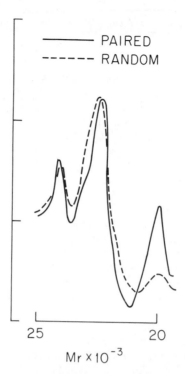

Figure 16.4. Densitometric scans of the 25,000- to 20,000-MW region of autoradi-ograms of eye samples from animals receiving paired (solid line) or random (dashed line) light and rotation.

Table 16.1. *Effect of paired, random, and unpaired stimulation on specific phosphoprotein bands in Hermissenda eyes*

Phosphoprotein band (MW)	Ratio: Experimental/untreated (mean ± *SEM*)		
	Paired (*n* = 16)	Random (*n* = 12)	Unpaired (*n* = 5)
72,000	0.83 ± .08	1.05 ± .23	1.30 ± .38
55,000	1.02 ± .04	1.04 ± .09	0.88 ± .07
44,000	0.88 ± .12	1.35 ± .10	1.04 ± .44
42,000	1.27 ± .14	1.18 ± .08	1.08 ± .13
38,000	1.11 ± .07	0.87 ± .10	1.23 ± .14
34,000	1.55 ± .21	1.22 ± .20	1.56 ± .32
31,000	1.12 ± .14	1.23 ± .18	0.91 ± .16
29,000	0.89 ± .11	1.24 ± .17	1.15 ± .41
24,000	1.12 ± .17	0.90 ± .15	1.10 ± .18
22,000	1.04 ± .10	1.16 ± .14	0.76 ± .07
20,000	2.16 ± .28[a]	1.13 ± .10	1.18 ± .14

[a] $F_{2,30} = 6.49$; p < .01.
Note: For experimental procedures, see Ref. 22, table 1.

resulting in the removal of phosphate group(s) from the 20,000-MW protein that could then be phosphorylated by protein kinase during the assay. Another possibility is that the process of acquisition could increase the rate of synthesis of a 20,000-MW acceptor protein.

Although we cannot exclude any of these possibilities at this time, the activation of protein kinase(s) may play a role in the increase in the incorporation of ^{32}P into the 20,000-MW phosphoprotein band since injection of two types of protein kinases into type B photoreceptors of untreated animals mimics some of the effects of associative learning on the electrophysiological properties of these photoreceptors. The effects of conditioning on type B photoreceptors following acquisition include a decrease in membrane potential, increase in input resistance, increase in spontaneous discharge frequency in the dark, and an enhanced long-lasting depolarization (LLD) in the tail of the generator potential in response to light [21]. When type B photoreceptors from trained animals are tested 1–2 days after the termination of conditioning, there is a reduction in I_A, a transient, rapidly inactivating K^+ current [26], an increase in input resistance in the dark, and an enhanced long-lasting depolarization response to light [27]. When the catalytic subunit of cAMP-dependent protein kinase is injected intracellularly into type B photoreceptors, increases in input resistance in the dark and in the LLD following a light step are observed [28,29]. Intracellular injection of a Ca^{2+}-dependent protein kinase, phosphorylase kinase, increases the input resistance and the LLD after Ca^{2+} influx is stimulated by light steps [30]. The increased input resistance (decreased conductance) can be explained by a reduction in K^+ conductance since the outward flow of K^+ normally acts to repolarize the membrane following depolarization. Voltage clamp studies have shown that both a Ca^{2+}-dependent protein kinase and the catalytic subunit of cAMP-dependent protein kinase reduce the early (I_A) and delayed (I_B) K^+ currents in the type B photoreceptors and that they exhibit a differential effect on these K^+ currents [29,30]. The catalytic subunit of cAMP-dependent protein kinase reduces I_B to a greater extent than I_A, whereas the Ca^{2+}–calmodulin–dependent protein kinase appears to have a greater effect on I_A. It may be that both Ca^{2+}–calmodulin- and cAMP-dependent protein kinases play a role in regulating the K^+ currents. Recent studies on phosphoproteins such as glycogen synthetase [31–34], phospholambden [35], and a neural-specific protein, protein I [36,37], have shown that Ca^{2+}–calmodulin- and cAMP-dependent protein kinases phosphorylate different sites on the same protein.

Since I_A is reduced in type B photoreceptors of trained animals [26], and since the aforementioned biochemical and electrophysiological studies suggest a role for protein phosphorylation in K^+ currents and associative learning, we have recently begun a series of experiments to investigate the effects of K^+ channel blockers on ^{32}P-phosphoproteins in eyes and ganglia of *Hermissenda* [38]. Previous studies in molluscan somata [39–42], including *Hermissenda* photoreceptors and other neurons [26; J. Shoukimas and D. L. Alkon pers. comm.], have shown that I_A can be blocked by 4-aminopyridine. In *Hermissenda*, I_B can be blocked by Ba^{2+} [43], and I_C, the Ca^{2+}-dependent K^+ current, can be blocked by Ni^{2+} [44]. When I_A is blocked by 4-aminopyridine, a 7- to 10-fold reduction in the level of phosphorylation in a 25,000-MW phosphoprotein band is observed in eyes, pedal ganglia, and the intact circumesophageal nervous system [38]. Reduction of I_A by high external K^+ (100 mM) also leads to a decrease in the level of phosphorylation in the same band. Blockade of the I_B current with Ba^{2+} or the I_C current with Ni^{2+} has no significant effect on the level of phosphorylation in this band. Recovery from I_A block by removal of 4-aminopyridine results in an increase in ^{32}P incorporation in the 25,000-MW band and in an additional phosphoprotein of 23,000 MW. These experiments suggest that a distinct type of K^+ current, I_A, may be related to the state of phosphorylation of neuronal proteins.

In order to correlate the evidence from our biochemical, electrophysiological, and behavioral studies, several questions need to be addressed. Firstly, since I_A is reduced in type B photoreceptors of trained animals and since 4-aminopyridine block of I_A results in a decrease in the level of phosphorylation of a 25,000-MW phosphoprotein band, one might expect to see a decrease in phosphorylation in the same band following acquisition. But as shown in Table 16.1, no significant effect on this band was observed following acquisition. (The protein band identified as a 25,000-MW phosphoprotein in the 4-aminopyridine studies comigrates with the band designated 24,000 in Table 16.1 in the biochemical–behavioral experiments.) The reason for this may be that the method used for detecting changes in levels of phosphoproteins does not have adequate sensitivity to detect a phosphoprotein change related to the 30 percent reduction in I_A that has been found in type B photoreceptors following training [26] but is sufficiently sensitive to detect the change in the 25,000-MW phosphoprotein brought about by the much greater block of I_A caused by 4-aminopyridine. Another explanation for the fact that the level of phosphorylation in the 25,000-MW phosphoprotein band was reduced by 4-aminopyridine block of I_A but not by training may involve the difference in phosphoprotein labeling procedures in the two different sets of experiments. In the 4-aminopyridine experiments, phosphoproteins were labeled with ^{32}P before I_A was blocked, whereas in the behavioral experiments phosphoproteins were labeled after I_A was reduced by training.

Another question that arises from an attempt to correlate the evidence from our electrophysiological and biochemical experiments concerns, on the one hand, reduction in I_A by injection of the protein kinases, which catalyze phosphorylation, and, on the other hand, the dephosphorylation effect observed following I_A block by 4-aminopyridine. One possible explanation for these two observations is that the injection of protein kinase may result in the phosphorylation of a transient phosphoprotein that can activate a protein phosphatase or inhibit a protein kinase thereby leading to a subsequent dephosphorylation. For example, it has recently been shown that an inactive form of protein-phosphatase-1 can be activated by phosphorylation [45,46]. Experiments in which phosphatases are injected into photoreceptors may help to clarify this point. Another explanation is the possibility that an open K^+ channel may be related to the titration of a specific number of phosphorylation sites on the channel protein or a protein associated with the channel, whereas a closed channel may reflect too few or too many phosphorylated sites on a protein. In this case, multisite phosphorylation, as induced by injected kinases, or multisite dephosphorylation, as induced by 4-aminopyridine, could both be related to a reduction in K^+ efflux through the I_A channel. Alternatively, the difference in the effect on phosphorylation in the electrophysiological and biochemical experiments may be related to differences in the experimental protocols. For example, in the electrophysiological experiments, protein kinase was injected intracellularly and depolarization was achieved with brief voltage pulses [29], whereas in the biochemical experiments reported here, 4-aminopyridine blocks externally and high external K^+ leads to prolonged rather than brief depolarization.

From the preceding discussion, it is clear that further studies are needed to provide a detailed understanding of the relationships between the phosphorylation/dephosphorylation changes observed in several phosphoproteins and associative learning and K^+ conductance in *Hermissenda*. Our early studies suggest that three phosphoproteins can be affected by associative learning and agents which reduce the I_A K^+ current. If the phosphorylation/dephosphorylation systems we have begun to investigate are as intricate as those that regulate glycogen synthesis and degradation (for recent reviews, see Refs. 47, 48), then as we continue our studies we may find that there are many phosphoproteins related to ionic conductance, neurotransmitter activity, and other processes utilized by cells to transmit, receive, store, and recall information.

Future plans include an attempt to determine if the 20,000-MW phosphoprotein or the other phosphoproteins are affected by the injection of the Ca^{2+}–calmodulin- and cAMP-dependent protein kinases. In addition, an investigation of the types of protein kinases – that is, cAMP-, Ca^{2+}–calmodulin-, cGMP-, and Ca^{2+}–lipid-dependent protein kinases – present in *Hermissenda* neural tissue is now underway. A cAMP-dependent protein kinase in the *Hermissenda* nervous system and protein kinase activity in the eye have been measured by an in vitro enzyme assay (personal observations with P. Kandel and D. L. Alkon). Attempts will be made to look for changes in protein kinase and phosphatase activities and in the synthesis of the 20,000-MW substrate protein in the eyes of trained and control animals.

Another question concerns the number of phosphoproteins present in the 20,000-MW band. In preliminary experiments ^{32}P-phosphoproteins from eyes of untrained *Hermissenda* have been analyzed by the technique of two-dimensional gel electrophoresis [49]. As shown in Figure 16.5, one major ^{32}P-phosphoprotein has been found in the 20,000-MW region of the two-dimensional gel; a second ^{32}P-spot with a similar isoelectric point may also be present. Analysis of eyes from trained animals should make it possible to determine if a new phosphoprotein appears in the 20,000-MW region following training or if the the 20,000-MW phosphoproteins(s) from untrained animals is increased following training.

Other questions of interest raised by our initial protein phosphorylation study [22] include: (1) What is the nature of the 20,000-MW phosphoprotein band? (2) Does the observed change in incorporation occur within the type B photoreceptors? (3) What is the intracellular location of the protein? (4) How many phosphorylation sites are present in the 20,000-MW protein, and, as mentioned above, are they regulated by different protein kinases – for example,

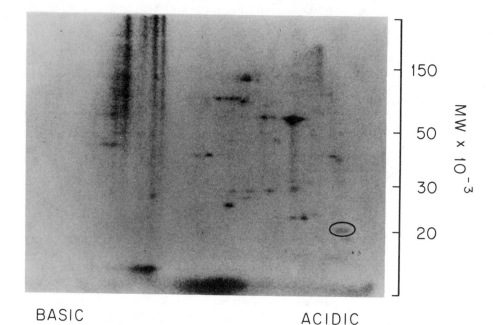

BASIC   ACIDIC

Figure 16.5. Two-dimensional gel autoradiogram of phosphoproteins from eyes of *Hermissenda*. Oval indicates the position of the 20,000-MW phosphoprotein. Two-dimensional gel procedure of O'Farrell [49] was employed.

cAMP-dependent and Ca^{2+}–calmodulin–dependent protein kinases? (5) Does the level of incorporation in the 20,000-MW band decrease as the animals' phototactic behavior returns to pretraining levels? (6) Is the incorporation of ^{32}P into the same protein band affected in other cells in the neural pathway underlying this example of associative learning? and (7) Do other stimuli that affect different behaviors affect the same or a different phosphoprotein?

As mentioned previously, more than one biochemical process is likely to be involved in associative learning, and studies are now in progress to investigate the possibility that other biochemical processes such as RNA synthesis or translation, phospholipid metabolism, lipid methylation, glycoproteion processing, as well as other types of posttranslational protein modifications, could be related to associative learning in *Hermissenda*.

In summary, the results of biochemical–behavioral experiments demonstrate that the level of incorporation of ^{32}P into a 20,000-MW phosphoprotein band in the eye of *Hermissenda* is altered in animals that exhibit a change in their photoactic behavior due to the conditioning procedure. This suggests that the phosphorylation of specific proteins is one of the biochemical processes utilized by cells in neural pathways involved in this example of associative learning. It will be of interest to examine vertebrate models of associative learning to determine if protein phosphorylation is related to learning in animals with more complex nervous systems.

ACKNOWLEDGMENTS

I am grateful to my colleagues, Dan Alkon, Terry Crow, Juan Acosta-Urquidi, Howard Rasmussen, Jim Olds, Greg Kuzma, and Paul Kandel for their contributions to the work described here. I also wish to thank Jeanne Kuzirian for secretarial support.

REFERENCES

1. Kandel, E. R., and Spencer, W. A. 1968. *Physiol. Rev. 48*:65–134.
2. Kandel, E. R. 1976. *Cellular Basis of Behavior*. San Francisco: Freeman.
3. Krasne, F. B. 1976. In *Neural Mechanisms of Learning and Memory* (M. R. Rosenzweig and E. L. Bennet, eds.), pp. 401–429. Cambridge, Mass.: MIT Press.
4. Willows, A. O. D. 1973. In *Invetebrate Learning*, vol. 2, (W. C. Corning, J. A. Dyal, and A. O. D. Willows, eds.), pp. 187–274. New York: Plenum Press.
5. Alkon, D. L., 1980. *Biol. Bull. 159*:505–560.
6. Williams, M., and Rodnight, R. 1977. *Progr. Neurobiol. 8*:183–250.
7. Greengard, P. 1976. *Nature 260*:101–108.
8. Browning, M., Dunwiddie, T., Bennet, W., Gispen, W., and Lynch, G. 1979. *Science 203*:60–62.
9. Bär, P. R., Schotman, P., Gispen, W. H., Tielen, A. M., and Lopes da Silva, F. H. 1980. *Brain Res. 198*:478–484.
10. Ehrlich, Y. H., Reddy, M. V., Keen, P., Davis, L. H., Daugherty, J., and Brunngraber, E. G. 1980. *J. Neurochem. 34*:1327–1330.
11. Levitan, I. B., and Barondes, S. H. 1974. *Proc. Nat. Acad. Sci. 71*:1145–1148.
12. Ram, J. L., and Erlich, Y. H. 1978. *J. Neurochem. 30*:487–491.
13. Jennings, K. R., Kaczmarek, L. K., Hewick, R. M., Dreyer, W. J., and Strumwasser, F. 1982. *J. Neurosci. 2*:158–168.
14. Paris, C. G., Kandel, E. R., and Schwartz, J. H. 1980. *Soc. Neurosci. Abstr. 6*:844.
15. Levitan, I. B., and Norman, J. 1980. *Brain Res. 187*:415–429.
16. Pant, H. C., Yoshioka, T., Tasaki, I., and Gainer, H. 1979. *Brain Res. 162*:303–313.
17. Neary, J. T. 1980. *Fed. Proc. 39*:2166.
18. Goy, M. F., Schwarz, T. L., and Kravitz, E. A. 1981. *Soc. Neurosci. Abstr. 7*:933.
19. Kandel, E. R., Brunelli, M., Byrne, J., and Castellucci, V. 1976. *Cold Spring Harbor Symp. Quant. Biol. 40*:465–482.
20. Crow, T. J., and Alkon, D. L. 1978. *Science 201*:1239–1241.
21. Crow, T. J., and Alkon, D. L. 1980. *Science 209*:412–414.
22. Neary, J. T., Crow, T., and Alkon, D. L. 1981. *Nature 293*:658–660.
23. Tyndale, C. L., and Crow, T. J. 1979. *IEEE Trans. Biomed. Eng. 26*:649–655.
24. Laemmli, U. K. 1970. *Nature 227*:680–685.
25. Laskey, R. A., and Mills, A. D. 1977. *FEBS Lett. 82*:314–316.
26. Alkon, D. L., Lederhendler, I., and Shoukimas, J. J. 1982. *Science 215*:693–695.
27. West, A., Barnes, E., and Alkon, D. L. 1982. *J. Neurophys. 48*:1243–1255.
28. Acosta-Urquidi, J., Alkon, D. L., Olds, J., Neary, J. T., Zebley, E., and Kuzma, G. 1981. *Soc. Neurosci. Abstr. 7*:944.
29. Alkon, D. L., Acosta-Urquidi, J., Olds, J., Kuzma, G., and Neary, J. T. 1983. *Science 219*:303–6.
30. Acosta-Urquidi, J., Neary, J. T., and Alkon, D. L. 1982. *Soc. Neurosci. Abstr. 8*:825.

31. Cohen, P. 1978. *Curr. Top. Cell Reg. 14*:117–196.
32. Roach, P. J., DePaoli-Roach, A. A., and Larner, J. 1978. *J. Cyclic Nucl. Res. 4*:245–257.
33. Soderling, T. P., Srivastava, A. K., Bass, M. A., and Khatra, B. S. 1979. *Proc. Nat. Acad. Sci. 76*:2536–2540.
34. Walsh, K. Y., Millikin, D. M., Schlender, K. K., and Reimann, E. M. 1979. *J. Biol. Chem. 254*:6611–6616.
35. LePeuch, C. J., Haiech, J., and Demaille, J. G. 1979. *Biochemistry 18*:5150–5157.
36. Sieghart, W., Forn, J., and Greengard, P. 1979. *Proc. Nat. Acad. Sci. 76*:2475–2479.
37. Huttner, W. B., and Greengard, P. 1979. *Proc. Nat. Acad. Sci. 76*:5402–5406.
38. Neary, J. T., and Alkon, D. L., 1983. *J. Biol. Chem. 258*:8979–8983.
39. Thompson, S. H. 1977. *J. Physiol* (London) *265*:465–488.
40. Adams, D. J., and Gage, P. W. 1979. *J. Physiol.* (London) *289*:115–141.
41. Byrne, J. H., Shapiro, E., Dieringer, N., and Koester, J. 1979. *J. Neurophysiol. 42*:1233–1250.
42. Thompson, S. H. 1982. *J. Gen. Physiol. 80*:1–18.
43. Alkon, D. L., Shoukimas, J., and Heldman, E. 1982. *Biophys. J. 40*:245–250.
44. Alkon, D. L. 1979. *Science 205*:810–816.
45. Yang, S. D., Vandenheede, J. R., Goris, J., and Merlevede, W. 1980. *J. Biol. Chem. 255*:11759–11767.
46. Hemmings, B. A., Yellowlees, D., Kernohan, J. C., and Cohen, P. 1981. *Eur. J. Biochem. 119*:443–451.
47. Rasmussen, H. 1981. *Calcium and cAMP as Synarchic Messengers*. New York: Wiley.
48. Cohen, P. 1982. *Nature 296*:613–620.
49. O'Farrell, P. H. 1975. *J. Biol. Chem. 250*:4007–4021.

17 · Dose-dependent effects of intracellular cyclic AMP on nerve membrane conductances and internal pH

PHILIP E. HOCKBERGER and JOHN A. CONNOR

INTRODUCTION

Shortly after its discovery in liver cells (Rall et al. 1957), detectable levels of the ribonucleotide adenosine 3′, 5′-cyclic monophosphate (cyclic AMP; cAMP) were found in other tissues, including brain (Rall and Sutherland 1958). In fact, the synthetic and degradative enzymes of cAMP, adenylate cyclase, and phosphodiesterase, respectively, were shown to have their highest activity in brain tissue (Butcher and Sutherland, 1962; Sutherland et al. 1962) Measurable levels of cAMP and its associated enzymes were subsequently found in brain slices, ganglia, neurons in culture, as well as cell-free homogenates (cf. Daly 1977). Kakiuchi and Rall (1968) demonstrated that the neurotransmitter norepinephrine increased the content of cAMP in cerebellar slices. This result initiated a search for additional stimulatory agents that showed that various hormones, neurotransmitters, psychoactive drugs, and electrical stimulation were capable of elevating cAMP levels in neurons of both vertebrate and invertebrate species.

Even though the conditions for changing cyclic nucleotide levels were being elucidated, there was still no clear-cut evidence of a subcellular role for this nucleotide in nervous tissue. Nevertheless, cAMP has been implicated in three physiological processes: neurotransmitter synthesis, microtubular function, and membrane conductance changes – for example, those underlying synaptic potentials and spontaneous pacemaker activity (for a review, see Nathanson 1977).

The first compelling evidence that cAMP may be involved in synaptic transmission was demonstrated by Greengard and coworkers using histochemical methods. Kebabian et al. (1975) used immunofluorescent techniques to demonstrate that dopamine elevated intraneuronal cAMP levels in bovine superior cervical ganglion cells. Then Nathanson and Greengard (1977) used autoradiographic staining to show that cAMP-dependent phosphodiesterases were concentrated next to the synaptic membrane of postganglionic neurons. Electrophysiological examinations of the effects of cAMP in these neurons have proved inconclusive, however. Intracellular injections of the nucleotide have resulted in only occasional membrane potential effects (Gallagher and Shinnick-Gallagher 1977; Kobayashi et al. 1978; Weight et al. 1978), and bath-applied analogs have yielded inconsistent results (for review see Libet 1979).

Injections of cAMP into the giant neurons of invertebrates, on the other hand, have been reported to induce a variety of electrophysiological changes. Liberman et al. (1975) found that iontophoretic injections of cAMP into the somata of *Helix* neurons routinely depolarized the cell membranes. Subsequent studies by Liberman et al. (1977, 1978, 1981) demonstrated that the duration of these responses was dose-related and potentiated in the presence of the phosphodiesterase inhibitors isobutylmethylxanthine (IBMX), papaverine, or SQ 20009. Other investigators have injected cAMP or associated agents into different gastropod neurons, and they too have reported excitatory (Drake and Treistman 1980; Pellmar 1981; Gillette et al. 1982) and inhibitory responses (Treistman and Levitan 1976a, b), as well as effects upon synaptically evoked potentials (Brunelli et al. 1976; Shimahara and Tauc 1977; Klein and Kandel 1978; Alkon 1979; Kaczmarek and Strumwasser 1981). Missing from these studies, however, is quantitative information regarding the dose dependency and duration of the elicited responses. Also missing is any indication as to how general the responses were between neurons as well as between species following injections of cAMP.

In this chapter we will describe evidence that has accumulated in our laboratory over the past several years that indicates that there are changes in neuronal properties that are dependent upon the dose of cAMP injected. Some of these changes were short-lived (i.e., several minutes in duration) whereas others were found to last much longer periods of time. In addition, we will report on the effects which cAMP injections had on the levels of two important cellular parameters, intracellular free Ca^{2+} (Ca_i) and pH (pH_i) measured directly using intracellular probes. Finally, we will comment on the generality of the nucleotide-induced changes obtained in 14 different *Archidoris* neurons, as well as in neurons from other gastropods.

METHODS

All experiments were performed on cells from the isolated brain of the marine slug *Archidoris montereyensis*. A diagrammatic representation of an isolated brain, sans buccal ganglia, is shown in Figure 17.1. Each brain contained 4 pairs of ganglia (pleural, pedal, cerebral, and buccal) with 14 large, reidentifiable neurons. Examination of intact specimens by tracing each nerve trunk projecting from the ring and buccal ganglia until it either innervated an abdominal organ or else became embedded in the body wall of the animal revealed no additional major ganglia within the body cavity. It is therefore likely that what has been termed the pleural ganglia in *Archidoris* (Blackshaw 1976; Connor 1979) is in fact a visceral–partial–pleural complex of ganglia. For most experiments, individual ganglia were used to maximize electrical isolation during voltage clamping and to facilitate positioning of the neurons over the incident light fiber with minimal obstruction of the light path by adjoining tissue (see below).

All experiments were run at 12°C. Physiological (normal) saline composition was 490 mM NaCl, 8 mM KCl, 20 mM $MgCl_2$, 30 mM $MgSO_4$, 10 mM $CaCl_2$, 5 mM $C_6H_{12}O_6$, and 10 mM MOPS at pH 7.6. Ion-substituted salines were osmotically balanced using one of the normal ingredients or a drug sub-

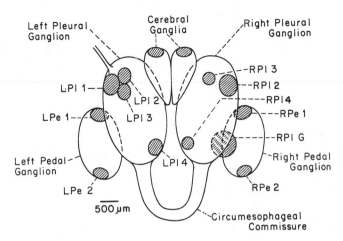

Figure 17.1. Diagram showing the approximate locations of the 14 identifiable neurons in the circumoesophageal ring ganglia of *Archidoris montereyensis*. R, right; L, left; Pl, pleural; Pe, pedal; Ce, cerebral; G, giant cell. (Modified from Connor, 1979)

stitute (e.g., tetramethylammonium chloride). Isolated brains stored at 9°C in normal saline were quite stable with regard to their cAMP responses for periods of 1 to 2 days. In several cases the same neuron was examined on 2 consecutive days.

Voltage clamp and dye absorbance techniques have been described elsewhere (Connor 1979; Ahmed and Connor 1979, 1980). The experimental chamber and optical recording system are shown schematically in Figure 17.2. The optical recording system was used for two separate purposes: monitoring changes in Ca_i and pH_i, and for quantifying the amount of cAMP pressure injected into cells. Both procedures employed the metallochromic indicator dye Arsenazo III (grade I, Sigma Chemical Company, St. Louis). In addition, both procedures employ single- as well as dual-wavelength absorbance measurements. The latter were preferable during intracellular pressure injections since they reduced distortion artifacts (due to changes in cell volume, light scattering, lamp intensity fluctuations, etc.) by subtracting a reference signal (e.g., 700 nm) from the signal of interest (e.g., 580, 610, or 660 nm). Photomultiplier outputs were low-pass-filtered with time constants of 50 msec or 1 sec for single- and dual-wavelength measurements, respectively.

Intracellular pH was also monitored using recessed-tip pH microelectrodes (Thomas 1978). The tip diameters were 1 to 2 μ, and the electrodes were calibrated in buffered internal saline solutions: 350 mM KCl; 50 mM NaCl; 30 mM MOPS. The output of the pH microelectrode was fed to a varactor bridge preamplifier and then into the noninverted input of a differential amplifier. The inverted input was attached to the intracellular voltage-recording electrode that served as the reference electrode. The slope of the pH microelectrode was measured for a 1-unit difference (6.7–7.7) in buffered standards, and only electrodes with slopes greater than 51 mV were used. The pH_i was calculated using

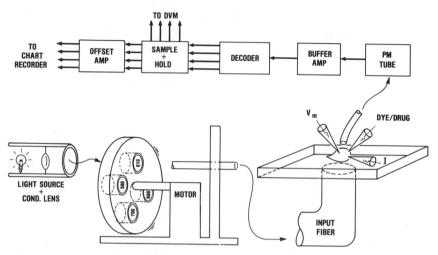

Figure 17.2. Rendering and block diagram of the optical recording system used in the present experiments. The input fiber under the prep dish is enlarged to approximate its size in relation to the soma diameter and to the output fiber leading to the photomultiplier (PM) tube.

the slope and by measuring the drop in voltage upon moving the electrode from the bath (pH = 7.6) into the cell. The pH microelectrode was considered to be inside the cell when a negative voltage step of 20 mV across the cell membrane produced a change of 1 mV or less in the differential pH electrode output.

RESULTS

Iontophoretic or pressure injection of cAMP into any one of the 14 identifiable nerve cells in *Archidoris* resulted in a rapid, transient depolarization that was generally sufficient to cause repetitive firing. The duration as well as the magnitude of the response depended upon the quantity of nucleotide injected. Figure 17.3 shows intracellular voltage records from a pedal neuron pressure-injected with three different doses of cAMP. The cell was spontaneously active, and the first injection induced a transient increase in firing rate lasting several minutes (Figure 17.3*A*). Larger doses resulted in further increased firing rates but with concomitant reduction in spike amplitude and undershoot (Figure 17.3*B*).

Neurons that underwent marked spike reduction following nucleotide injections often displayed subsequent modifications in their firing patterns. The most severe modification occurred almost simultaneously with spike reduction, and it involved a period lasting 1–10 min during which action potentials were absent (Figure 17.3*C*). A more interesting type of modification occurred in some cells following 1–10 min of increased firing behavior. These cells subsequently displayed the development of slow, rhythmic bursts of action potentials

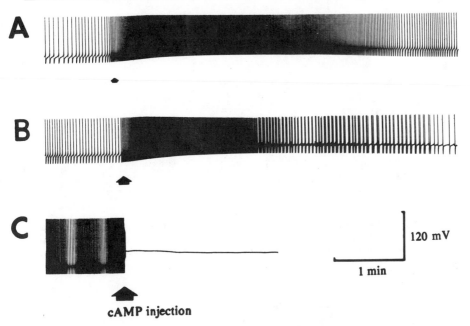

Figure 17.3. Intracellular recordings from neuron LPe 2 following three different pressure injections of cAMP. *A*: Small injection caused a transient increase in firing frequency with a slight reduction in spike undershoot. *B*: Medium injection also resulted in a period of increased firing but with concomitant reduction in both spike overshoots and undershoots. A slow-burst firing pattern developed within 2 min and persisted for several more minutes. *C*: Large injection induced spike reduction followed immediately by the absence of spike activity. Action potentials could not be evoked for several minutes thereafter.

(slow-bursting) frequently containing a uniform number of spikes per burst (Figure 17.3*B*). Closer examination, seen in Figure 17.4, revealed that each burst was usually preceded by subthreshold membrane oscillations and followed by a depolarizing after-potential. The subthreshold oscillations often exhibited a progressive increase in size prior to a burst, and once bursting and oscillatory behaviors were induced, they often persisted for hours. However, since bursting and oscillatory behaviors were only seen in cells where resting potentials were near threshold, sometimes current stimulation was needed to demonstrate the persistence of these behaviors.

A quantitative examination of the dose dependence of these responses was made by pressure-injecting a mixture of cAMP plus Arsenazo III (20:1) and monitoring the amount of dye injected. The dose of nucleotide was estimated by measuring the net change in absorbance baseline at the isosbestic wavelength, 580 nm. The absorbance change was converted to a dye concentration change using Beer's law and the molar extinction coefficient of Arsenazo III ($3 \times 10^4 \text{ m}^{-1} \text{ cm}^{-1}$). The injected mixture was assumed to have the same proportions as that in the barrel. Figure 17.5 shows the firing pattern of a pedal

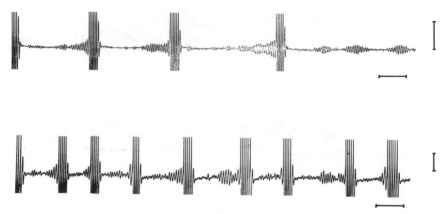

Figure 17.4. Two different examples of subthreshold membrane oscillations that occur both spontaneously as well as after cAMP injections. Top: Recording of typical waveform in which oscillations grew progressively in size until a burst was elicited. Bottom: Recording less frequently seen exhibiting a random correlation between the size of the oscillations and initiation of a burst.

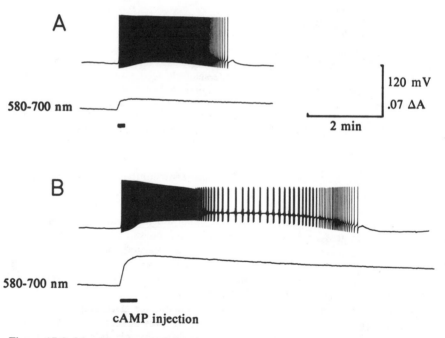

Figure 17.5. Membrane potential and transcellular absorbance records during two successive pressure injections of cAMP plus Arsenazo III mixture into the same neuron (LPe 1). *A*: This relatively small dose (0.4 mM) of cAMP, quantified using the net change in dual wavelength absorbance, evoked a transient depolarization lasting about 2 min. *B*: A large dose (1.2 mM) of cAMP initiated intense firing with reduction in spike size followed by slow-bursting.

neuron following two successive injections of the drug–dye mixture. The smaller dose, indicated by the smaller dual-wavelength absorbance increase (Figure 17.5*A*), resulted in a transient period of spike activity during which a slight reduction in spike amplitude and undershoot occurred. Doses estimated between 0.1 and 7.6 mM cAMP (mean 3.3 mM, $n = 32$) resulted in this type of response. A larger injection (Figure 17.5*B*) resulted in more pronounced spike reduction with subsequent induction of slow-bursting. Injected doses between 1.5 and 17 mM cAMP (mean 9.0 mM, $n = 13$) induced this type of response. It should be noted that although larger doses were needed to induce slow-bursting, no single dose was found that could routinely induce this firing pattern.

These estimated doses do not take into account either the endogenous phosphodiesterase (PDE) activity or the subcellular localization of the injected dose. As a result, the concentrations reported here only reflect relative increases (or "effective" doses) of cellular cAMP concentration. In spite of this shortcoming, a lower limit of cAMP efficacy was estimated. Figure 17.6 shows the effect on membrane potential of three barely detectable doses of cyclic nucleotide injected into a pleural neuron. The first two injections (Figure 17.6*A*) exhibited no clear change in the absorbance measured between the beginning and end of the record, making these doses below the minimum detectable dose (i.e., 0.1 mM). Nevertheless the second dose was enough to induce five action potentials. The third injection (Figure 17.6*B*) is an example of a 0.1-mM dose of cAMP that evoked 15 action potentials. The rates of injection and cellular responses indicated that the third dose was in the neighborhood of threefold larger than the second dose, arguing that even one-third the minimum detectable dose was effective in eliciting action potentials. Also, it can be seen that even at the higher injection rate the cell began firing long before the total dose was delivered. Upon termination of the second and third injections the membrane potential recovered much more rapidly than the dye absorbance signal, suggesting that loss of cAMP from the local injection site was assisted in large part by hydrolysis of the molecule as well as diffusion.

A variety of control solutions, including Arsenazo III, 5′AMP, Arsenazo III plus 5′AMP, and buffered internal saline were pressure-injected into these neurons without effects on membrane potential. Relatively large doses of 5′AMP plus Arsenazo III (mean $= 5.7$ mM) did not change the resting potential in 15 out of 17 cells. Injections of either 5′AMP or internal saline that increased cell volume by 5 percent or less were without effect on cellular excitability. Saline injections greater than 20 percent by volume resulted in transient depolarizatons. In contrast, cAMP injections that caused comparable depolarizations were achieved with as little as 0.2 percent increases in cell volume.

Ion-substituted salines and voltage clamping were used to investigate the nature of the conductance changes induced by cAMP. Injecting cAMP plus dye into a cell voltage-clamped at its resting potential evoked a dose-dependent increase in the membrane holding current (Figure 17.7*A*). Voltage-clamping neurons at potentials between -30 and -100 mV did not greatly affect the size of the induced current, although it was difficult to match injections exactly. Replacing sodium in the superfusion solution with Trisma (base) or tetramethylammonium chloride decreased the holding current baseline and blocked

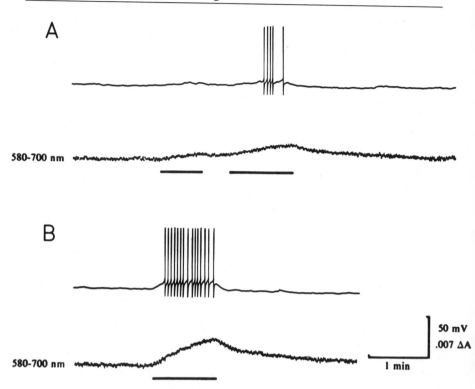

Figure 17.6. Membrane potential and transcellular absorbance records of RPl 4 during three successive pressure injections of cAMP-plus-dye mixture. Horizontal bars indicate periods during which light pressure was applied. *A*: The second injection elicited five action potentials although the dose was below the minimum quantifiable amount, 0.1 mM. *B*: This injection evoked 15 spikes with the minimum detectable dose of 0.1 mM cAMP. (Reproduced from Connor and Hockberger in press)

the nucleotide-induced response even when larger doses were administered than in normal saline (Figure 17.7*B*). Bathing cells in 5×10^{-4} M ouabain did not affect the response, nor did perfusing the cells in potassium-free saline or 40-mM TEA saline (Figure 17.7*C*). Calcium replacement with magnesium also did not affect the cAMP response. Taken together these results indicate that cAMP injections induce a reversible increase in the membrane sodium permeability of *Archidoris* neurons.

Examination of possible voltage-dependent conductance changes induced by cAMP was more difficult, although two general patterns of results emerged. The clearest, most consistent effects were observed in spontaneously firing cells where the current–voltage *(I–V)* relationship has a negative slope resistance (NSR) region for long clamp pulses (see Partridge et al. 1979). That is, for small positive voltage steps (5 to 20 mV) from resting potential, membrane current became increasingly inward instead of outward (Figures 17.8*A*, *C*). Cyclic AMP injections into these neurons always increased the net inward cur-

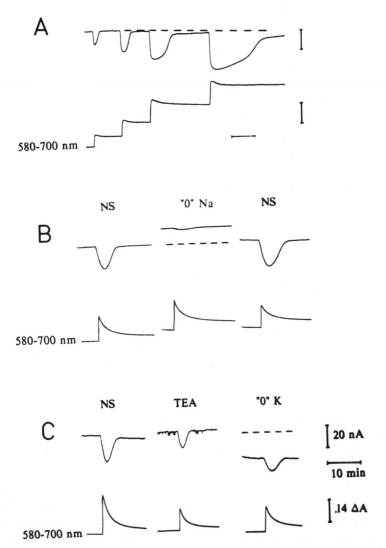

Figure 17.7. Injections of cAMP-plus-dye mixtures into neurons under voltage clamp resulted in reversible, dose-dependent increases in membrane holding current. *A*: Four successively larger injections showing eventual saturation of the current response. *B*: Removing sodium from the perfusion solution blocked the current response even when a larger dose was administered. This effect was reversed upon returning the cell to normal saline (NS). *C*: Bathing the cell in 40 mM TEA or in K-free saline did not block the current response.

rent in the NSR region (Figures 17.8A, C, after AMP). This enhancement often persisted beyond the induced inward current at -40 mV (e.g., Figure 17.7). Ion substitution and pharmacological studies have demonstrated that the inward current underlying the NSR region is primarily carried by sodium ions. Thus, cAMP enhanced both the resting Na^+ permeability and the voltage-dependent sodium conductance underlying the NSR region in these neurons.

About one-half the total neuron population did not exhibit an NSR region; rather, the membrane current quickly became outward following a small, positive voltage step. In these cells the net effect of cAMP was to increase the outward current (Figure 17.8B, D, after cAMP). We have not yet sorted out the more basic changes in current underlying this effect. That is, even in cells where the total I_m records are like those of Figure 17.8B, one can demonstrate the presence of a rather constant inward current that is simply of smaller size than the outward current carried by other channels (see Connor 1979). We have tentatively concluded that the characteristics of both inward and outward current-carrying channels may be altered as a result of cAMP injection, and where outward current dominates the slow voltage clamp records, any effect on inward current is masked.

We found that the four pedal ganglion neurons studied almost invariably showed an NSR region, whereas LPl 4, LPl 1, and RPl 4 generally did not. The remaining neurons were observed to fall into either category an appreciable fraction of time. Since the effect of cAMP seemed to be a strengthening of either type of characeristic (NSR or non-NSR) rather than a conversion from one characteristic to another, it appears unlikely that differing endogenous levels of the nucleotide are directly involved in establishing the variability seen in the individual neurons. However, since cells capable of displaying a slow-burst firing pattern were always characterized by a non-NSR $I–V$ relation, we are inclined to believe that enhancing the outward current with cAMP injections may underlie the conversion of a spontaneously firing neuron into one exhibiting slow-bursting.

Although calcium-free saline had no noticeable effect on the cAMP-induced current response, it was still possible that a fraction of the induced current was carried by Ca^{2+}. The metallochromic dye Arsenazo III was used to further investigate this possibility. Figure 17.9 shows two different dual-wavelength absorbance recordings of a voltage-clamped cell filled with 0.3 mM Arsenazo following intracellular injections of Ca^{2+} and H^+. Calcium ion injections caused a greater increase in absorbance of the 660- to 700-nm signal, and the subsequent return to baseline has been shown to reflect cellular calcium regulation (Connor and Nikolakopoulou 1982). In contrast, injections of hydrogen ions resulted in a greater decrease in absorbance of the 610- to 700-nm signal. Here, again, the return of the signal to baseline corresponded to a pH recovery confirmed by simultaneously monitoring pH_i with a pH microelectrode. Note that the pH electrode recorded a slower onset and recovery for the change compared to the dye signals. This difference stems from the slower electrical properties of the pH electrode, as well as from diffusion of the ions to the electrode.

Pressure injection of cAMP (buffered with MOPS, pH $= 7.5$) into an Arsenazo-filled neuron, as shown in Figure 17.10A, resulted in a membrane

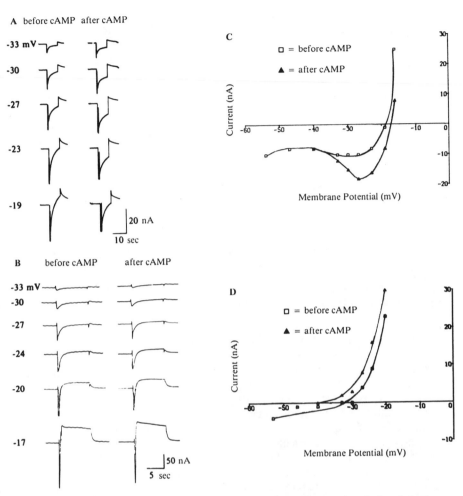

Figure 17.8. *A*: Current records during voltage clamp steps before and after injection of cAMP into a neuron exhibiting an NSR region. After the injection there was an increase in the quasi-steady-state inward current (plotted in *C*), which persisted even after the holding current recovered. *B*: Current records similarly taken before and after cAMP injection from a neuron not exhibiting an NSR region. The injection into this cell enhanced the steady-state outward current (plotted in *D*) upon recovery of the holding current.

depolarization and a transient bout of action potentials. Concurrently the 660- to 700-nm signal increased, indicating an elevation in Ca_i during the bout. Afterward the dye signal decreased to a level below the original baseline. An intracellular pH microelectrode showed that a pH_i decrease occurred during and after the bout, and thus the shift in 660- to 700-nm baseline was likely the result of this pH_i decrease. Both these effects, however, were produced by a

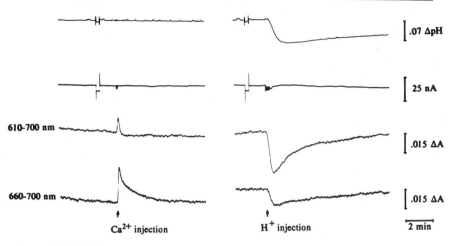

Figure 17.9. Simultaneous measurements of pH_i, current, and absorbance at two different dual wavelengths of an Arsenazo-filled cell (LPl 1) under voltage clamp during injections of Ca^{2+} and H^+. See text for details. A voltage step prior to each injection was performed to ensure that the pH microelectrode was completely iside the cell. (Reproduced from Hockberger and Connor 1983)

similar spike train elicited with current stimulation (Figure 17.10*A*). Therefore, cAMP was injected into this same neuron under voltage clamp and examined in a similar fashion. Figure 17.10*B* shows the characteristic change in holding current following an injection, but notice that there was no increase in dye absorbance at 660–700 nm. However, the dye signal decreased following the injection, and the pH microelectrode confirmed that this reflected an intracellular acidification.

The dye Arsenazo III was also used to examine possible changes in the voltage-dependent calcium conductance during and after cAMP injections. Figure 17.11*A* shows the characteristic increase and recovery in the 660-nm absorbance signal for a voltage pulse to +20 mV. During the following several minutes a total of five intracellular iontophoretic injections of cAMP were administered under voltage clamp, of which three are shown in Figure 17.11*B*. When the voltage was unclamped following an injection, the cell showed the characteristic depolarization that sometimes elicited action potentials. Immediately after the fifth injection (not shown) the voltage was stepped to +20 mV as in the control. The absorbance, recorded in Figure 17.11*C*, showed no observable change in either the characteristics of calcium influx or its subsequent regulation as reflected by the recovery of the dye signal ($n = 10$).

As previously mentioned, intracellular pH microelectrodes were used to verify the pH_i decrease induced by cAMP. Sixty-one cells were examined in this manner, exhibiting a range in resting pH_i of 7.05–7.75 (mean = 7.39, *SD* ± .16). Figure 17.12*A* shows the change in pH_i following pressure injection of buffered cAMP (pH = 7.2) plus Arsenazo III into a neuron with a resting pH of 7.31. The cell was voltage-clamped at rest to eliminate pH_i changes that would have occurred if the cell had been allowed to fire action potentials

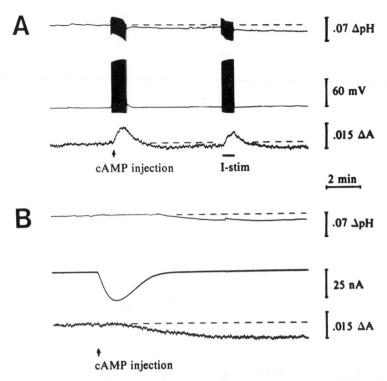

Figure 17.10. *A*: Simultaneous measurements of pH_i, voltage, and absorbance (660–700 nm) of Arsenazo-filled cell (LPl 1) during pressure injection of a buffered solution of cAMP (pH = 7.5) and during a brief period of current stimulation (I-stim). Each procedure elicited a train of action potentials with a corresponding increase in dye absorbance (i.e., rise in Ca_i) and cellular acidification, the latter persisting even after the membrane potential recovered. *B*: Similar measurements done while the cell was voltage-clamped. A larger injection of cAMP induced the typical inward current response, but no increase in absorbance (660–700nm) was detected. However, a subsequent internal acidification was detected. Resting pH_i was 7.46. (Reproduced from Hockberger and Connor 1983)

(Ahmed and Connor 1980; see also Figure 17.10*A*). The ΔpH_i of Figure 17.12*A* showed a reversible acidification of approximately 0.14 pH unit following the injection. A comparable injection of 5′AMP plus Arsenazo III (pH-buffered to 7.5) into the same cell exhibited a slight alkalinization. Additional experiments have shown that the small initial transient responses in Figures 17.12*A* and *B* are stimulus artifacts due to the mismatch between cell resting pH and the pH of the injected solution.

The magnitude and time course of the cAMP-induced pH_i response were somewhat variable, but in general they appeared to be dose-dependent and reversible. The latency before onset of the pH change ranged from 1 to 5 min and the maximum pH_i response, which always occurred subsequent to the peak

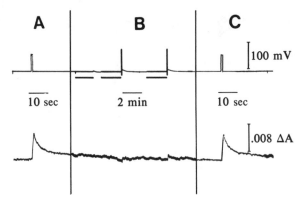

Figure 17.11. *A, C:* Transcellular absorbance changes at 660 nm (lower trace) of an Arsenazo-filled cell (LPl 1) generated by identical voltage clamp pulses (upper trace) before and after iontophoretic injections of cAMP. The two records are nearly identical, indicating no change in Ca^{2+} influx or regulation characteristics. *B:* Absorbance and voltage records during cAMP injections. Bars underlying the voltage trace indicate three periods of iontophoretic injection. Following each injection period, voltage was discontinued. The second and third injections were followed by a train of action potentials. Record *B* is contiguous with *A*, whereas two additional injections were interposed between the end of record *B* and the start of record *C*. (Reproduced from Hockberger and Connor 1983)

current response, reached its peak within 12 to 25 min after injection. The amplitude and duration of both the inward current and the pH_i responses were dramatically enhanced if the cell was bathed in 10^{-3} M IBMX, a phosphodiesterase inhibitor. Figure 17.13*B* shows this enhancement occurring even when the administered dose was smaller than the control (Figure 17.13*A*).

Two additional observations were made using cells bathed in IBMX. First, no change in resting pH_i was evident prior to a cAMP injection even though IBMX depolarized most cells examined (seen as a downward drift in the current record of Figure 17.13*B*). If IBMX were acting to raise endogenous nucleotide levels rather than working through some presynaptic element, then raising endogenous levels did not produce a pH_i decrease. A second troublesome observation was that the current response following a cAMP injection often recovered in IBMX even though its peak value was markedly enhanced. This result may reflect imperfect inhibition of phosphodiesterases by 10^{-3} M IBMX.

One notable variation in the typical pH_i response occurred in a few cells. Following cAMP injections into these cells an initial alkalinization occurred that could not be accounted for as a stimulus artifact. Figure 17.14 shows such a record in which the alkalinization was followed by the usual acidification. This type of pH response was most often seen in the two giant white cells (LPl 2 and 3), although it was not unique to them. The peak alkalinization occurred between 5 and 10 min postinjection depending upon the degree of acidification that followed. The subsequent acidification seemed to run a similar time course as seen in cells without this initial alkalinization. It is of course possible that

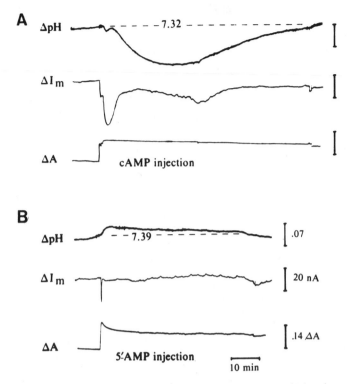

Figure 17.12. *A*: Simultaneous measurements of pH$_i$, current, and absorbance (580–700 nm) of cell RPe 2 under voltage clamp during an injection of cAMP plus dye (pH = 7.2). An injection of 11.1 mM cAMP resulted in a large current response as well as an intended acidification that was reversible. *B*: Same measurements during a comparable injection of 5′AMP plus dye (pH = 7.5) showed no change in holding current. A small alkalinization was elicited that was reversible. Calibrations are the same for *A* and *B*. (Reproduced from Connor and Hockberger in press)

both responses were always present if one assumes that in most cells the acidification predominated.

Intracellular pH measurements were also performed in sodium-free saline that blocked the cAMP-induced current response. Figure 17.15 shows the current and pH records during successive cAMP injections in normal saline and Na-free saline. Although the current response was blocked in Na-free saline, an internal acidification was evoked nevertheless. The slightly smaller pH$_i$ response in Na-free saline reflected the smaller injected dose of cAMP (0.6 mM) compared with the one in normal saline (0.8 mM). Notice that a slight internal acidification occurred when Na-free saline was applied, reflected by the prevention of a complete pH$_i$ recovery following the first injection. This acidification did not always appear, although the cellular hyperpolarization (seen as an upward deflection in the current record) routinely occurred in Na-free saline.

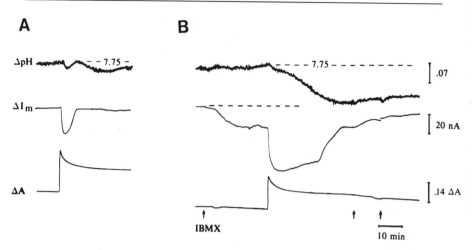

Figure 17.13. *A*: Simultaneous measurements of pH$_i$, current, and absorbance (580–700 nm) of cell LPl 4 under voltage clamp during an injection of cAMP plus dye in normal saline. *B*: Slightly smaller injection after 20-min exposure to 10^{-3} M IBMX. The smaller injection (4.2 mM) in IBMX yielded much larger current and pH$_i$ responses than those induced with the first injection (4.7 mM). The downward drift in the current record upon adding IBMX was consistently obtained in voltage-clamped cells. Cells not clamped became depolarized during the exposure period. Arrows denote normal saline washes. Calibrations apply to *A* and *B*. (Reproduced from Connor and Hockberger in press)

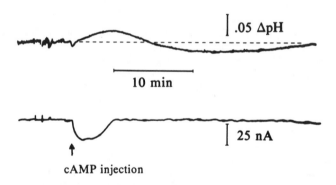

Figure 17.14. Internal pH (upper trace) and membrane current (lower trace) of a neuron under voltage clamp during an injection of cAMP (3.5 mM) plus dye (pH = 7.4). This biphasic type of response was most often seen in the white cells LPl 2 and 3, although it was not exclusive to those cells. (Reproduced from Connor and Hockberger in press)

DISCUSSION

Injections of cAMP into the somata of *Archidoris* neurons induced a rapid depolarization in all 14 identifiable cells tested. Sufficient injection always trig-

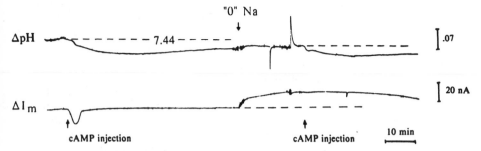

Figure 17.15. Internal pH and membrane current of cell RPl 4 under voltage clamp during comparable injections of cAMP in normal saline (left) and in Na-free saline (right). The current response was blocked in Na-free saline whereas the pH$_i$ response was not affected. The upward drift in the current record while the cell was exposed to Na-free saline was probably due to blockage of a high resting sodium permeability. All neurons that were examined responded to Na-free saline in this manner. (Reproduced from Connor and Hockberger in press)

gered repetitive firing and occasionally induced a "slow-burst" pattern, a term used here to distinguish this pattern from parabolic-bursting (e.g., R15 in *Aplysia*) or fast-bursting (e.g., cyberchron neurons in *Helisoma*) recorded from other gastropod neurons. Similar cAMP-induced results have been noted in *Helix* (Liberman et al. 1975, 1977).

Using the Arsenazo–nucleotide injection technique, we have found that relatively small doses of cAMP were capable of transiently increasing the neuronal firing rate, whereas larger injections were necessary to initiate a sustained depolarization or slow-burst firing pattern. Although attempts at quantifying the cAMP elevations were hampered by the action of endogenous PDE activity, we believe that the data presented in Figure 17.7 yield the best estimate currently available of a concentration-dependent effect of cAMP in situ. By gradually increasing the cellular cAMP level using constant pressure injection, we found that doses in the range of 30–40 μM injected over a period of several minutes, were sufficient to cause repetitive firing in an otherwise quiescent neuron.

The larger and longer-lasting effects of cAMP on membrane conductances and pH$_i$ were observed with estimated doses ranging up to several millimolar. Gallagher and Shinnick-Gallagher (1977) reported that intracellular iontophoretic injections of cAMP depolarized rat superior cervical ganglion cells when given in doses estimated to be 0.3 to 3 mM. Although it is still unclear whether or not intracellular cAMP concentrations ever reach millimolar levels under physiological conditions, it is likely that high endogenous PDE activity reduces the actual elevation of an injection below the estimated concentration. As a result, the absence of effect reported following injections of cAMP into certain vertebrate (Weight et al. 1978) and invertebrate (Treistman and Levitan 1976a) neurons may reflect doses that were inadequate for obtaining membrane conductance changes.

Cyclic AMP levels measured in isolated gastropod neurons using biochemical techniques have been shown to increase following exposure to certain neurotransmitters. Cedar and Schwartz (1972) determined the total cellular content of cAMP in the two giant neurons, R2 and LPG, in *Aplysia* using the isotope displacement method of Gilman (1970). They found that each cell contained approximately 0.6 pmol of cAMP. Assuming cell body diameters of 500 μm (100-g animals), the baseline concentrations of cAMP were about 10 μM. These authors then showed that bathing either cell in 0.2 mM serotonin (a transmitter known to depolarize R2) for 5 min doubled the intracellular concentration of cAMP. Levitan and Drummond (1980) have reported resting cAMP levels to be about one-half the values reported by Cedar and Schwartz for cells R2 and LPG (0.3 pmol per cell) and even smaller amounts in cell R15 (0.03 pmol per cell). However, the latter value probably reflects the smaller size of R15, typically one-third the diameter of R2 (Frazier et al. 1967). Levitan and Drummond also reported doubling (R2 and LPG) and tripling (R15) of the neuronal cAMP levels in the presence of 0.01 mM serotonin (which also depolarized R15), which raised the concentration of cAMP to about 10 μM in the giant cells and 5 μM for cell R15 (assuming soma diameters of 500 and 150 μm, respectively). If one allows that the resting concentration of cAMP in gastropod neurons might be twice the amount measured using biochemical methods (a rather conservative estimate considering the effects of endogenous PDE activity and the likelihood of subcellular compartmentalization), then neurotransmitter-stimulated levels may reach 20–40 μM in situ. In addition, within the synaptic neuropil where neurotransmitter receptors are more commonly located in gastropod ganglia, the locally stimulated levels of cyclic nucleotides may even be greater than the values obtained from isolated somata (see Cedar et al. 1972).

In the present experiments cAMP elevation induced two dose-dependent and reversible changes in cellular properties: an increase in membrane sodium permeability and an increase in intracellular [H$^+$]. Two lines of evidence indicated that these processes were not coupled to one another: (1) The pH$_i$ response lasted much longer then the sodium influx, and (2) the pH$_i$ response persisted in Na-free saline. In addition, the initial alkalinization evoked in some neurons during the depolarization indicates that the pH$_i$ change probably reflects more than one process being induced by cAMP. At present we have no additional data to suggest what process in particular these might be. Since the buffering capacity of these neurons for [H$^+$] is considerable, that is, in the neighborhood of 10–20 mEq/unit pH (Ahmed and Connor 1980; see also Thomas 1976), a pH$_i$ change of 0.1 unit reflects a rather large change in total [H$^+$] evolution. Whether cAMP levels are ever elevated sufficiently in situ to produce similar pH$_i$ changes is still an open question. However, if such a substantial rise in [H$^+$]$_i$ were to occur, it could serve to alter various cellular functions (Nuccitelli and Deamer 1982).

The indicator dye data further support the idea that the cAMP-induced current is carried by Na in that we were unable to detect changes in Ca$_i$ during the time when this current was flowing. These measurements were performed while the cell was voltage-clamped to prevent calcium influx accompanying action potentials. The cAMP-induced currents measured here were 20–50 nA for 300- to 500-μm cells and lasted several minutes. Ahmed and Connor (1979) have shown that voltage steps to $+20$ mV in these neurons results in a calcium current of approximately 400 nA. A flux of this magnitude for only 400 msec produces a large indicator absorbance change. Therefore, if only 10 percent or so of the cAMP-induced current were carried by calcium ions, the dye should have detected it.

Although this result can mean that there are truly no direct effects of cAMP on either Ca_i regulation or Ca_i transport systems in these particular neurons, we would emphasize that increased cAMP levels will have a large effect on internal calcium levels under physiological conditions. That is, the sodium current induced by cAMP could depolarize the membrane of unclamped neurons and thereby generate large Ca_i changes due to calcium influx during action potentials (see Figure 17.10A). This would, for example, result in an especially important functional consequence of cAMP elevation within presynaptic terminals.

Finally, we have also examined the effects of cAMP on the membrane potential of several neurons in gastropod species other than *Archidoris montereyensis* (Figure 17.16). Cyclic AMP routinely depolarized most neurons and induced slow-bursting in a few, although a small percentage showed no response even with doses estimated to be greater than 1 mM. Where examined, the induced depolarization was invariably dose-dependent and reversible. Most notable was the finding that cells R2 and LP1 in *Aplysia* behaved in a fashion similar to *Archidoris* neurons following cAMP injections. That is, cAMP induced a dose-dependent increase in the membrane sodium permeability. Cell R2 was particularly sensitive and exhibited loss of membrane potential with doses of cAMP estimated to be below 0.1 mM. We are encouraged by these results to speculate that the cAMP-induced processes described in this chapter for *Archidoris* neurons may occur under physiological conditions (e.g., response to neurotransmitters) and that they are generalizable to neurons of other gastropod species.

SUMMARY

Pressure or iontophoretic injection of cyclic AMP into any one of the 14 identified neurons in *Archidoris montereyensis* routinely resulted in the following changes in cellular properties: (1) a dose-dependent, reversible increase in membrane sodium permeability; (2) long-lasting changes in voltage-dependent membrane conductances that sometimes produced a slow-burst firing pattern; and (3) a dose-dependent, reversible intracellular acidification that outlasted the sodium influx and persisted in the absence of extracellular sodium. The phosphodiesterase inhibitor IBMX (10^{-3} M), externally applied, potentiated these cAMP-induced responses. Comparable injections of 5'AMP did not elicit any of these responses. No change was found in the resting level of internal free Ca^{2+} nor in the characteristics of calcium influx or regulation during electrical activity following injections of cAMP. Finally, injections of cAMP into neurons from nine other gastropod species (including *Aplysia californica*) yielded similar membrane conductance changes, indicating that these induced effects may be generalized to neurons of species other than *Archidoris montereyensis*.

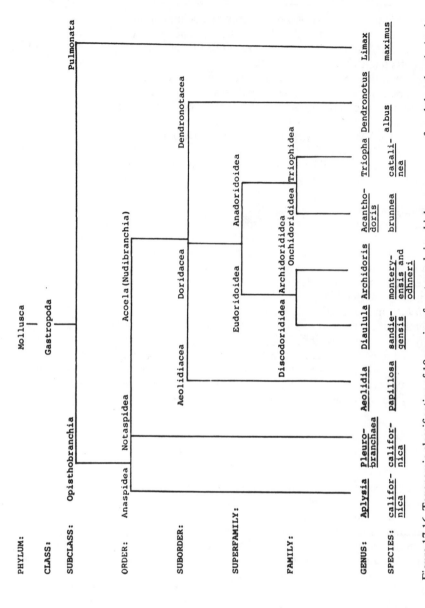

Figure 17.16. Taxonomic classification of 10 species of gastropods in which neurons were found that depolarized following cAMP injections.

REFERENCES

Ahmed, Z., and Connor, J. A. 1979. Measurement of calcium influx under voltage clamp in molluscan neurons using the metallochromic dye Arsenazo III. *J. Physiol. 286*:61–82.

1980. Intracellular pH changes induced by calcium influx during electrical activity in molluscan neurons. *J. Gen. Physiol. 75*:403–426.

Alkon, D. 1979. Voltage-dependent calcium and potassium ion conductances: A contingency mechanism for an associative learning model. *Science 205*:810–816.

Blackshaw, S. E. 1976. Dye injection and electrophysiological mapping of giant neurons in the brain of *Archidoris*. *Proc. Roy. Soc. B. 192*:393–419.

Brunelli, M., Castellucci, V. and Kandel, E. 1976. Synaptic facilitation and behavioral sensitization in *Aplysia:* possible role of serotonin and cyclic AMP. *Science 194*:1178–1181.

Butcher, R., and Sutherland, E. 1962. Adenosine 3′,5′-phosphate in biological materials. I. Purification and properties of cyclic 3′,5′-nucleotide phosphodiesterase and use of this enzyme to characterize adenosine 3′,5′-phosphate in human urine. *J. Biol. Chem. 237*:1244–1250.

Cedar, H., Kandel, E., and Schwartz, J. 1972. Cyclic adenosine monophosphate in the nervous system of *Aplysia californica*. I. Increased synthesis in response to synaptic stimulation. *J. Gen. Physiol. 60*:558–569.

Cedar, H., and Schwartz, J. 1972. Cyclic adenosine monophosphate in the nervous system of *Aplysia californica*. II. Effect of serotonin and dopamine. *J. Gen Physiol. 60*:570–587.

Connor, J. A. 1979. Calcium current in molluscan neurons: Measurement under conditions which maximize its visibility. *J. Physiol. 286*:41–60.

Connor, J. A., and Hockberger, P. In press. Intracellular pH changes induced by injection of cyclic nucleotides into gastropod neurons. *J. Physiol.*

Connor, J. A., and Nikolakopoulou, G. 1982. Calcium diffusion and buffering in nerve cytoplasm. In *Lectures on Mathematics in the Life Sciences,* Vol. 15 (R. Miura, ed.), pp. 79–101. Providence, R.I.: American Mathematical Association.

Daly, J. 1977. *Cyclic Nucleotides in the Nervous System,* New York: Plenum Press.

Drake, P., and Treistman, S. 1980. Alterations of neuronal activity in response to cyclic nucleotide agents in *Aplysia*. *J. Neurobiol. 11*:471–482.

Frazier, W., Kandel, E., Kupferman, I., Waziri, R., and Coggeshall, R. 1967. Morphological and functional properties of identified neurones in the abdominal ganglion of *Aplysia*. *J. Neurophys. 30*:1288–1351.

Gallagher, J., and Shinnick-Gallagher, P. 1977. Cyclic nucleotides injected intracellularly into rat superior cervical ganglion cells. *Science 198*:851–852.

Gillette, R., Gillette, M., and Davis, W. J. 1982. Substrates of command ability in a buccal neuron of *Pleurobranchaea*. II. Potential role of cyclic AMP. *J. Comp. Physiol. B. 146:461–70.*

Gilman, A. G. 1970. A protein binding assay for adenosine 3′,5′-cyclic monophosphate. *Proc. Nat. Acad. Sci. 67*:305–307.

Hockberger, P., and Connor, J. A. 1983. Intracellular calcium measurements with Arsenazo III during cyclic AMP injections into molluscan neurons. *Science 219*:869–871.

Kaczmarek, L., and Strumwasser, F. 1981. The expression of long-lasting afterdischarge by isolated *Aplysia* bag cell neurons. *J. Neurosci. 1*:626–634.

Kakiuchi, S. and Rall, T. 1968. The influence of chemical agents on the accumulation of adenosine-3′,5′-phosphate in slices of rabbit cerebellum. *Mol. Pharmacol. 4*:367–378.

Kebabian, J., Bloom, F., Steiner, A., and Greengard, P. 1975. Neurotransmitters increase cyclic nucleotides in postsynaptic neurons: Immunocytochemical demonstration. *Science 190*:157–159.

Klein, M., and Kandel, E. 1978. Presynaptic modulation of voltage-dependent Ca current: Mechanism for behavioral sensitization in *Aplysia californica*. *Proc. Nat. Acad. Sci. 75*:3512–3516.

Kobayashi, H., Hashiguchi, T., and Ushiyama, N. 1978. Postsynaptic modulation of excitatory process in sympathetic ganglia by cyclic AMP. *Nature 271*:268–270.

Levitan, I., and Drummond, A. 1980. Neuronal serotonin receptors and cyclic AMP: Biochemical, pharmacological, and electrophysiological analysis. In *Neurotransmitters and their Receptors* (U. Littaver, Y. Dudai, I. Silman, V. I. Tiechberg, and Z. Vogel, eds.), pp. 163–176. London: Wiley.

Liberman, Y., Minina, S., and Golubtsov, K. 1975. Study of the metabolic synapse. I. Effect of intracellular microinjection of 3′,5′-AMP. *Biophysics* (USSR) *20*:457–463.

1977. Study of the metabolic synapse. II. Comparison of the effects of cyclic 3′,5′-AMP and 3′,5′-GMP. *Biophysics* (USSR) *22*:73–80.

Liberman, Y., Minina, S., and Shklovskii, N. 1978. Depolarization of the neural membrane on exposure to cyclic 3′,5′-adenosine monophosphate and its possible role in the work of the molecular computer (m.c.) of the neurone. *Biophysics* (USSR) *23*:308–314.

Liberman, Y., Minina, S., and Shklovskii-Kordi, N. 1981. Study of the diffusion modelling system of the molecular computer of the neurone. *Biophysics* (USSR) *25*:470–477.

Libet, B. 1979. Which synaptic action of dopamine is mediated by cyclic AMP? *Life Sci. 24*:1043–1057.

Nathanson, J. 1977. Cyclic nucleotides and nervous system function. *Physiol. Rev. 57*:157–256.

Nathanson, J., and Greengard, P. 1977. Second messengers in the brain. *Sci. Am. 237*:108–119.

Nuccitelli, R., and Deamer, D. 1982. *Intracellular pH: Its Measurement, Regulation, and Utilization in Cellular Functions.* New York: Liss.

Partridge, L., Thompson, S., Smith, S., and Connor, J. A. 1979. Current–voltage relationships of repetitively firing neurons. *Brain Res. 164*:69–80.

Pellmar, T. 1981. Ionic mechanism of a voltage-dependent current elicited by cyclic AMP. *J. Cell. Mol. Neurobiol. 1*:87–97.

Rall, T. and Sutherland, E. 1958. Formation of cyclic adenine ribonucleotide by tissue particles. *J. Biol. Chem. 232*:1065–1076.

Rall, T., Sutherland, E., and Berthet, J. 1957. The relationship of epinephrine and glucagon to liver phosphorylase. IV. Effect of epinephrine and glucagon on the reactivation of phosphorylase in liver homogenates. *J. Biol. Chem. 224*:463–475.

Shimahara, T., and Tauc, L. 1977. Cyclic AMP induced by serotonin modulates the activity of an identified synapse in *Aplysia* by facilitating the active permeability to calcium. *Brain Res. 127*:168–172.

Sutherland, E., Rall, T., and Menon, T. 1962. Adenyl cyclase. I. Distribution, preparation, and properties. *J. Biol. Chem. 237*:1220–1227.

Thomas, R. C. 1976. The effect of carbon dioxide on intracellular pH and buffering power of snail neurones. *J. Physiol. 255*:715–735.

1978. *Ion-Sensitive Intracellular Microelectrodes.* New York: Academic Press.

Treistman, S., and Levitan, I. 1976a. Alteration of electrical activity in molluscan neurons by cyclic nucleotides and peptide factors. *Nature 261*:62–64.

1976b. Intraneuronal guanylyl–imdodiphosphate injection mimics longterm synaptic hyperpolarization in *Aplysia. Proc. Nat. Acad. Sci. 73*:4689–4692.

Weight, F., Smith, P., and Schulman, J. 1978. Postsynaptic potential generation appears independent of synaptic elevation of cyclic nucleotides in sympathetic neurons. *Brain Res. 158*:197–202.

18 · Patterns of intracellular information transfer

HOWARD RASMUSSEN

INTRODUCTION

A question might be raised as to why someone who has spent most of his investigative career in the study of hormone action in nonexcitable cells should contribute to a volume devoted to neural substrates of learning and behavior. This question is legitimate only as long as one persists in accepting the concept that stimulus–response coupling in nonexcitable cells is fundamentally different from stimulus–response coupling in excitable cells. Although initially this appeared to be the case with Ca^{2+} serving as a major coupling factor in excitable cells and cyclic AMP (cAMP) in nonexcitable ones, it is no longer permissible to make such a distinction [1, 2]. Ca^{2+} serves as coupling factor or messenger in the action of numerous hormones, and cAMP in the action of neurotransmitters. Furthermore, these two messengers appear to convey qualitatively similar types of information from cell surface to cell interior. A dramatic validation of this statement comes from the studies of the control of exocrine secretion in pancreas and parotid gland. Both glands secrete fluid and electrolytes, and digestive enzymes by a process of exocytosis. The intracellular messenger that couples neural stimulation to enzyme secretion in the pancreas is Ca^{2+} and to fluid secretion, cAMP. The converse operates in the parotid: Ca^{2+} couples stimulus to fluid secretion, and cAMP couples stimulus to enzyme release [1, 3, 4].

Rather than separate cell types employing distinctly different intracellular messengers, there is a single universal system employing the dual messengers, Ca^{2+} and cAMP, that couples a particular type of extracellular stimulus to cell response. The characteristics of this type of extracellular messenger are (1) it combines with receptor(s) on the cell surface, (2) in doing so it causes a change in the concentration of an intracellular messenger (Ca^{2+} and/or cAMP), and (3) it thereby induces the differentiated cell to perform its work function.

In an effort to call attention to this universal system, the term "synarchic messengers" has been introduced to characterize the general phenomenon of dual messenger control of cellular response [1]. This concept will be developed more fully in the latter part of this chapter, and the various particular patterns of Ca^{2+}–cAMP interaction within this universal system will be presented.

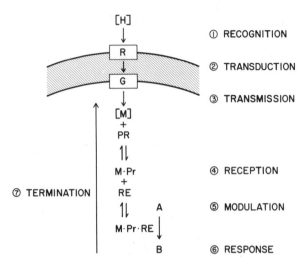

Figure 18.1. A model of the steps in information flow during peptide hormone or neurotransmitter action. H, hormone; R, surface receptor; G, signal generator; M, intracellular messenger; PR, receptor protein for this messenger; M·Pr, complex of messenger with its receptor protein; RE, response element; M·Pr·RE, activated response element; A → B, metabolic step(s) being regulated.

INTRACELLULAR INFORMATION TRANSFER

From the point of view of the present discussion, the most important developments have been the great strides made in defining the molecular events that underlie stimulus–response coupling. Because of these developments, it is possible to consider hormone and neurotransmitter action as processes of information transfer [1, 2] and to define at least seven distinct steps in such information transfer processes: (1) *recognition* of the extracellular messenger by a receptor on the cell surface; (2) *transduction* of this messenger into a new intracellular messenger; (3) *transmission* of this message from cell surface to cell interior; (4) *reception* of the intracellular messenger by specific cytosolic (or membrane-bound) receptor proteins; (5) *modulation* of the activity of other proteins, called response elements, by the messenger–receptor protein complex; (6) *response* as a consequence of a change in response element function; and (7) *termination* of messenger generation and/or response (Figure 18.1).

Within this conceptual framework, it is possible to discuss the messenger function of either Ca^{2+} or cAMP. However, I shall consider only that of Ca^{2+} in detail because it presents several unique features and because recent work has gone further in defining particularly the steps of intracellular receptor, modulation, and response. Nevertheless, from what is known, it seems evident that there are many operational similarities between the Ca^{2+} messenger system and the cAMP messenger systems.

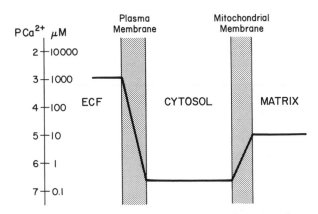

Figure 18.2. A schematic representation of cellular calcium metabolism showing the concentration of free calcium in extracellular fluid (ECF), cell cytosol, and mitochondrial matrix space. The concentration in the matrix space is not known with certainty, so the value depicted is only an estimate.

Calcium as messenger

One way in which to approach the question of how Ca^{2+} serves its intracellular messenger function is to consider this question within the context of cellular calcium metabolism and the calcium-binding properties of the calcium receptor protein, calmodulin. In doing so, three assumptions will be made: (1) that when Ca^{2+} serves as messenger it conveys its information by binding to and thereby altering the conformation of specific intracellular protein(s) whose sole function is that of binding Ca^{2+} and thence interacting with one or more response elements to change their (the response elements') behavior; (2) that all calcium receptor proteins of this type are members of a homologous class of low-molecular-weight proteins, of which calmodulin is representative [5]; (3) that it is the free calcium ions in the cell cytosol that are involved in this information transfer process [2, 6].

The most compelling fact about cellular calcium metabolism is that the calcium ion concentration in the cytosol is approximately 0.2 μM, but that of the extracellular fluids 1000 μM (Figure 18.2). This 5000-fold calcium concentration gradient existing across the plasma membrane of the cell [6] is maintained by three processes in the plasma membrane. First, the rate of influx of calcium into the cell is quite small because the membrane is relatively impermeable to this cation. Second, external Na^+ is exchanged for internal Ca^{2+}, a process of secondary active transport driven by the Na^+ gradient across the membrane, which is maintained, in turn, by the activity of the Na^+–K^+–ATPase. Third, the active extrusion of Ca^{2+} across the membrane is driven by a specific Ca^{2+}–ATPase or calcium pump.

One of the most interesting properties of this Ca^{2+} pump is that its function is modulated by Ca^{2+}–calmodulin [7–9]; that is, it is regulated by the calcium

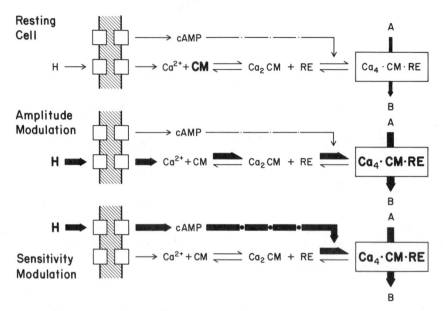

Figure 18.3. Models of modulation in the calcium messenger system.

ion concentration of the cell cytosol. This property is of critical importance in understanding the way in which the calcium messenger system operates within the cell (see below).

Because there are circumstances under which calcium influx into the cell increases considerably, there are also intracellular systems whose functions supplement those of the Na^+–Ca^{2+} exchange mechanism and the calcium pump. They help control the calcium ion concentration in the cell cytosol $[Ca^{2+}]_c$ [6]. These are principally a poorly defined passive buffer system in the cell cytosol and an active (energy-dependent) buffering of cytosolic calcium ion concentration by a calcium pump in the inner mitochondrial membrane [10, 11]. In other words, the mitochondria serve as a critically important sink for Ca^{2+} in time of excessive cellular Ca^{2+} uptake (Figure 18.2). This mitochondrial function is essential for cellular calcium homeostasis. It is also important to the maintenance of cellular integrity because excess cytosolic calcium is extremely toxic to the cell [12–14]. High $[Ca^{2+}]$ alters energy metabolism, activates proteases and phospholipases, and if sustained for any length of time leads to cell death. As might be anticipated, the mitochondrial capacity to store excess calcium is large but finite, and if it is exceeded, cell death ensues.

With this information as way of background, it is now possible to consider the role of Ca^{2+} in cellular information transfer. In doing so, it is convenient to focus on the question of how reception and modulation occur. The generally assumed model as to how these events occur is one that can be labeled as an example of amplitude modulation (Figure 18.3): An increase in the concentration (amplitude) of the messenger (message) is perceived by the calcium receptor protein. As a consequence, the concentration of the Ca^{2+}–protein complex

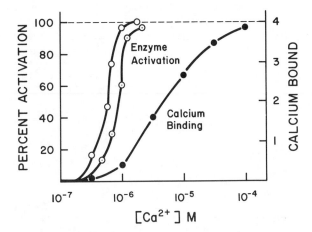

Figure 18.4. A comparison of the binding of calcium to free calmodulin (closed-circle curve) with the activation of either the red cell plasma membrane calcium pump (dotted-circle curve) or of phosphodiesterase (open-circle curve) in the presence of calmodulin as a function of the calcium ion concentration.

increases. This complex (calcium–calmodulin) binds to a response element, and in doing so modulates its function, that is, alters its conformation. If we consider the binding of calcium to calmodulin in terms of this model, and relate these considerations to our knowledge of cellular calcium metabolism, a problem arises. The binding of Ca^{2+} to isolated calmodulin (CM) displays an average K_d of about 10^{-5} M [15–19]. Hence, practically no Ca^{2+} is bound to CM below 10^{-6} M, and saturation of binding sites on CM occurs at 10^{-4} M (Figure 18.4). If we assume that the CM must be fully saturated ($Ca_4 \cdot CM$) in order to modulate response element function, then a $[Ca^{2+}]_c$ of at least 10^{-5} M must be achieved within the cell cytosol when a cell displays amplitude modulation following neurotransmitter or hormone action. However, if we consider the behavior of typical mitochondria in isolation and in situ [10, 11], a prediction can be made as to the rate of Ca^{2+} uptake by the mitochondria in a cell in which the $[Ca^{2+}]_c$ is 10^{-5} M. This prediction shows that the rate would be enormous (Table 18.1) and if sustained the mitochondrial capacity to accumulate calcium would be exceeded in 1.5 hr or less. Yet many cells (that employ Ca^{2+} as a message) can maintain an activated state for longer than 1.5 hr; hence the system must operate differently than just discussed.

It does operate differently, even in vitro, as soon as a response element is included along with Ca^{2+} and calmodulin [17]. For example, as shown in Figure 18.3, the Ca^{2+}–CM-dependent activation of the Ca^{2+} pump has a quite different Ca^{2+} dependency than that of Ca^{2+} binding to isolated CM. There are at least three factors that could contribute to this difference: (1) CM has four Ca^{2+} binding sites per mole and it is possible that only one or two need be occupied in order for CM to exert its modulatory function; (2) there is a large excess of CM over response element; or (3) there are cooperative interactions between Ca^{2+}, CM, and response element.

Table 18.1. *Cellular calcium metabolism as a function of cytosolic calcium concentration*

	Flux rate (μM/kg cell H_2O/min)				
	(R) 0.2 μM	10 μM	1.0 μM	0.4 μM	(A) 0.3 μM
Cell					
Influx	4	100	30	20.5	20.0
Efflux	4	40	24	18.5	18.2
Net	0	60	6.0	1.5	1.8
Mitochondria					
Influx	0.5	40.5	5.5	1.8	2.0
Efflux	0.5	0.5	0.5	0.5	0.5
Net	0	40.0	5.0	1.3	1.5
Cell death (hr)	—	1.5	20	75	55

Note: Values are of the estimated rates of cellular and mitochondrial calcium flux when the steady-state concentration of ionized calcium is changed in the cell cytosol. The first column (R) is what is seen in a resting cell, the last column in a hormone activated cell (A), and the intervening ones the hypothetical situations when the cytosolic calcium is maintained at 10, 1.0, or 0.4 μM.
Source: Estimates based on data from Borle [6] and Nicholls [10].

There is no evidence to support the first possibility. All recent data indicate that $Ca_4 \cdot CM$ is the modulatory species of calmodulin [17, 19]. The second possibility is supported by the observation that both in vitro and in situ there is a considerable (100-fold or greater) molar excess of CM over response element (RE) [5, 17]. Hence, part of the difference seen in Figure 18.3 is due to [CM] $>$ $>$ [RE]. However, this is only part of the explanation. Another part is due to the fact that the interactions between the reactants Ca^{2+}, CM, and RE occur in an ordered and highly cooperative sequence [15–18, 20]. This sequence is depicted schematically in Figure 18.5. There are three major steps. The first is the cooperative binding of two calciums to calmodulin. This leads to a conformational change in CM so that it now interacts with a RE with a very high affinity. The binding of $Ca_2 \cdot CM$ to RE leads to another conformational change in CM so that it now binds the second two calciums with a very high affinity. When the last two calciums bind to $Ca_2 \cdot CM \cdot RE$, the RE undergoes a conformational change from an "inactive" to an "active" conformation.

This ordered and highly cooperative sequence that involves Ca^{2+} as a ligand in both the first and third steps has a unique property. It takes a larger bolus of Ca^{2+} to shift it from the inactive to the active state than to maintain it in the active state. (An appropriate analogy is the concept of activation energy in a chemical reaction.) Because of this property, the system possesses the potential of displaying an oscillatory pattern of behavior.

This property is particularly important to the problem of cellular calcium toxicity discussed above. If we consider how this system operates in a cell that has been activated to produce a sustained response, there are data showing that the characteristic pattern of change in the $[Ca^{2+}]_c$ is one of an initial rise from

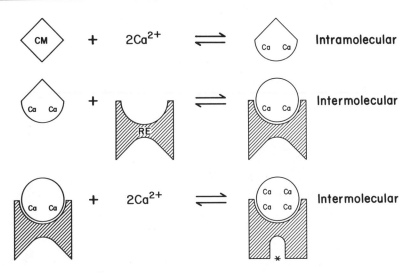

Figure 18.5. A three-step model of the activation of a calcium–calmodulin regulated enzyme. CM, calmodulin; RE, response element. See text for description.

0.2 to 1.0 μM followed by a fall to a steady-state value in the range 0.5–0.6 μM [21, 22]. This is coupled to sustained higher rates of calcium influx and efflux accross the plasma membrane [23, 24]. The sustained high rates of calcium efflux indicate that the calcium pump is associated with and being regulated by $Ca_4 \cdot CM$ (pump). If the system does indeed have the kinetic properties discussed above, then the coupling of $Ca_4 \cdot CM$ to the plasma membrane Ca^{2+} pump is an essential element in the calcium messenger system. The $[Ca^{2+}]_c$ must fall below its original value in order for the response to be terminated. This means that in order to achieve a sustained response, the rate of calcium influx must remain high to balance both the higher rate of efflux from the cell and the uptake of Ca^{2+} by the mitochondria when $[Ca^{2+}]_c$ shifts from 0.2 to 0.6 μM. An estimate of the magnitude of this uptake is shown in Table 18.1 (see Refs. 6 and 23). The kinetic behavior of the $Ca \cdot CM \cdot RE$ system allows the cell to employ Ca^{2+} as a messenger while simultaneously minimizing the danger of cellular calcium intoxication.

The process of amplitude modulation in the calcium messenger system is considerably more complex and elegant than originally considered. Even so, the above discussion touches on only one facet of its behavior within the cell. In addition to amplitude modulation, the system displays another type of modulation, *sensitivity modulation* (Figure 18.3).

By definition, sensitivity modulation is a process in which there is an increase in the concentration of $Ca_4 \cdot CM \cdot RE$ (where RE stands for a single specific response element) without a change in $[Ca^{2+}]_c$. This could come about by one of several mechanisms: (1) a change in calmodulin structure so that binding of the first two calciums is enhanced; (2) an increase in the affinity of the RE element for $Ca^2 \cdot CM$; or (3) an increase in calmodulin concentration – for example, the release of CM from a membrane-bound pool.

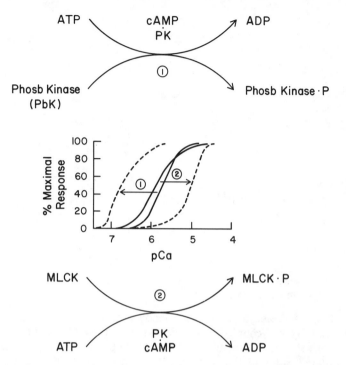

Figure 18.6. Sensitivity modulation of phosphorylase b kinase and myosin light-chain kinase.

The most clearly defined examples of sensitivity modulation in the calcium messenger system are ones in which cAMP is involved. In none of these does cAMP alter the binding of Ca^{2+} to CM by changing the structure of CM. There are examples in which cAMP alters the binding of CM to plasma membranes, and these may be examples of a special type of sensitivity modulation. However, the type of sensitivity modulation of most interest is the one in which a cAMP-dependent phosphorylation of a response element alters its affinity for $Ca_2 \cdot CM$. This change in affinity can be either an increase or a decrease, so it is convenient to talk of either positive or negative sensitivity modulation. Operationally, this type of change in affinity of response element for calmodulin leads to a change in the calcium activation profile of the enzyme. This is illustrated in Figure 18.6 for both phosphorylase b kinase and myosin light-chain kinase [25–32]. Each is a Ca^{2+}-dependent, calmodulin-regulated enzyme, and each can serve as a substrate for the cAMP-dependent protein kinase. Phosphorylation of either enzyme by this cAMP-dependent kinase causes a shift of the calcium activation profile of the enzyme. In the case of phosphorylase b kinase, the shift is to the left, indicating an increase in the sensitivity of the enzyme to activation by Ca^{2+}, that is, positive sensitivity modulation. In the case of myosin light-chain kinase, the shift is to the right, indicating a decrease in sensitivity of the enzyme to activation by Ca^{2+}, that is, negative sensitivity modulation.

Although these examples are both cases in which a covalent modification of a response element leads to a shift in the apparent sensitivity of the system to activation by Ca^{2+}, there is no reason to believe that this is the exclusive means by which sensitivity modulation is achieved within the cell. It is already known, for example, that an increase in the substrate concentration (cAMP) of the calcium-dependent phosphodiesterase causes a positive shift in the activation profile of the enzyme [33]. In a sense, this is a specialized example of an allosterically mediated change in calcium sensitivity. From this example, one should be able to generalize and propose that other allosteric modifiers (other than substrates) will be found to bring about changes in the calcium activation profiles of calmodulin-regulated enzymes. The key point to emphasize is that this is likely to be an important mechanism of control in the calcium messenger system because it allows a single type of response element to be turned on or off selectively.

Plasticity of cellular control systems

The remarkable feature of the nervous system is that in spite of a rather stereotyped mechanism of information transfer from one cell to the next via an action potential and its eventual consequence, synaptic secretion, it exhibits enormous plasticity in behavioral responses. The same is true of the flow of information within the cell in the apparently rather stereotyped calcium messenger system. The single most important element conferring a potential for this type of plastic behavior is the diversity of $CM \cdot RE$ complexes. Rather than displaying a single calcium activation profile, they display a remarkably broad range of individual profiles. This diversity of sensitivities of various calmodulin-dependent response elements to activation by Ca^{2+} means that in addition to quantitative differences in cellular responses to quantitative differences in the amount of neurotransmitter or hormone they are exposed to, there can be qualitative differences in response. These could be achieved simply if the strength (concentration) of the intracellular message (messenger) is a direct function of the strength of the extracellular stimulus. Under this circumstance, a small stimulus would give rise to a small change in $[Ca^{2+}]_c$ that would be sufficient to activate only a few of the most sensitive calmodulin-regulated response elements. A large stimulus by causing a large change in $[Ca^{2+}]_c$ would lead to the activation of many of the less sensitive response elements. Furthermore, the particular sensitivity of any given response element would be determined by the prior history of the cell. Hence, a qualitatively different response of the same cell type to the same stimulus would occur if two different groups of that particular cell type had been exposed to different hormonal or environmental influences that alter the calcium sensitivity of specific response elements. There is then a potentially broad range of cellular responses to a given stimulus.

Patterns of calcium–cAMP interactions

This feature of plasticity within a single messenger system provides the cell with only one element of its remarkable potential for plastic behavior. Another element is that provided by using a dual rather than a single messenger system to control cell function. There are now available data from an extremely wide variety of different cell types from different organisms attesting to the fact that the calcium and cAMP messenger systems nearly always interact during their functioning in stimulus–response coupling [1, 2]. These two messenger systems convey a qualitatively similar kind of information, as can best be seen in their control of exocrine secretion from pancreas and parotid. As discussed in the Introduction, the roles of Ca^{2+} and cAMP appear interchangeable in regulating the secretion of the two major components, enzymes and fluids, of the gland product. In both cases, this dual control provides a considerable plasticity of response by modulating fluid and protein secretion separately.

Not only do the two messenger systems carry the same type of information, they interact at several levels of intracellular information transfer from transduction to termination [1, 2]. Thus, a rise in $[Ca^{2+}]_c$ can either activate or inhibit adenylate cyclase, and increase or decrease the activity of phosphodiesterase [1, 2, 5, 6, 34]. It is also possible that Ca^{2+} may influence the interaction of cAMP with its receptor proteins. Conversely, an increase in $[cAMP]_c$ can increase or decrease the entry of Ca^{2+} into the cell across the plasma membrane, can increase Ca^{2+} uptake by endoplasmic reticulum, can increase Ca^{2+} pumping out of the cell, and may decrease net calcium uptake by mitochondria [1, 2, 6]. A rise in $[cAMP]_c$ can also increase or decrease the sensitivity of various response elements to activation by Ca^{2+}. Depending on which of these multiple interactions operates in a particular cell type, any one of a number of distinct patterns of Ca^{2+}–cAMP is seen. At least five general patterns of their interactions are identifiable [1].

Patterns of synarchic regulation

The five identifiable patterns of synarchic regulations are coordinate, hierarchical, redundant, antagonistic, and sequential (Figure 18.7). Representative examples of each type are given in Table 18.2.

A particularly well-defined example of coordinate control is the control of fluid and electrolyte secretion from the blowfly salivary gland by serotonin (5HT) [1, 2, 21, 24, 35, 36]. When 5HT acts, there is both an increase in Ca^{2+} entry into the gland and an activation of adenylate cyclase. These two effects are presumably mediated by two different populations of 5HT receptor, although this point remains to be established. In any case, the two messengers interact within the cell in two ways: (1) They exert control at sequential points in the secretory process by regulating the activity of a K^+ pump in the luminal membrane (cAMP) and controlling the $Cl-$ permeability of this membrane (Ca^{2+}); and (2) they regulate each other's concentrations by modulating either generation and/or termination of the other messenger. A point of considerable interest is the fact that, experimentally, fluid secretion is activated when the concentration of only one of the two messengers increases within the cell. Nonetheless, under physiological circumstances both are clearly involved in stimulus–response coupling in this tissue. The value of the arrangement is that after the initial transient rise in $[Ca^{2+}]_c$ and $[cAMP]_c$ a minimal perturbation in both $[Ca^{2+}]_c$ and $[cAMP]_c$ is sufficient to maintain these secretory cells in their activated state for hours.

A particularly interesting example of a hierarchical control pattern is that

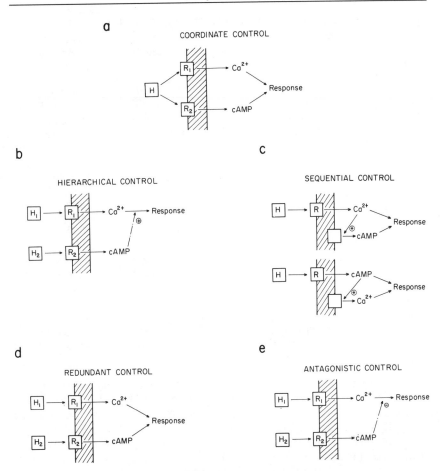

Figure 18.7. Patterns of synarchic regulation.

found in the beta (insulin-secreting) cells of the endocrine pancreas [2, 37–40]. These cells are electrically excitable in the classic sense and display trains of action potentials when stimulated by a rise in extracellular glucose concentration. The rise in glucose concentration is coupled to the calcium messenger system so that this endocrine gland behaves operationally like a neural synapse: Calcium couples stimulus to secretion, and a considerable part of the calcium employed comes from the extracellular pool. However, the secretory response of the cell to a given change in extracellular glucose concentration depends upon the degree of activation of the cAMP messenger system. Agents such as glucagon, which activate adenylate cyclase, enhance cAMP concentrations within the beta cell and enhance the secretory response to glucose, whereas agents such as norepinephrine, which inhibit adenylate cyclase, diminish cAMP concentrations and thereby diminish the secretory response to glucose. Similar patterns are seen in the control of neurosecretion at synapses. The

Table 18.2 *Examples of different patterns of synarchic regulation*

Patterns	Tissue or cell	Intracellular messenger	Hormone neurotransmitter
Coordinate	Fly salivary gland	Ca^{2+}– cAMP	Serotonin
	Renal tubules	Ca^{2+}–cAMP	Parathyroid
Hierarchical	Beta cell of pancreas	Ca^{2+}–cAMP	Glucose
	Adrenal fasiculata	Ca^{2+}–cAMP	Glucagon
	Aplysia muscle	Ca^{2+}–cAMP	ACTH
			ACTH
			Acetylocholine
			Serotonin
Redundant	Liver	cAMP–Ca^{2+}	Glucagon
	Adrenal glomerulosa	cAMP–Ca^{2+}	Epinephrine (α)
			ACTH
			Angiotensin
Antagonistic	Smooth muscle	Ca^{2+}–cAMP	Acetylocholine
	Platelet	Ca^{2+}–cAMP	Epinephrine (β)
			Thrombin
			PGE_2
Sequential	Heart	cAMP $\rightarrow Ca^{2+}$	Epinephrine
	Adrenal medulla	$Ca^{2+} \rightarrow$ cAMP	Acetylcholine

question yet to be answered is how cAMP acts within the cell to change its response to glucose. There is incomplete evidence that it acts both to expand the temporal and spatial domain of the calcium messenger and to alter the sensitivity of calcium response elements to activation by the calcium message.

The counterpart of this type of pattern in a neuromuscular system is found in the buccal muscle of *Aplysia* [2, 41]. The cells of accessory radula closer muscle do not display action potentials when stimulated to contract by cholinergic neurons. Only a membrane deplorization is seen. Nevertheless, acetylcholine, acting via the calcium messenger system, causes the muscle to contract. Serotonergic neurons also innervate this muscle. Stimulation of these neurons alone causes no electrical or contractile responses in the resting muscle, but does increase its content of cAMP. However, if these serotonergic neurons are stimulated just prior to cholinergic stimulation, the resulting contractile response is enhanced. This pattern is operationally similar to that seen in the endocrine pancreas. At present the mechanism by which cAMP acts to enhance this response is not known.

Not surprisingly, some cell types that perform functions of critical survival value have a redundant pattern of control. Liver cells that provide glucose to the blood that is vital to brain function is the best-studied example [2, 11, 42–44]. Adrenal glumerulosa cells that secrete aldosterone that is essential for Na^+ conservation is another [2]. In the case of the hepatocyte, glucagon acting via the cAMP system, and α-adrenergic agents, vasopressin, and/or angiotensin acting via the Ca^{2+} system, all stimulate both glycogenolysis and gluconeogenesis, that is, activate those reactions responsible for both short-term (glycogenolysis) and long-term (gluconeogenesis) glucose production. If we consider the dozen or so reactions involved in converting lactate or alanine to glucose, there are several regulated steps: (1) glucose-6-phosphatase, (2) fructose diphosphatase, (3) PEP-carboxykinase, (4) pyruvate kinase, (5) pyruvate carboxylase, and (6) phosphofructokinase. Steps 4 and 6 are inhibited, the others activated. In considering how redundant control might operate in such a complex metabolic sequence, one of at least two possibilities exists. Different messengers (Ca^{2+} or cAMP) control different key steps, or these messengers regulate the same steps. Since integration of cellular response is a prime attribute of intracellular messenger function, the likelihood is that the two messengers regulate the functions of the same key enzymes in this metabolic sequence. Support for the supposition is provided by the observations that in hepatocytes prelabeled with [^{32}p]-phosphate and then treated with either norepinephrine (Ca^{2+}) or glucagon (cAMP), the pattern of phosphoprotein labeling is nearly identical. For this result and observations in other systems, it appears likely that in this hepatocyte response, an enzyme such as pyruvate kinase is a substrate for either a cAMP-dependent or a calcium-dependent protein kinase, that each enzyme catalyzes the phosphorylation of a different site on the pyruvate kinase molecule, yet these separate phosphorylations decrease the activity of the enzyme. It is possible to suggest as a general rule that in redundant control systems the point of convergence will be at the level of modulation of response element structure.

One of the most interesting patterns of synarchic regulation is the antagonistic one. To date these systems are all ones in which Ca^{2+} is the activator of the particular cellular response, and the rise in cAMP concentration, as a consequence of cyclase activation, leads to a blunting or inhibition of the calcium-mediated response. The prime example of this type of pattern is that seen in several types of smooth muscle [2, 28–32, 45]. There is considerable evidence in support of the view that the calcium-dependent phosphorylation of myosin light chain by myosin light-chain kinase is the event that regulates the contractile response in these muscle cells. The myosin light-chain kinase is a calcium–calmodulin-modulated enzyme. When acetylcholine stimulates these cells, a rise in $[Ca^{2+}]_c$ leads to an activation of the enzyme and hence a contractile response. Addition of epinephrine brings about a relaxation of the acetylcholine-mediated contraction. The effect of epinephrine is mediated via the cAMP messenger system. The rise in cAMP acts in two ways to antagonize the acetylcholine effect. As already discussed above, one effect of cAMP is to catalyze the phosphorylation of myosin light-chain kinase, making it thereby less sensitive to activation by calcium ion. The second effect of cAMP is that of stimulating the removal of Ca^{2+} from the cytosol by increasing both calcium efflux from the cell and calcium uptake into the sarcoplasmic reticulum.

No examples are known of systems in which cAMP mediates the basic cellular response and calcium produces an antagonistic effect. Whether this represents an evolutionary asymmetry in the development of synarchic regulation or is merely a lack of information about such systems is not known presently.

The final pattern is that of sequential control. There are examples in which Ca^{2+} is the initial intracellular messenger that initiates a rise in the concentration of cAMP, the subsequent messenger; and others in which cAMP is initial and Ca^{2+} is subsequent [2].

An example of a system in which Ca^{2+} is an initial messenger in a sequential pattern is the adrenal medulla [2, 46–48]. When acetylcholine stimulates catecholamine amine release from this tissue, calcium is responsible for coupling stimulus to response. However, it is also responsible for the sequential activation of adenylate cyclase, causing a rise in $[cAMP]_c$ and then of phosphodiesterase, causing a subsequent fall in $[cAMP]_c$. Even so, the rise in $[cAMP]_c$ is essential in activating the synthesis of the enzyme tyrosine hydroxylase, a major enzyme of the catecholamine biosynthetic pathway. Thus, by employing this sequential type of messenger pattern, the adrenal medullary cell activates two different functions within the cell with completely different time constants and metabolic functions. Nonetheless, these two functions both serve the specific differentiated function of the cell, the secretion of catecholamines. Thus, as is always the rule, synarchic regulation in this system regulates cell function in an integrative sense.

The converse type of sequential regulation is seen in the heart when catecholamines induce an inotropic effect [2, 49, 50]. The immediate effect of the catecholamines is to activate adenylate cyclase and thereby cause a rise in $[cAMP]_c$. The rise in $[cAMP]_c$ has two effects in cellular calcium metabolism: (1) It increases the time during which calcium enters the cell via the voltage-dependent calcium channel; and (2) it enhances the rate of calcium accumulation by the sarcoplasmic reticulum. Both of these effects bring about a larger calcium signal during subsequent systoles and therefore an increase in contractile force.

Undoubtedly, other examples of sequential regulation, particularly in the nervous system, will be found.

CONCLUSION

Not only is the recognition that cAMP and Ca^{2+} serve as synarchic messengers important from an experimental point of view, but an attempt to develop a conceptual framework within which to consider the functioning of these messengers has heuristic value in redefining the framework within which all cell regulation operates. The sorting out of the different patterns of synarchic regulation is useful from a conceptual and didactic point of view. However, the examples given have all been simplified. Cell regulation must not simply be thought of as a single chord, but as a symphony. Many of the systems discussed display a major theme, but most also display one or more minor ones. For example, in the case of the endocrine pancreas, in addition to the major hierarchical pattern delineated above, there is a minor one, a sequential one, as well. When glucose concentration rises, the calcium messenger system is activated, and this leads to a secretory response. Under appropriate conditions, the rise in $[Ca^{2+}]_c$ activates adenylate cyclase, leading to a rise in $[cAMP]_c$, which in turn enhances the secretory response to the rise in $[Ca^{2+}]_c$.

The most important conclusion to be reached is that stimulus–response coupling in so-called excitable tissues involves the same synarchic messengers as does such coupling in nonexcitable tissue. This system is universal. As such, the patterns seen in many hormonally regulated tissues must have their counterpart in the central nervous system, and these play a part in the learning process.

ACKNOWLEDGMENT

This work is supported by a grant (AM19813) from the National Institute for Arthritic, Metabolic, and Digestive Diseases.

REFERENCES

1. Rasmussen, H. 1981. *Calcium and cAMP as Synarchic Messengers.* New York: Wiley.
2. Rasmussen, H., and Waisman, D. M. 1981. The messenger function of calcium in endocrine systems. *Biochem. Actions Horm. 8*:1–115.
3. Schramm, M., and Selinger, Z. 1975. The function of α- and β-adrenergic receptors and a cholinergic receptor in the secretory cell of rat parotid gland. *Adv. Cytopharmacol. 2*:29–32.
4. Pearson, G. T., Singh, J., Daoud, M. S., Davison, J. S., and Petersen, O. H. 1981. Control of pancreatic cyclic nucleotide levels and amylase secretion by noncholinergic, nonadrenergic nerves. *J. Biol. Chem. 256*:11025–11031.
5. Scharff, P. 1981. Calmodulin and its role in cellular activation. *Cell Calcium 2*:1–28.
6. Borle, A. 1981. Control, modulation, and regulation of cell calcium. *Rev. Physiol. Biochem. Pharmacol. 90*:14–152.
7. Penniston, J. T., Graf, E., and Itano, T. 1980. Calmodulin regulation of the Ca^{2+} pump of erythrocyte membranes. *Ann. N.Y. Acad. Sci. 356*:245–257.
8. Vincenzi, F. F., Hinds, T. R., and Raess, B. V. 1980. Calmodulin and the plasma membrane calcium pump. *Ann. N.Y. Acad. Sci. 256*:233–244.
9. Waisman, D. M., Gimble, J., Goodman, D. B. P., and Rasmussen, H. 1981. Studies of the Ca^{2+} transport mechanism of human inside-out plasma membrane vesicles. I. Regulation of the Ca^{2+} pump by calmodulin. *J. Biol. Chem. 256*:409–414.
10. Nicholls, D. G. 1978. The regulation of extramitochondrial free calcium ion concentration by rat liver mitochondria. *Biochem. J. 179*:511–522.
11. Murphy, E., Catt, K., Rich, T. L., and Williamson, J. R. 1980. Hormonal effects on calcium homeostasis in isolated hepatocytes. *J. Biol. Chem. 255*:6600–6608.
12. Schanne, F. A. X., Kane, A. B., Young, E. E., and Farber, J. L. 1979. Calcium dependence of toxic cell death: A final common pathway. *Science 206*:206–208.
13. Fleckenstein, A. 1974. Drug-induced changes in cardiac energy. *Adv. Cardiol. 12*:183–197.
14. Wrogemann, K., and Pena, S. D. J. 1976. Mitochondrial calcium overload: A general mechanism for cell necrosis in muscle disease. *Lancet 1*:672–673.
15. Crouch, T. H., and Klee, C. B. 1980. Positive cooperative binding of calcium to bovine brain calmodulin. *Biochemistry 19*:3692–3698.
16. Wang, J. H., Sharma, R. K., Huang, C. Y., Chau, V., and Chock, P. B. 1980. On the mechanism of activation of cyclic nucleotide phosphodiesterase by calmodulin. *Ann. N.Y. Acad. Sci. 356*:190–204.
17. Haung, C. Y., Chau, V., Chock, P. B., Wang, J. H., and Sharma, R. K. 1981. Mechanism of activation of cyclic nucleotide phosphodiesterase: Requirement of the binding of four Ca^{2+} to calmodulin for activation. *Proc. Nat. Acad. Sci. 18*:871–874.
18. Harech, J., Klee, C. B., and Vemaille, J. G. 1981. Effects of cations on affinity of calmodulin for calcium: Ordered bindings of calcium ions allows for the specific activation of calmodulin-stimulated enzymes. *Biochemistry 20*:3890–3897.
19. Blumenthal, D. K., and Stull, J. T. 1980. Activation of skeletal muscle myosin light chain kinase by calcium and calmodulin. *Biochemistry 19*:5608–5614.

20. Cheung, W. Y., Lynch, T. J., Wallace, R. W., and Tallant, E. C. 1981. cAMP renders Ca^{2+}-dependent phosphodiesterase refractory to inhibition by a calmodulin-binding protein (calcineuria). *J. Biol. Chem. 256*:4439–4443.

21. Berridge, M. J. 1975. The interaction of cyclic nucleotides and calcium in the control of cellular activity. *Adv. Cyclic Nucl. Res. 6*:1–98.

22. O'Doherty, J., Youmans, S. J., Armstrong, W. McD., and Stark, R. J. 1980. Calcium regulation during stimulus–secretion coupling: Continuous measurement of intracellular calcium acitivties. *Science 209*:510–513.

23. Borle, A., and Uchikawa, T. 1978. Effects of parathyroid hormone on the distribution and transport of calcium in cultural kidney cells. *Endocrinology 102*:1725–1732.

24. Berridge, M. J., and Lipke, H. 1979. Changes in calcium transport across *Calliphora* salivary glands induced by 5-hydroxytryptamine and cyclic nucleotides. *J. Exp. Biol. 78*:137–138.

25. Brostrom, C. O., Hunkeler, F. L., and Krebs, E. G. 1971. The regulation of skeletal muscle phosphorylase kinase by Ca^{2+}. *J. Biol. Chem. 246*:1961–1967.

26. Shenolikar, S., Cohen, P. T. W., Cohen, P., Nairn, A. C., and Perry, S. V. 1979. The role of calmodulin in the structure and regulation of phosphorylase kinase from rabbit skeletal muscle. *Eur. J. Biochem. 100*:327–329.

27. Walsh, K. X., Millikin, D. M., Schendler, K. K., and Reimann, E. M. 1980. Stimulation of phosphorylase b kinase by the calcium-dependent regulator. *J. Biol. Chem. 255*:5036–5042.

28. Hartshorne, D. J., and Persechini, A. J. 1980. Phosphorylation of myosin as a regulatory component in smooth muscle. *Ann. N.Y. Acad. Sci. 356*:130–141.

29. Adelstein, R. S., Conti, M. A., and Pato, M. D. 1980. Regulation of myosin light chain kinase by reversible phosphorylation and calcium-calmodulin. *Ann. N.Y. Acad. Sci. 356*:142–150.

30. Kerrick, W. G. L., Hoar, P. E., and Cassidy, P. A. 1980. Calcium-activated tension: The role of myosin light chain phosphorylation. *Fed. Proc. 39*:1558–1563.

31. Silver, P. J., Holroyde, M. J., Solaro, J., and Disalvo, J. 1981. Ca^{2+}, calmodulin, and cyclic AMP–dependent modulation of actin-myosin interactions in aorta. *Biochim. Biophys. Acta 674*:65–70.

32. Conti, M. A., and Adelstein, R. S. 1980. Phosphorylation by cyclic adenosine 3′:5′-monophosphate–dependent protein kinase regulates myosin light chain kinase. *Fed. Proc. 39*:1569–1573.

33. Smith, S. B., White, H. D., Siegel, J. B., and Krebs, E. G. 1981. Cyclic AMP–dependent protein kinase. I. Cyclic nucleotide binding, structural changes, and release of catalytic subunits. *Proc. Nat. Acad. Sci. 78*:1591–1595.

34. Clayberger, C., Goodman, D. B. P., and Rasmussen, H. 1981. Regulation of cAMP metabolism in the rat erythrocyte during chronic adrenergic stimulation: Evidence for calmodulin-mediated alteration of membrane-bound phosphodiesterase activity. *J. Memb. Biol. 58*:191–201.

35. Prince, W. T., Rasmussen, H., and Berridge, M. 1973. The role of calcium in fly salivary gland secretion analyzed with the ionophore A23187. *Biochim. Biophys. Acta 329*:98–107.

36. Prince, W. T., Berridge, M., and Rasmussen, H. 1972. Role of calcium and adenosine 3′:5′-cyclic monophosphate in controlling fly salivary gland secretion. *Proc. Nat. Acad. Sci. 69*:553–557.

37. Grill, U., and Cerasi, E. 1974. Stimulation by D-glucose of cyclic adenosine 3′:5′-monophosphate accumulation and insulin release in isolated pancreatic islets of the rat. *J. Biol. Chem. 249*:4196–4201.

38. Karl, R. C., Zawalich, W. S., Farrendell, J. A., and Machinsky, F. M. 1975. The

role of Ca^{2+} and cyclic adenosine 3':5'-monophosphate in insulin release induced *in vitro* by the divalent cation ionophore A23187. *J. Biol. Chem. 250*:4575–4579.

39. Mathews, E. K. 1975. Calcium and stimulus–secretion coupling in pancreatic islet cells. In *Calcium Transport in Contraction and Secretion* (E. Carofoli, F. Clementi, W. Drabikowsky, and A. Margreth, eds.), pp. 203–210. Amsterdam: North Holland.

40. Malaisse, W. J. 1973. Insulin secretion multifactorial regulation for a single process of release. *Diabetologia 9*:167–173.

41. Kupfermann, I., Cohen, J. G., Mandelbaum, D. E., Schonberg, M., Susswein, A. J., and Weiss, K. R. 1979. Functional role of serotonergic neuromodulation in *Aplysia. Fed. Proc. 38*:2095–2102.

42. Tolbert, M. E. M., and Fain, J. M. 1974. Studies on the regulation of gluconeogenesis in isolated liver cells by epinephrine and glucagon. *J. Biol. Chem. 249*:1162–1166.

43. Exton, J. H., and Harper, S. C. 1975. Role of cyclic AMP in the actions of catecholamines on hepatic carbohydrate metabolism. *Adv. Cycl. Nucl. Res. 5*:519–532.

44. Garrison, J. C. 1978. The effects of glucagon, catecholamines, and the calcium ionophore A23187 on the phosphorylation of rat hepatocyte cytosolic proteins. *J. Biol. Chem. 253*:7091–7100.

45. Bolton, T. B. 1979. Mechanism of action of neurotransmitters and other substances on smooth muscle. *Physiol. Rev. 59*:609–718.

46. Douglas, W. W., and Rubin, R. P. 1961. The role of calcium in the secondary response of the adrenal medulla to acetylcholine. *J. Physiol.* (London) *159*:40–57.

47. Guidotti, A., Hanbauer, I., and Costa, E. 1975. Role of cyclic nucleotides in the induction of tyrosine hydroxylase. *Adv. Cycl. Nucl. Res. 5*:619–640.

48. Guidotti, A., Chuang, D. M., Hollenbeck, R., and Costa, E. 1978. Nuclear translocation of catalytic subunits of cytosol cAMP-dependent protein kinase in the transsynaptic induction of medullary tyrosine hydroxylase. *Adv. Cycl. Nucl. Res. 9*:185–207.

49. Tsien, R. W. 1977. Cyclic AMP and contractile activity in heart. *Adv. Cycl. Nucl. Res. 8*:363–420.

50. Chapman, R. A. 1979. Excitation–contraction coupling in cardiac muscle. *Progr. Biophys. Mol. Biol. 35*:1–52.

Index

electrical stimulation
  comparison of techniques of, 107, 109
  extracellular microampere technique,
    105–6
  extracellular nanoampere technique, 109,
    111–13
electric currents
  applied intracellularly, 102
  tachycardia from stimulation of right
    cardiac nerve by, 132
electroconvulsive shock (ECS), 154–5, 267,
  269
electromyographic (EMG) response, 105,
  107, 118–19
epileptic seizures, 165
EPSP, 173, 176, 186
excitatory postsynaptic potential (EPSP),
  173, 176, 186
excitors, inhibitors and, 40–1
eye blink response
  cerebellum and, 87, 89–90, 93–4
  elicited by brain stimulation, 102, 118
  high decerebration and delay conditioned,
    75
  latency of, 77, 86, 119
  see also nictitating membrane (NM)
    reponse
eyes of *Hermissenda,* 295, 297, 328–9, 331–4

Farley, J., 292
fear
  conditioned, 47
  fourth ventricle and learned, 86
  mammalian expressions of, 50, 55–8
  see also disgust
feeding
  defense and, 48, 49
  see also consumption
feeding motor program *(Limax),* 230–1, 233–
  4
FMP, *see* feeding motor program *(Limax)*
food odor, 247, 248, 249
Frankstein, S. I., 158
Fudim, O. K., 37

Galef, J., 53
Garcia, J., 49
garden slug, *see Limax maximus*
gastropods, *see* mollusks; *names of specific*
  gastropods
geese, color–illness associations in, 53
giant neurons, 338, 339, 340–1, 343–4, 346–
  55
Giurgea, C., 155–6
gravity-detecting systems, *see* statocysts

habituation, behavioral model for study of,
  130
handling effects, reducing, 207

hawks, taste potentiation of color in, 53
heart rate CR
  associative and nonassociative training and,
    134
  behavioral model for visually conditioned
    changes in, 129–30
  brain and, 82, 86
  central processing time and, 135–6
  experimental variables and, 130
  innervations influencing, 132–4
  mediation of, 133–4
  retinal output and, 134–5
Hebb, D. O., 178–9, 180
  lack of support for postulate of, 201
Hebb synapses, 201
Helisoma, 353
Helix, 353
Hermissenda (Pacific nudibranch), 69
  associative learning in, 205–25, 294, 299,
    300–6, 311, 316–17, 319–22
  behavioral strategy to train, 205–9, 211,
    304–310
  biochemical explanation of associative
    learning in, 319–20
  biophysical basis of learning in, 308–16
  causal role of type B photoreceptor
    changes in, 305–7
  cellular analysis of associative learning in,
    205–25, 308–16
  cellular correlates of learning in, 217, 219–
    20, 222–4, 308–17, 319, 328–34
  chemosensory information in, 297
  CS specificity in, 213–14
  CS and US pathways in, 211–17, 308–16
  different responses to same stimulus in, 291
  electrophysiologic changes after associative
    training in, 302
  eyes of, 214–15, 217, 295, 297
  interneurons of, 297–9, 301–2
  ion flow across membranes in, 316–17,
    319–20
  laboratory-reared, 209
  light-pairing with statocyst stimulation in,
    206, 308–16
  modification of primary sensory neurons in
    CS pathways in, 217, 219–20, 308–16
  neural membrane changes with
    conditioning in, 316
  neural systems in, 294–5, 291–322
  photoreceptors of, 295, 297–9, 302–8, 311,
    313–15, 319–21
  phototaxic behavior in, 206, 208–9, 213–
    14, 220, 222, 225, 298, 328–34
  retention of learned behavior in, 311, 312–15
  rotation as a US in, 206–9, 291–322
  spatial orientation in, 297
  statocysts of, 211, 297, 302
  tentacles of, 297
  visual information in, 297

unconditioned stimulus (UCS *or* US)
  associative learning studies and, 6
  convergence with CS, 151–2
  food shock as, 49
  input of, 150–1
  retinal ganglion cells that do not respond to, 147
  rotation as a, 206, 291–322
  second-order conditioning and, 25
  visual neurons and, 147
  x-rays as, 49
US, *see* unconditioned stimulus

verbal learning, primacy and recency effects in, 265
vision, sensory gates for, 53–5
visual clues potentiated by taste, 53

visual CS information, pathways of, 137–9
visual learning, interruption of visual pathways to cortex and, 138–9
visual neurons and pathways, 134–5, 137-41, 291–322
  modifiability of, 139–47
  movement-sensitive neurons and, 281–4
  sensitization and response changes in, 280–4
visual signal processing in brain of honeybee, 279
visual stimuli
  pathways and, 137–41
  in study of conditioned heart rate response, 129–30, 132, 134
Vives, J. L., 47